Environmentally Friendly Catalysts for Energy and Water Treatment Applications

Environmentally Friendly Catalysts for Energy and Water Treatment Applications

Guest Editors

José Ignacio Lombraña
Hugo Olvera-Vargas
Sandra Yazmin Arzate Salgado

MDPI

Basel • Beijing • Wuhan • Barcelona • Belgrade • Novi Sad • Cluj • Manchester

Guest Editors

José Ignacio Lombraña
Department of Chemical Engineering
University of the Basque Country (UPV/EHU)
Bilbao
Spain

Hugo Olvera-Vargas
Institute of Renewable Energies
National Autonomous University of Mexico (IRE-NAUM)
Temixco
Mexico

Sandra Yazmin Arzate Salgado
Department of Earth Resources
Universidad Autónoma Metropolitana—Lerma
Lerma
Mexico

Editorial Office
MDPI AG
Grosspeteranlage 5
4052 Basel, Switzerland

This is a reprint of the Special Issue, published open access by the journal *Catalysts* (ISSN 2073-4344), freely accessible at: https://www.mdpi.com/journal/catalysts/special_issues/0YVJV9S48L.

For citation purposes, cite each article independently as indicated on the article page online and as indicated below:

Lastname, A.A.; Lastname, B.B. Article Title. *Journal Name* **Year**, *Volume Number*, Page Range.

ISBN 978-3-7258-5801-9 (Hbk)
ISBN 978-3-7258-5802-6 (PDF)
https://doi.org/10.3390/books978-3-7258-5802-6

Contents

About the Editors

José Ignacio Lombraña

José Ignacio Lombraña completed his PhD in Industrial Chemistry at the University of the Basque Country (UPV/EHU). Since then, he has held various teaching and research positions at UPV/EHU, and in 1997, he was appointed as Professor of Chemical Engineering. He has taught subjects including Fluid Mechanics, Water Treatment, and Energy and Sustainability in the BEng and MEng Chemical Engineering programmes.

As a researcher, José Ignacio Lombraña has led numerous publicly funded projects and research contracts for industry, including recent work on controlling the formation of disinfection by-products in drinking water. He has published over 90 papers in leading peer-reviewed journals (JCR), with a particular focus on advanced oxidation technologies for the removal of pollutants from water. He has supervised and directed 17 doctoral theses and coordinated the PhD Programme in Chemical Engineering for more than 15 years.

Research interests: water treatment, advanced oxidation, catalytic ozonation, adsorption, photocatalytic oxidation.

Hugo Olvera-Vargas

Hugo Olvera-Vargas holds a degree in Chemistry and a Master's in Chemical Sciences with honors from the Autonomous University of the State of Mexico. He earned his Ph.D. in Chemistry and Environmental Sciences through joint programs at the University of Paris-Est (France), the UNESCO Institute for Water Education (Netherlands), and the University of Cassino and Lazio (Italy). He completed postdoctoral research at the University of Toulouse III – Paul Sabatier and the National University of Singapore (NUS). Hugo Olvera-Vargas is currently an Associate Professor at the Institute for Renewable Energies, National Autonomous University of Mexico (UNAM). His research focuses on the water–energy–environment nexus, particularly on (photo)electrochemical technologies for the treatment and valorization of industrial wastewater. He also investigates the degradation and environmental impact of emerging contaminants. Hugo Olvera-Vargas has secured funding for several research projects and has published over 50 peer-reviewed articles in high-impact journals. He also holds international patents and has an h-index of 35 (Google Scholar). He has mentored undergraduate and graduate students in Mexico and abroad, evaluated research proposals nationally and internationally, and serves as a reviewer for various scientific journals. In 2015, he received the best PhD Theses Award in Science of Engineering from the University of Paris-Est, and in 2024, he was honored with the Early Career Researcher Award at the Iberoamerican Conference on Advanced Oxidation Technologies (CIPOA) held in Brazil. He is currently a member of Mexico's National Researchers of the Ministry of Science and an Editor of the *Journal of Environmental Chemical Engineering* (Elsevier, IF 7.2).

Sandra Yazmin Arzate Salgado

Sandra Arzate works as an Associated Professor at Autonomous Metropolitan University (2024-present). She has been involved in projects related to the use of catalytic technology in environmental applications. She has participated in Research, Development, and Innovation projects related to the catalytic process for drinking water, water treatment, waste valorization, and energy production. Her academic production comprises almost 40 communications in national and international congresses and 13 published papers, of which she is the main author of 6; 8 of these papers are published in Q1 journals. So, considering the citations, according to the Web of Science, she has an h impact factor of 7. She has been recognized as National Researcher Level I (2021-2030) by the National Research System (SNI, Mexico).

Preface

Water and the quality of water resources—increasingly threatened by the overexploitation of the planet's natural reserves and by industrial intensification—require urgent protection. Such protection is only achievable through the development of technologies that ensure the safe use of this essential resource in accordance with established environmental quality standards. As specialists in water treatment technologies, the Guest Editors of this Special Issue (SI) have sought to highlight research that is focused on the development and application of environmentally friendly catalysts (EFCs), that are aligned with the Sustainable Development Goals (SDGs).

The incorporation of efficient catalytic materials has fostered the emergence of a new generation of so-called advanced oxidation processes (AOPs). These processes, while chemical in nature, deliver substantial environmental benefits and are of critical importance for the removal of non-biodegradable pollutants. AOPs can function as pretreatments for industrial effluents—facilitating their acceptance at wastewater treatment plants (WWTPs)—or as tertiary and even quaternary treatments for water reuse.

From the catalytic perspective, AOPs can generally be divided into two main groups depending on whether catalysis is homogeneous or heterogeneous. Classical homogeneous catalytic systems, such as Fenton-type processes ($H_2O_2/Fe^{2+}/UV$), have been complemented by catalytic processes based on solid catalysts derived from semiconductor materials. These include transition metal oxides (MnO_2, TiO_2, Al_2O_3), as well as granular activated carbon (GAC). Such catalytic and photocatalytic materials, often endowed with high adsorption capacities, are able to interact effectively with pollutants at very low concentrations, achieving high removal efficiency.

This SI gives prominence to studies on catalytic materials and their integration into AOPs, illustrated with examples targeting the removal of microcontaminants and emerging pollutants. The results presented reveal highly promising perspectives, providing consistent alternatives for pollutant removal considering the stricter requirements set forth in recent EU directives.

In view of these characteristics, this collection of contributions aspires to serve as a timely and valuable reference in the development of catalytic materials for water treatment. The encouraging findings reported—both in terms of pollutant removal efficiency and energy performance—point to the likely implementation of these technologies in commercial AOP applications in the few next years.

José Ignacio Lombraña, Hugo Olvera-Vargas, and Sandra Yazmin Arzate Salgado
Guest Editors

catalysts

MDPI

Editorial

Environmentally Friendly Catalysts for Energy and Water Treatment Applications

José Ignacio Lombraña [1,*], Hugo Olvera-Vargas [2] and Sandra Yazmin Arzate Salgado [3]

1 Department of Chemical Engineering, Faculty of Science and Technology, University of the Basque Country (UPV/EHU), P.O. Box 644, 48080 Bilbao, Spain
2 Institute of Renewable Energies, National Autonomous University of Mexico (IRE-NAUM), Priv. Xochicalco S/n, Col. Centro, Temixco C.P. 62580, Morelos, Mexico; huolv@ier.unam.mx
3 Department of Earth Resources, Division of Basic Sciences and Engineering, Universidad Autonoma Metropolitana-Lerma, Av. de las Garzas 10, Lerma C.P. 52005, State of Mexico, Mexico; s.arzate@correo.ler.uam.mx
* Correspondence: ji.lombrana@ehu.eus

Water and energy are two essential factors for the sustainable development of human activities, and are increasingly driven by continuous industrial growth and societal development. Water resources and their quality, which are increasingly threatened by the overexploitation of the planet's natural reserves, require appropriate protection to ensure their universal availability for current and future generations. Only a very small fraction (<0.03%) of the Earth's water mass is potable. The lack or inefficiency of water treatment processes make the development of novel technologies, in many cases catalytic, necessary for the treatment of wastewater and drinking water.

While catalysts are useful for many environmental applications, some of them may pose negative environmental or health impacts. Therefore, ensuring their safety and sustainability should be a primary consideration in catalyst research and design. In this Special Issue (SI), priority has been given to the development and application of environmentally friendly catalysts (EFCs). As a result, a wide range of materials have been proposed, including single-metal catalysts and mixed metal oxides supported on alumina, titania, activated carbon, zeolites, or immobilized directly in the reactor. This approach facilitates controlled usage and enables the efficient recycling of the catalysts at the end of their service life.

First, the publications included in this SI can be broadly classified into two main thematic areas:

Water treatment as target. The primary objective of all the publications is to present the latest trends in the design and application of environmentally friendly catalysts (EFCs) for the development of advanced oxidation processes (AOPs). In water treatment, it is important to distinguish between processes aimed at wastewater and those focused on the production of drinking water. For wastewater treatment, there is growing concern over the removal of so-called emerging contaminants before the treated water is discharged back into the environment. Several contributions in this SI address this challenge, including a review article (Contribution 1) that explores various treatment strategies. Specific studies report the degradation of pharmaceuticals and their mixtures, such as dexamethasone, naproxen, and ketorolac (Contribution 2), diclofenac (Contribution 3), paracetamol (Contribution 4), and the antibiotic levofloxacin (Contribution 5). The removal of other emerging pollutants, such as polystyrene nanoplastics (Contribution 6), is also emphasized in response to the increasing demand for water reuse systems. In the context of drinking water production, notable examples include the use of photocatalytic materials for pathogen removal (Contribution 7) and the degradation of trace organic matter (Contribution 8), with the latter also being addressed using non-photocatalytic graphene-based materials (Contribution 9).

Energy concern. The energy requirements associated with treatment processes can be a determining factor in assessing their viability and overall efficiency. In this context, Contribution 2 stands out for presenting the development of an iron-based ligand derived from metallurgical slag, which demonstrates high efficiency in harnessing solar energy. Contribution 1 further explores the remarkable potential of graphitic carbon in the design of highly effective photocatalytic materials, with applications extending to solar energy conversion and hydrogen production. Additionally, Contributions 4 and 9 highlight other advanced catalytic systems designed for efficient energy capture and utilization. These include materials capable of storing energy to enhance the activation of oxidants, such as peroxymonosulfate (Contribution 9), or generating singlet oxygen (Contribution 4), both of which are essential for advancing modern non-photolytic catalytic processes.

Second, the publications in this SI have been classified according to the type of catalytic system employed, with an additional category created to highlight the most relevant contributions focused on the treatment of microcontaminants, as detailed below.

Homogeneous Catalysis. This section refers to Contributions 2 and 3, both of which employ the photo-Fenton process using iron catalysts in solution for the removal of pharmaceutical contaminants. In both studies, elimination efficiency serves as the basis for selecting the most suitable operational strategy. Contribution 3 investigates the degradation of diclofenac using the conventional photo-Fenton process with iron salts. The study places particular emphasis on the degradation pathway, identifying intermediates and by-products, and assessing water quality indicators such as turbidity, color, and aromaticity. Interestingly, the findings reveal that UV irradiation is not always beneficial. In the case of diclofenac, UV exposure appears to promote the formation of aromatic compounds during degradation, ultimately reducing the overall removal efficiency compared to other pharmaceuticals. In Contribution 2, the authors propose a novel circular economy approach by using metallurgical slag as a source of iron to form an iron–citrate complex, which then serves as a homogeneous catalyst in the photo-Fenton process. This strategy integrates waste valorization into water treatment. The study demonstrates the feasibility of this approach for the degradation of a mixture of three pharmaceuticals and shows the system's resilience to variations in both liquid depth and radiation intensity. Furthermore, the investigation offers valuable insight into the influence of key operational parameters, contributing to the process's potential for scale-up.

Heterogeneous Catalysis. It is worth highlighting Contributions 4, 7 and 5, which present materials that promote electron transfer from pollutant to oxidant without the need for light irradiation. In Contribution 4, the Fe-N-$C_{1.5}$ catalyst was synthesized by pyrolysis in a mixed solvent, generating Fe–N_x active sites, graphitic nitrogen, and carbonyl (C=O) groups. This Fe-N-$C_{1.5}$ catalyst activates peroxymonosulphate (PMS) through non-radical pathways, leading to the production of singlet oxygen ($^1O_2$). Notably, it achieves high efficiency (100% degradation of paracetamol in 7 min) and stable reusability (94% degradation) after several cycles. This makes it highly recommended for treating water contaminated with persistent organic microcontaminants, where high selectivity is required. With a similar reactive behavior, Contribution 5 reports the synthesis of a MnO_2/CN@SiO_2 catalyst, an amorphous compound containing Si–OH groups that enhance the rapid adsorption of PMS, thereby improving its reactivity. Its main advantages include a high degradation efficiency (80.8% removal of levofloxacin) and excellent stability after multiple cycles, achieved through more selective, non-radical pathways based on electron transfer and the generation of $^1O_2$, maintaining effectiveness even in complex water matrices. Contribution 7 presents a catalyst based on ZnO nanoparticles (~20 nm) with a wurtzite structure, synthesized via a green method using leaf extracts. Its advantages include a higher photocatalytic efficiency (up to 74% degradation of methylene blue), notable antibacterial activity against both Gram-positive and Gram-negative bacteria, and an environmentally friendly synthesis process. This makes it especially suitable for sustainable disinfection and the efficient degradation of contaminants.

Heterogeneous Photocatalysis. In this field, Contribution 1 highlights the potential of graphitic carbon nitride (gCN) as a promising, metal-free photocatalyst for sustainable water decontamination. Key strategies to enhance its photocatalytic performance include morphological modifications (e.g., 0D–2D hybrid structures), improved visible-light absorption (via doping), and better charge separation (through defects and functional groups). These approaches significantly boost gCN's ability to degrade emerging pollutants in a cost-effective and environmentally friendly manner, supporting its scalability for large-scale photocatalytic applications. Contribution 10 investigates the interaction of different strontium titanate photocatalysts with alkaline (solonetz) soil solutions. After exposure, most samples showed improved phenol degradation activity despite surface changes and the accumulation of organic residues. Although the crystalline structures remained unchanged, the samples exhibited alterations in the bandgap and crystallite size. The findings underline the importance of assessing the environmental interactions of photocatalysts, especially for solar-driven water purification applications.

Microcontaminants. This section includes Contributions 8, 6, and 9, which describe the use of highly adsorptive catalytic materials capable of interacting with low concentrations of microcontaminants, thus enabling their removal. Contribution 8 reports the synthesis of a catalyst consisting of carbon nanotubes supporting silica–alumina nanoparticles, forming mesoporous 3D structures with a high surface area and a notable bandgap reduction to 1.65 eV. This promotes light absorption and the generation of reactive species that accelerate photodegradation. The catalyst shows excellent performance in the complete removal of persistent dyes at low concentrations under UV/visible light, with much faster degradation rates than conventional TiO_2 and ZnO catalysts. Contribution 6 is based on natural pyrolusite (n-MnO_2) combined with oxalic acid, which enhances the generation of reactive oxygen species (ROS) during ozonation, facilitating both chemical degradation and the physical fragmentation of polystyrene nanoparticles (PSNPs). The material exhibits a stable, reusable structure without Mn leaching, achieving up to 75% removal of total organic carbon and a reduction in turbidity within just 30 min. Contribution 9 presents the synthesis of the β/N-rGO catalyst (nitrogen-doped graphene combined with β-cyclodextrin), which features a high specific surface area and good dispersion. This design enhances PMS reactivity and the generation of oxidizing species derived from $^1O_2$. Its high stability across a wide pH range makes it particularly suitable for the rapid removal of trace antibiotics in water.

Conflicts of Interest: The authors declare no conflict of interest.

List of Contributions:

1. Barreiro-Xardon, X.; Rosales, E.; Sanromán, M.Á. Current strategies to improve the properties of graphitic carbon nitride for effective and scalable wastewater pollutant removal: A Critical review. *Catalysts* **2025**, *15*, 523. https://doi.org/10.3390/catal15060523.
2. Arzate, S.Y.; Yañez-Aulestia, A.; Ramírez-Zamora, R.M. The direct formation of an iron citrate complex using a metallurgical slag as an iron source for micropollutant removal via the Photo-Fenton process. *Catalysts* **2024**, *14*, 426. https://doi.org/10.3390/catal14070426.
3. Villota, N.; Echevarria, B.; Duoandicoechea, U.; Lombraña, J.I.; De Luis, A.M. Kinetic Study of the Water Quality Parameters during the Oxidation of Diclofenac by UV Photocatalytic Variants. *Catalysts* **2024**, *14*, 580. https://doi.org/10.3390/catal14090580.

4. Zhuo, Y.; Meng, H.; Zhang, Y.; Chen, Y.; Cui, J. Peroxymonosulfate activation by FE/C composites for paracetamol degradation: Performance evaluation and mechanism insight. *Catalysts* **2025**, *15*, 217. https://doi.org/10.3390/catal15030217.
5. Xia, L.; Wang, X.; Li, J.; Xu, D. Amorphous MNO2 supported on CN@SIO2 for levofloxacin degradation via a Non-Radical pathway by PMS activation. *Catalysts* **2025**, *15*, 419. https://doi.org/10.3390/catal15050419.
6. Mello, V.; Nieto-Sandoval, J.; Dezotti, M.; Sans, C. Natural Pyrolusite-Catalyzed ozonation for nanoplastics degradation. *Catalysts* **2025**, *15*, 502. https://doi.org/10.3390/catal15050502.
7. Redjili, S.; Ghodbane, H.; Tahraoui, H.; Abdelouahed, L.; Chebli, D.; Ola, M.S.; Assadi, A.A.; Kebir, M.; Zhang, J.; Amrane, A.; et al. Green Innovation: Multifunctional Zinc Oxide Nanoparticles Synthesized Using Quercus robur for Photocatalytic Performance, Environmental, and Antimicrobial Applications. *Catalysts* **2025**, *15*, 256. https://doi.org/10.3390/catal15030256.
8. Saber, O.; Alshehab, A.; Shaalan, N.M.; Hegazy, A.M.; Aljasem, F.K.; Osama, A. Fabrication of a novel Silica–Alumina-Based photocatalyst incorporating carbon nanotubes and nanofiber nanostructures using an unconventional technique for Light-Driven water purification. *Catalysts* **2025**, *15*, 452. https://doi.org/10.3390/catal15050452.
9. Yao, M.; Wu, N. β-Cyclodextrin functionalization of Nitrogen-Doped graphene to enhance dispersibility and activate persulfate for trace antibiotic degradation in water. *Catalysts* **2025**, *15*, 541. https://doi.org/10.3390/catal15060541.
10. Ariya, B.; Chagu, J.; Solymos, K.; Gyulavári, T.; Tóth, Z.-R.; Kukovecz, Á.; Kónya, Z.; Veréb, G.; Pap, Z. Effect of alkaline soil solution on the material characteristics and photocatalytic activity of strontium titanate nanomaterials. *Catalysts* **2025**, *15*, 608. https://doi.org/10.3390/catal15070608.

catalysts

MDPI

Review

Current Strategies to Improve the Properties of Graphitic Carbon Nitride for Effective and Scalable Wastewater Pollutant Removal: A Critical Review

Xan Barreiro-Xardon , Emilio Rosales and María Ángeles Sanromán *

CINTECX, Universidade de Vigo, Bioengineering and Sustainable Processes Group, Department of Chemical Engineering, Campus Universitario As Lagoas-Marcosende, 36310 Vigo, Spain; xan.barreiro@uvigo.gal (X.B.-X.); emiliorv@uvigo.es (E.R.)

* Correspondence: sanroman@uvigo.es; Tel.: +34-986-812-383

Abstract: Heterogeneous photocatalysis has emerged in recent years as a promising and sustainable decontamination method. However, several drawbacks limit the effective usage of this process up to date, such as photocatalysts' limited properties, difficulty in modifying and improving their properties, as well as the environmental impact and cost associated with the use of the metals on which conventional photocatalysts are based. Graphitic carbon nitride (gCN), a new carbon-based photocatalyst, offers the possibility of easy modification and improvement of their properties. There are several strategies to improve the properties of these derivatives, such as increasing the surface area (modifying morphology into 0D, 1D, 2D, or porous structures), increasing the absorption in the visible (doping), and improving the separation and mobility of photogenerated charges (introducing defects, vacancies, functional groups, and doping). In this review, a compilation of these modifications, the associated improvements in its properties, and its derivatives was carried out with focus on the degradation of emerging pollutants (EPs). The property modifications enhance their behavior and efficiency of degradation, all in a cheaper and more sustainable way. Thus, improved gCN derivatives offer real possibilities for the upscaling of heterogeneous photocatalytic processes as an effective alternative for decontaminating water bodies.

Keywords: AOPs; heterogeneous photocatalysis; surface area; band gap; valence and conduction bands; excitons; graphitic carbon nitride

1. Introduction

Graphitic carbon nitride (gCN) has emerged as a highly promising photocatalytic material, particularly for applications under visible light irradiation. Its advantages include the use of low-cost, widely available, and environmentally friendly precursors, along with its straightforward synthesis and ease of structural and electronic modification. It offers suitable band gaps (valence band (VB) $\approx$ [(+1.6)–(+1.4)] eV; conduction band (CB) $\approx$ [(−1.1)–(−1.3)] eV) as well as appropriate redox potentials (suitable mainly for the production of superoxide, $O_2(g)/O_2^{\bullet-} \approx -0.33$ eV, or perhydroxyl radicals if the medium is acid, $O_2/HO_2^{\bullet} \approx +0.07$ eV, but not so much for the direct production of hydroxyl radicals, $HO^{\bullet}/H_2O \approx +2.3$ eV). Additionally, gCN shows high thermal stability (up to 600 °C) in diverse reaction media (acidic media, basic media, and organic solvents such as alcohols, DMF, THF, diethyl ether, toluene), and demonstrates excellent resistance to oxidation in air. In addition, gCN is inherently non-toxic, which makes it a compelling alternative to conventional metal-based photocatalysts [1–7].

The history of this compound is almost 200 years old. It was Jöns Jakob Berzelius who, in 1822, first described the synthesis of gCN (heptazines synthesized by pyrolysis of mercury thiocyanate, $Hg(SCN)_2$), and it was Justus von Liebig, in 1834, who ended up calling it 'melon'. Its empirical composition was detailed by Franklin in 1922, who determined that it varied according to the conditions of synthesis. In 1937, Linus Pauling studied the crystal structure of heptazines; it was in 1940 that Redemann and Lucas determined 'melon' has a lamellar structure like that of graphite [8,9]; and it was in 1966 that Teter and Hemley elucidated the five different crystalline phases of C_3N_4 (graphitic, cubic, pseudo-cubic, α, and β phases) [10]. However, it was not until 2009 that Wang et al. [9,11] described the first documented use of gCN in photocatalysis (production of H_2 from H_2O).

The applications of bulk gCN and its modifications in photocatalysis have been documented extensively in the literature, with over 8619 publications (limited to articles, reviews, and book chapters with g-C_3N_4 (gCN) in the article title) in the last five years focusing on gCN and its derivatives, particularly in fields such as physical chemistry, materials science, and chemical engineering (Scopus, 8 April 2025). The application areas of these modified gCN materials include not only environmental remediation, but also hydrogen production, CO_2 reduction, and solar energy conversion, as inferred from the keyword co-occurrence map (Figure 1), based on the total referenced articles ($\geq$250 rep.).

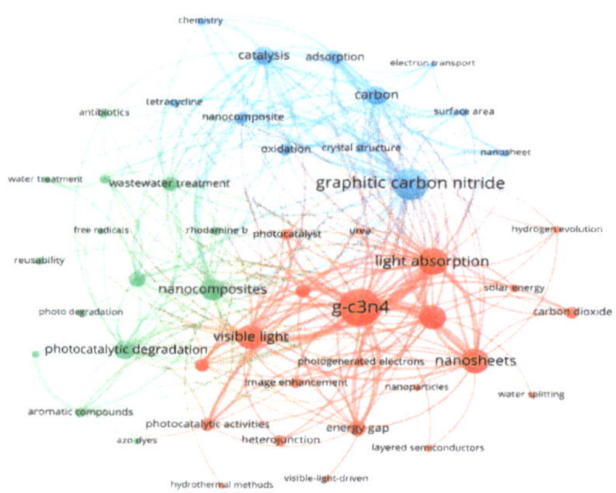

Figure 1. Keyword co-occurrence map for gCN generated using VOSviewer with data from Scopus.

The main reported synthesis methods include thermal condensation and polymerization (in inert environment and in aerobic environment, respectively), hydrothermal condensation, reflux techniques, the ionothermal method, chemical and physical vapor deposition (CVD and PVD), and electrochemical deposition [1,6,7,12]. Among the methods mentioned, thermal condensation and polymerization are the most widely used [1,9,12,13], but this synthesis has poor yield and purity [1,5]. Additionally, some of the main precursors used in the synthesis of gCN include cyanamide, dicyandiamide, urea, thiourea, and melamine, but also other nitrogen-rich carbon compounds such as melamine, cyanurate, ammonium thiocyanate, hexamethylenetetramine, guanidine carbonate or guanidine hydrochloride, etc. [4–7,9,14].

Thus, two main phases are possible within the gCN family: those based on triazines, and those based on heptazines (Figure 2). It has been found that heptazine-based structures are more thermodynamically stable (30 kJ/mol more stable) [15,16]. For example, precursors such as guanidinium, cyanamide, and melamine; those used with eutectic molten salts;

or those copolymerized with trimesic acid, as well as use of inert environments, favor these triazine structures [17–20].

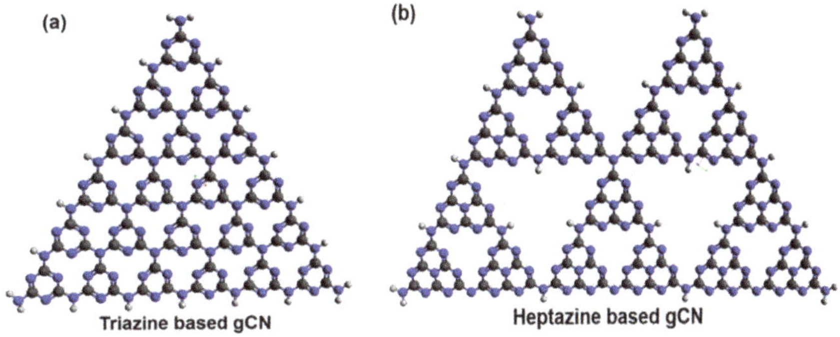

Figure 2. (**a**) Triazine-based gCN and (**b**) heptazine-based gCN. Carbon atoms: grey spheres, Hydrogen atoms: white spheres, and Nitrogen atoms: blue spheres.

Triazine-based gCN tends to have a wider band gap (2.8–3.1 eV) and tends to have a faster recombination of photogenerated charges due to the smaller extent of the π cloud (less conjugation) [21–24]. It is usually easy to differentiate between the two structures; the diffractogram of a triazine-based gCN (with A-B stacking) usually shows a peak corresponding to the (1,0,0) plane at $2\theta = 20–21°$ (CCDC 2217016) [19], and heptazine-based gCN (with A-B stacking) at around $2\theta = 15–17°$ (CCDC 1714809) [25], both for the hexagonal crystalline system (P-6m2 space group, Figure 3a). In the FTIR spectra, the peaks corresponding to the out-of-plane bending of the rings, around $\approx$810 cm^{-1} in the heptazine-based gCN, are usually sharp and pronounced, while the triazine-based peaks are usually either attenuated or absent; also, heptazine-based gCNs show higher intensity in N-H stretches (3000–3500 cm^{-1}) due to the higher content of primary and secondary amines (Figure 3b) [10,17,26].

The choice of precursor and reaction conditions of the thermal treatment (Figure 4) directly influence the specific surface area, band gap, and overall photocatalytic efficiency of gCN [6,12,14,27–30].

Regarding reaction conditions, it is worth noting that higher condensation temperatures result in a larger surface area and a smaller band gap, although they also lead to reduced yield. When synthesized under the same conditions, the yield decreases according to the following preference order: melamine $\approx$ dicyandiamide > cyanamide > thiourea > urea [6,12,31]. Conversely, a lower heating rate enhances crystallinity and promotes charge mobility but significantly reduces the surface area [31,32].

Under the same thermal synthesis conditions, urea as a precursor yields a significantly larger surface area (especially in air, in a non-inert environment) although slightly wider band gap than with other precursors [27,29,33]. Among common precursors, dicyandiamide results in the smallest band gap [30].

The precursors urea and thiourea are particularly noteworthy. Urea is one of the most used precursors due to its low cost and ease of handling. During thermal treatment, urea undergoes a series of transformations (Figure 5), starting with the formation of isocyanic acid ($\approx$175 °C), followed by condensation into melamine (>200 °C), which, after condensing as melam (390 °C), melem (>390 °C), and melon (500 °C), undergoes final condensation as gCN at 550 °C [1,5,34]. However, precursors such as cyanamide or dicyandiamide, apart from being expensive reagents, are difficult reagents to handle due to their explosive nature [1]. It is worth mentioning a common misconception found in the literature, namely, that the transformation of urea into cyanamide (by dehydration of the urea) then into dicyandiamide (by dimerization of two cyanamide units) gives rise to melamine: this route is impossible; it would only be possible in inert, anhydrous environments, and at low pressure [15,35–38].

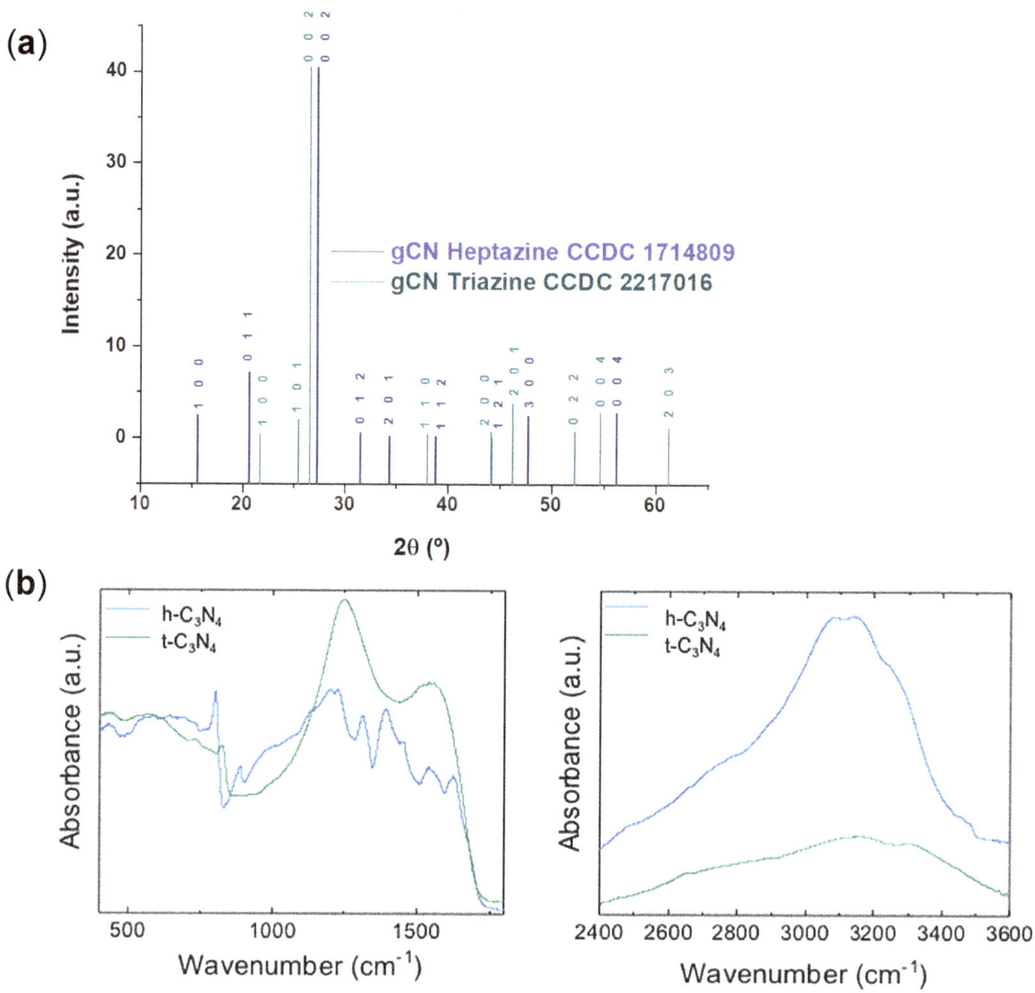

Figure 3. (**a**) Diffractograms of triazine-based gCN (green line) and heptazine-based gCN (blue line) from CCDC data [19,25], and (**b**) FTIR spectra of triazine-based (green line) and heptazine-based gCN (blue line) [17].

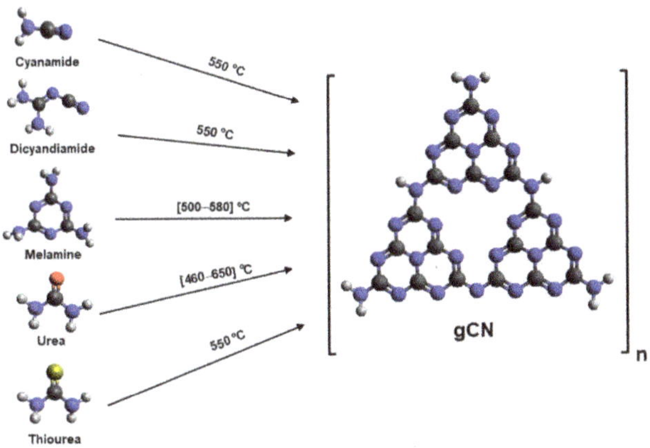

Figure 4. Reaction conditions of thermal treatment of different precursors for synthetizes heptazine-based gCN. Carbon atoms: grey sphere, Nitrogen atoms: blue spheres, Oxigen atom: red spheres, Hydrogen atoms: white spheres, and Sulfur atom: yellow sphere.

Currently, advanced synthesis methods are being used to overcome the limitations of traditional thermal synthesis; alternative methods such as microwave-assisted synthesis and solvothermal methods have been explored. Microwave-assisted synthesis, for example, not only reduces synthesis time but also increases the number of active sites on the gCN surface, leading to improved photocatalytic performance [7,9]. In the case of solvothermal methods, although more complex and requiring toxic solvents, they offer better control over the morphology and crystallinity of the synthesized gCN, making them suitable for specific high-performance applications [13,39].

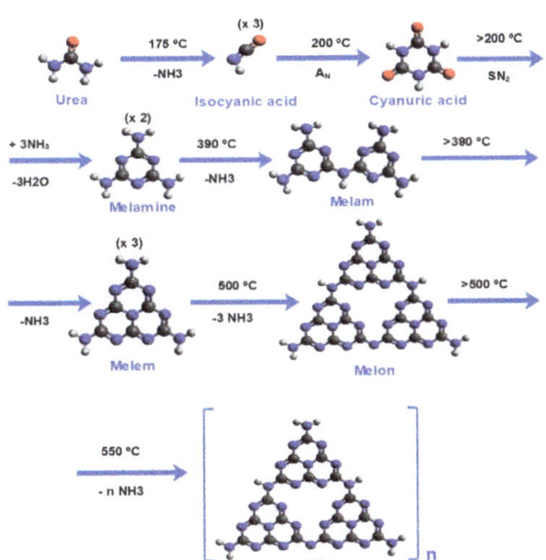

Figure 5. Transformations and conditions for thermal synthesis from urea of gCN. Redrawn with data from Refs. [1,5,34]. Carbon atoms: grey spheres, Oxigen atoms: red spheres, Hydrogen atoms: white spheres, and Nitrogen atoms: blue spheres.

Despite the aforementioned advantages, pristine gCN suffers from several intrinsic limitations, such as a relatively large band gap (~2.75 eV, or 450 nm; only absorbs in the blue region of visible spectra), low specific surface area, rapid recombination of photogenerated electron–hole pairs, low conductivity (high internal resistance), low electronic mobility, and slow surface reactions [1,7,40]. These factors significantly limit their photocatalytic efficiency, particularly in environmental applications such as the degradation of emerging pollutants (EPs) in water bodies.

To address these challenges, various modification strategies have been developed to enhance the photocatalytic properties of gCN (Figure 6). These modifications can be broadly categorized as follows:

- **Morphological modifications**: Adjusting the morphology of gCN to create structures like 0D quantum dots (QDs); 1D nanorods, nanowires, or nanotubes (NRs, NWs, and NTs, respectively); 2D nanosheets (NSs); and 3D porous networks. These modifications mainly enhance the surface area as well as improve the mobility of the charge carriers (their diffusion to the surface) and may even reduce the band gap of the material by introducing defects in its morphology [4,6,13].
- **Surface functionalization**: Introducing functional groups like amino, imino, cyano, hydroxyl, and carboxyl groups to the gCN surface increases the specific surface area and thereby increases the number of active sites in the gCN [4,15,41,42].
- **Modification of band structure**:
  Vacancies and defects: The introduction of defects such as nitrogen and carbon vacancies in gCN can significantly influence its photocatalytic performance. Nitrogen vacancies (Nv) can enhance charge separation and extend the lifetime of photogenerated excitons [4,6], while carbon vacancies (Cv) can serve as electron reservoirs, facilitating the generation of superoxide radicals ($O_2^{\bullet-}$) [43,44].
  Metal, non-metal, and self-doping: Doping with metal and non-metal elements influences the electronic structure of gCN by not only improving the absorption range (due to the introduction of intermediate energy levels in the band gap, as well as vacancies) but also by increasing the lifetime of the charge carriers and improving their separation [1,13]. This doping can also correct the increased gap produced in nanostructures (e.g., nanosheets) by the quantum confinement effect [1,13].

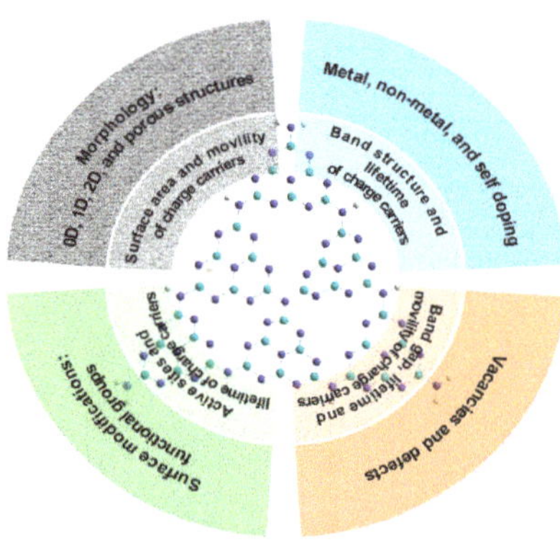

Figure 6. Methods of gCN modification and improvements in its properties. Grey: morphology modifications, blue: modifications by doping, beige: modifications by vacancies and defects, and green: surface modifications.

These modifications, applied individually or synergistically to gCN, improve its photocatalytic efficiency to degrade emerging pollutants in aquatic environments; however, applied individually, they can also lead to disadvantages such as possible increases in band gap (nanoarchitecture), the use of toxic reagents and reduced exciton mobility (vacancies), accelerated exciton recombination (non-metal doping), or secondary contamination due to leaching of metal ions (metal doping) (Table 1).

Table 1. Advantages and drawbacks of the gCN different modifications.

Structure	Photocatalyst	Advantages	Refs.	Drawbacks	Refs.
0D, 1D, 2D and porous 3D	0D: gCNQDs; 1D: gCNNW, gCNNRs, gCNNF, or gCNNTs; 2D: nanosheets; or porous 3D	Favor the migration of charge carriers. Enlarge the active surface. Improve visible light absorption (3D porous). 0D, 1D, and 2D solubility in water. Non-toxic.	[4,6,13]	Quantum confinement effect by size reduction in 0D, 1D, and 2D structures. Gap broadening.	[6,13,14,45, 46]
Functional groups	gCN with amino, imino, cyano, ureido, hydroxyl, carboxyl, and aromatic groups.	Amino and imino groups improve the anchoring of metal oxides during their support, as well as their dispersion on the surface. Improved visible light absorption. Improved separation of charges. Increased specific surface area and active sites number (-OH) may act as h^+ trapping centers enhancing charge separation.	[4,41,42,47, 48]	Cyano, ureido, carboxyl and aromatic groups often require toxic reagents that can cause contamination. Mobility of photogenerated charges restricted by the presence of these groups.	[1,42,49,50]
Vacancies	Cv-gCN	Serve as a reservoir for photogenerated electrons. Inhibit the recombination of h^+ and e^-. Serve as electron transfer centers for the adsorbed molecular oxygen favoring the production of superoxide radicals. Reduce the effective band gap.	[43,44]	Mobility of photogenerated loads is restricted by the presence of these vacancies.	[50]
	Nv-gCN	Improve the separation of charges. Increase the lifetime of excitons. Reduce the effective band gap.	[4,6,50,51]		
Doping	Non-metallic doping	Introduction of intermediate energy levels in the gap. Reduction in effective gap. Increased exciton lifetime. Corrects for the increased gap due to the quantum confinement effect.	[1,2,13,40]	Non-metallic species do not participate in the transport of charges; they act as exciton recombination centers.	[12]
	Self-doping (C-gCN and N-gCN)	Introduction of intermediate energy levels in the gap; reduction in effective gap. Increased exciton lifetime. Corrects the increased gap due to quantum confinement effect. C-doping increases the number of delocalized π-bonds, improving conductivity, charge transfer, and their separation. Significant gap reduction.	[52–54]		
	Metal doping	Introduction of intermediate energy levels in the gap; reduction in effective gap (especially with alkalis). Increased exciton lifetime. Correction for the increased gap due to the quantum confinement effect. With noble metals, the effect of surface plasmon resonance (SPR) is introduced; the excitation of electrons to the conduction band is enhanced (greater number).	[1,13]	Causes secondary contamination due to leaching of metal ions. Excess metal ions act as exciton recombination centers. Serious pollution is associated with the mining of metals and their refining.	[12]

The modification strategies may be helped by the use of computational tools. It is worth highlighting the versatility of ab initio computational methods, such as Hartree–Fock theory and density functional theory (DFT), in the design and optimization of nanomaterials. While Hartree–Fock provides high accuracy for small systems (such as triazine or heptazine units), DFT stands out for its balance between accuracy and computational cost, making it ideal for studying complex nanostructures. These techniques enable the prediction of properties like band gaps, electronic density of states, and adsorption energies. The choice of exchange–correlation functional in DFT (e.g., B3LYP, HSE06) is critical, as it directly affects key parameters such as the band gap and band positions in DOS calculations [55,56].

In recent years, machine learning has emerged as a powerful complement, allowing for the prediction of complex material properties based on theoretical or experimental data [57,58]. In the case of g-C_3N_4 and its derivatives, these in silico techniques have been essential to understanding their structure [59,60], electronic and optical properties [59–62], and the effects of modifications such as morphology, vacancies, functionalization, and doping in this photocatalyst [63–69].

These computational tools not only guide nanoarchitecture design (Section 2), the introduction of functional groups (Section 3), and defect engineering, but also allow for synergy prediction in multi-element doping (Section 4), thereby reducing costly experimental trials. They provide researchers with a series of advantages that not only help optimize experimental design but also contribute to aligning such designs with the 12 principles of green chemistry [70].

Without disregarding the potential of catalyst combinations to enhance their electronic and optical properties—such as homojunctions or heterojunctions (between gCN derivatives or between gCN and other non-gCN-based photocatalysts, respectively); their immobilization, which facilitates recovery from the reaction medium (both of which are also key for developing effective and scalable photocatalytic pollutant degradation systems in water bodies); or the stability and toxicity (both also key aspects for the viability of heterogeneous photocatalysis as a decontamination method) of gCN and its derivatives—it should be noted that each of these aspects requires extensive, stand-alone monographic analysis owing to the diversity and complexity in these fields. It is for all these reasons that the main objective of this bibliographic review is solely to provide a comprehensive overview of current experimental strategies used to improve the intrinsic properties of gCN (morphology, defects, functionalization, and doping) for effective and scalable photocatalytic water remediation.

Thus, in this review, the synthesis methods, advantages, and limitations of various gCN modifications are examined, focusing on their impact on photocatalytic efficiency and the degradation of emerging pollutants.

2. Nanoarchitecture Design of gCN

Regarding modifications related to nanoarchitecture (morphology), this review refers to those aimed at altering not only the state of aggregation (0D, 1D, and 2D), but also the porosity in 3D structures. As previously noted, these modifications primarily contribute to enhancing surface area, improving charge carrier mobility, and even reducing the band gap. Thus, there are two main routes to this target: bottom-up, like thermal treatment for 3D structures from precursors, solvothermal-assisted microwave and chemical or physical vapor deposition (CVD and FVD, respectively) from precursor for 0D structures, and supramolecular self-assembly or template methods from precursor for 1D structures; top-down strategies like chemical, ultrasound, microwave, and thermal exfoliation for

2D structures from 3D structures and hydrothermal and sonication treatments from 2D structures for 0D structures (Figure 7) [4,6,13].

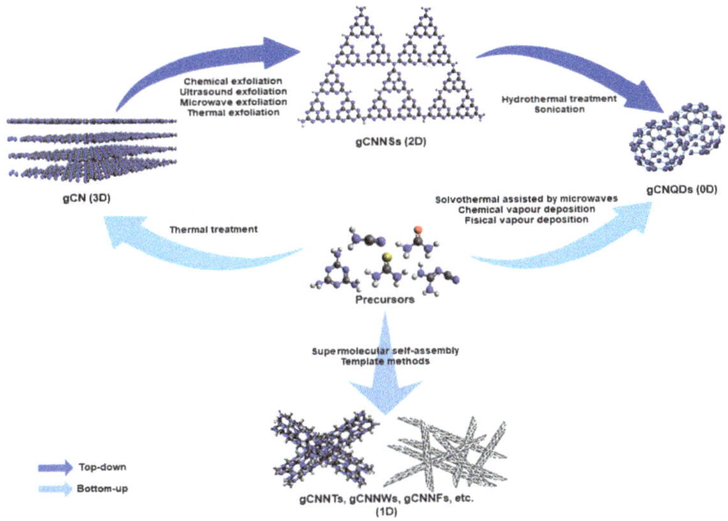

Figure 7. Techniques for gCN morphology modification (3D, 2D, 1D, and 0D). Carbon atoms: grey sphere, Nitrogen atoms: blue spheres, Oxigen atom: red spheres, Hydrogen atoms: white spheres, and Sulfur atom: yellow sphere.

2.1. Zero-Dimensional Structures (0D)

Zero-dimensional nanomaterials are structures confined to the nanoscale in all three spatial dimensions. They are particles smaller than 10 nanometers (<10 nm), typically consisting of a few thousand atoms such as quantum dots. These materials lack defined length, width, or height, and their properties are highly dependent on size and quantum effects. Typical 0D nanomaterials used in photocatalysis are metal oxides (TiO_2, ZnO), MXenes, chalcogenides (Cds), carbon quantum dots (CQDs), fullerenes (C_{60}), and gCNQDs, among others [71–73].

Due to their small size, gCNQDs exhibit a morphology that favors the migration of charge carriers [13], enlarging the active surface and enhance light absorption [13,45]. All these properties can be modulated by controlling their size. As their dimensions decrease, these structures display the so-called quantum confinement effect—a specific manifestation of quantum effects arising from size reduction—which, according to the Brus equation, leads to the band gap. In addition, gCNQDs also exhibit photoluminescence and water solubility and are considered non-toxic [6,13,14,45,74–76]. There are two different approaches to synthesize these types of nanostructures, as mentioned above, i.e., top-down and bottom-up techniques.

The most cited top-down methods include ultrasonic synthesis (sonication) and hydrothermal treatment, starting from exfoliated gCN (2D). Among these, it can be noted that sonication in a single step, although simple and easy to handle, requires a significant amount of time and ultimately results in a poor yield. However, sonication coupled with chemical oxidation, despite high yield and purity, is an expensive and complicated process. Similar observations apply to one-step and two-step hydrothermal treatments (coupled to oxidation): simple and cheap, with low yield for the one-step process, and complicated and expensive, but with good yield, for the two-step process [45,74–76].

Among the bottom-up processes, the microwave-assisted solvothermal method, the solid-phase process, and chemical vapor deposition (CVD) stand out. For the first one, there is an easy synthesis, with few steps, but with low yield; for the second one, there is high yield, good quantum yield of the material, and possibility to adjust the photoluminescence, but it is a long process, and post-treatments are necessary. As for chemical vapor deposition, it is only mentioned that it is an expensive process because of the high energy consumption, and it also requires post-treatment. Hydrothermal is the simplest, most sustainable, and most economical of the methods discussed [74–76].

Due to the main drawbacks of QDs, fast recombination of charge carriers, and wide band gap (owing to the effect of quantum confinement) [6,13,14,45,74–76], their main applications are in sensor fabrication, hydrogen production, and imaging.

The documented use of gCNQDs in decontamination involves both of the following:

- The fabrication of heterojunctions, mainly Z-type and II type (semiconductor/semiconductor);
- Heterojunction of the Schottky type (semiconductor/non-noble metal) [1,52].

As previously noted, since the aim of this work is not to provide an in-depth discussion of homojunctions or heterojunctions, only heterojunction types involving 0D gCN-based structures that demonstrate photocatalytic degradation of pollutants in aqueous environments are referenced in this subsection.

Examples of this type of compound in the degradation of pollutants in water bodies can be seen the work of Li et al. [77], in the degradation of Rhodamine B (RhB) (5 ppm) with a Z-type heterojunction of $rTiO_2$/gCNQDs (1:20; 1000 ppm) under visible light, prepared by heat treatment of TiO_2 (P25) and melamine at 500 °C for 4 h (2.3 $°C \cdot min^{-1}$), which degraded 95% of the RhB in 240 min (the pristine gCN degraded only 35% in the same time) (Table 2), attributing the improvements in degradation to a longer exciton lifetime (better charge separation; Figure 8a,b), and a slight increase in surface area, where the main species responsible for degradation are $HO^{\bullet}$ and h^+ radicals.

Table 2. Parameters of photodegradation of pollutants in water with different morphologies of gCN (0D, 1D, and 2D).

Morphology	Photocatalyst	Area ($m^2 \cdot g^{-1}$)	Gap (eV)	Pollutant (ppm)	Radiation	Efficiency (%)	Time (min)	k (min^{-1})	Refs.
0D	gCN (1000 ppm)	10.9	-	RhB: 5	500 W; Xe lamp (Vis.)	35	240	0.002	[77]
	$rTiO_2$/gCNQD 15% (1000 ppm)	43.9	-			95		0.012	
	gCN (600 ppm)	-	-	OTC: 40	500 W; Xe lamp (UV)	20	150	0.0017	[78]
	gCNQDs-$CoTiO_3$/$CoFe_2O_4$ (600 ppm)	-	-			88		0.0141	
	gCNQDs (600 ppm)	25.34	-	NOR: 30	1000 W; Hg lamp (UV)	35	120	0.0035	[79]
	gCNQDs/Ni_5P_4 8% (600 ppm)	83.61	-			92		0.022	
	1D MCN (200 ppm)	29.07	-	RhB: 10	300 W; Xe lamp (Vis.)	99.5	30	0.12	[80]
	0D UCN (200 ppm)	44.27	-			70	30	0.036	
	0D/1D MUCN (200 ppm)	57.24	-			99.96	20	0.26	

Table 2. *Cont.*

Morphology	Photocatalyst	Area ($m^2 \cdot g^{-1}$)	Gap (eV)	Pollutant (ppm)	Radiation	Efficiency (%)	Time (min)	k (min^{-1})	Refs.
1D	gCN (400 ppm)	9.9	2.65	SMX: 10	300 W; Xe lamp (Vis.)	15	140	0.001	[81]
	gCNNT (400 ppm)	100.4	2.31			100	120	0.035	
	gCN (1000 ppm)	5.3	2.62	MO: 10	350 W; Xe lamp (Vis.)	71.1	120	0.009	[82]
	gCNNWs (1000 ppm)	74.25	1.52			98.5		0.02	
	gCNNFs (1000 ppm)	60.16	1.61			90.9		0.01	
	TGCN (1000 ppm)	8.52	2.8	PhOH: 20 MO: 20	SLB-300A, 300 W (Simulated sunlight)	PhOH: 50 MO: 62	PhOH: 180 MO: 150	0.004 0.006	[83]
	SGCN (1000 ppm)	21.13	2.4			PhOH: 60 MO: 65		0.005 0.007	
	MGCN (1000 ppm)	31.84	2.45			PhOH: 85 MO: 92		0.01 0.013	
2D	Bulk (550 CN)	25	2.54	CIP: 5 RhB: 5	500 W; Xe lamp (Vis.)	CIP: 8 RhB: 58	CIP: 60 RhB: 40	0.001 0.017	[84]
	Ultrasounds (550 ul CN)	12	2.56			CIP: - RhB: 75		– 0.027	
	Thermic (475 CN)	29	2.42			CIP: - RhB: 50		– 0.009	
	Acid (550 H CN)	114	2.73			CIP: 17 RhB: 85		0.002 0.023	
	Thermic + ultrasounds (475 ul CN)	21	2.46			CIP: 7 RhB: 43		0.001 0.012	
	Acid + ultrasounds (550 ul H CN)	74	2.74			CIP: - RhB: 80		– 0.03	
	Thermic + acid (475 H CN)	71	284			CIP: - RhB: 80		– 0.018	
	Thermic + acid + ultrasounds (475 ul H CN)	57	2.67			CIP: 57 RhB: 92		0.006 0.037	
	CN-B	18	2.42	Rh6G: 5	400 W; Xe lamp (Vis.)	67	30	0.031	[85]
	CN 500-4	31	2.56			78		0.046	
	CN 550-4	107	2.68			92		0.085	
	CN 550-5	165	2.73			96		0.102	
	CN 550-6	295	2.89			98		0.139	
	gCN	22.2	2.8	CIP: 20	Sunlight	45	60	0.009	[86]
	E-gCN	63.8	2.94			78		0.023	

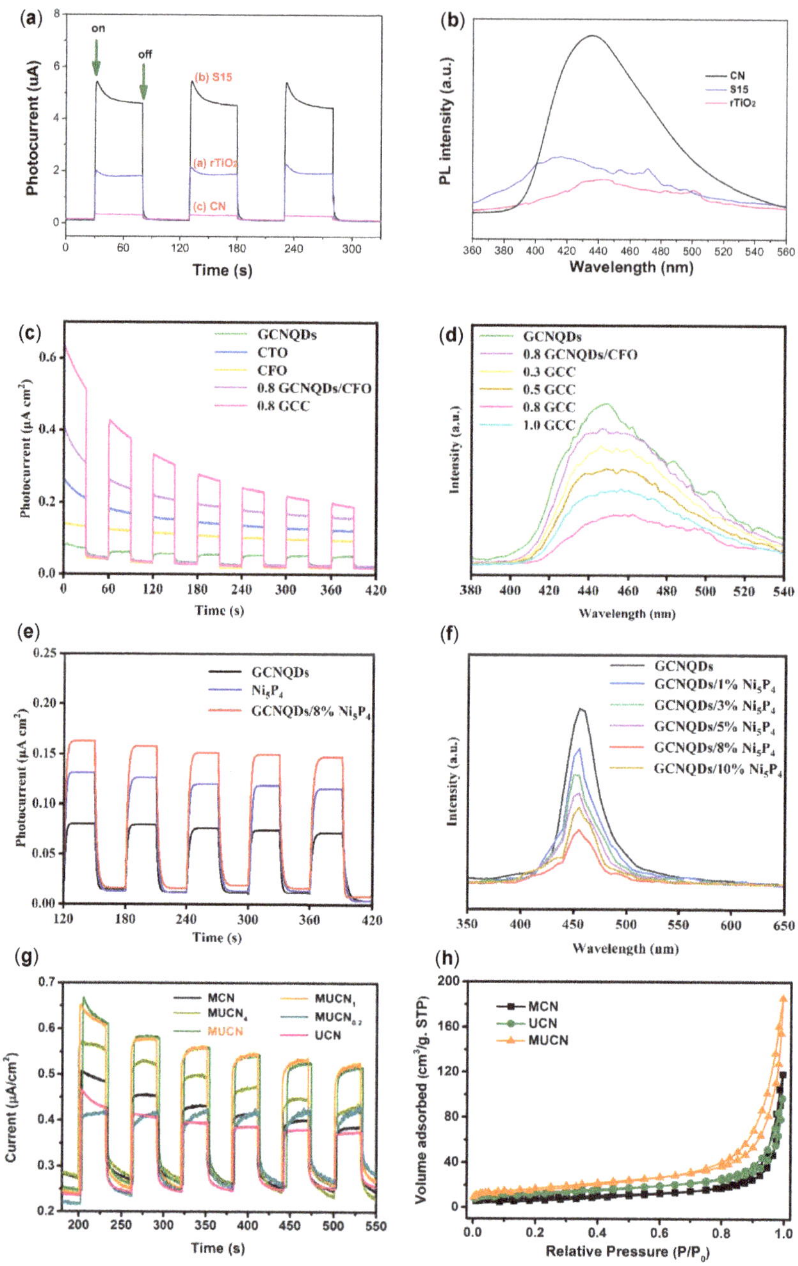

Figure 8. (**a**,**b**) Transient photocurrent, (**left**), and photoluminescence spectra (PL), (**right**), for gCN, $rTiO_2$, and $rTiO_2$/gCNQDs (15%) [77]; (**c**,**d**) transient photocurrent, left, and photoluminescence spectra (PL), right, for gCNQDs, cobalt titanate (CTO), cobalt ferrite (CFO), gCNQDs (0.8% in mass) with CFO, and dual Z-Scheme with 0.8 in mass of CTO (0.8 GCC) [78]; (**e**,**f**) transient photocurrent, (**left**), and photoluminescence spectra (PL), (**right**), for gCNQDs and gCNQDs/Ni_5P_4 (8% in mass) [79]; (**g**,**h**) transient photocurrent (**left**) and adsorption isotherms (**right**) for UCN, MCN, and MUCN [80].

Another example of a magnetically recoverable Z-type heterojunction based on gCN-QDs is given by Feng et al. [78]. These researchers prepared a dual Z-type heterojunction of gCNQDs-$CoTiO_3/CoFe_2O_4$, by hydrothermal treatment and microwave-assisted impregnation, for the degradation of oxytetracycline (OTC). The authors were able to degrade OTC (40 ppm) with the compound gCNQDs-$CoTiO_3/CoFe_2O_4$ (0.8% in $CoTiO_3$; 600 ppm) by 88% in 150 min under UV light, where pristine gCN degraded only 20%, and gCNQDs degraded 30% within the same time and under the same reaction conditions (Table 2). The removal rate of OTC can remain above 85% after four cycles. They claim that the improvement in OTC photodegradation is mainly due to an improvement in the separation of the photogenerated charges (longer lifetime of the photogenerated charges; Figure 8c,d).

An example of a Schottky-type heterojunction could be the study reported by Feng et al. [79], in which gCNQDs/Ni_5P_4 were prepared by impregnation of gCNQDs on Ni_5P_4 nanoflowers, in EtOH and with ultrasound. With this Schottky-type gCNQDs/Ni_5P_4 (8% by mass of Ni_5P_4; 600 ppm), 92% of norfloxacin (NOR) (30 ppm) was degraded in 120 min under UV radiation compared to 35% achieved with the gCNQDs and 20% achieved by the unmodified gCN within the same time (Table 2). The NOR removal rate remained above 85% after four cycles. The improvement in degradation efficiency is attributed by the authors to better migration and separation of the photogenerated charges (Figure 8e,f).

Finally, an interesting homojunction (type II) between 0D and 1D structures of gCN (sample named as MUCN) was prepared by Zheng et al. [80] from melamine and cyanuric acid as supramolecular assemblies (gCNNTs; 1D) plus urea (gCN 0D nanoparticles) in a 2:1 ratio, and subsequent heat treatment at 520 °C 2 h, for RhB degradation (10 ppm). With this homojunction (MUCN; 200 ppm), 99.96% degradation was achieved in 20 min under visible radiation, where the urea-derived gCN (UCN) achieved only 70% and the melamine-derived gCN (MCN) 99.5% in 30 min under the same conditions (Table 2). As for the previous examples, the improvements in efficiency over the bulk materials are mainly due to improvements in the separation of the photogenerated charges (longer exciton lifetimes) and to a larger surface area (Figure 8g,h).

2.2. One-Dimensional Structures (1D)

As far as 1D materials are concerned, it can be noted that this category includes structures such as nanowires (NWs) [82,87], nanorods (NRs) [88,89], or the better-known nanotubes (NTs) [81,90], to cite only a few examples. These structures can be synthesized via hard-, soft-, self-templating, or template-free methods [13]. Examples include the synthesis of gCN nanorods with templates, e.g., by thermal condensation of cyanamide using an AAO oxide membrane as a template, which improved the orientation and crystallinity of the material with respect to the bulk, or without templates, e.g., by infrared-controlled heating of dicyandiamide, a synthesis that improves charge separation [1]. These structures show improvements both in surface area (more active sites) and in the increased migration speed of the charge carriers (shorter distances for diffusion of the photogenerated charges), but not only in [1,13,91].

An example of a synthesis of this type of morphology without templates (e.g., molecular assembly) is reported by Zhang et al. [81], on porous gCNNTs for sulfamethoxazole (SMX) degradation. In this work, the authors prepared the nanotubes, in two steps, by molecular assembly (mixture of aqueous solution of melamine and 1,5-naphthalene disulfonic acid, in different amounts, plus aqueous solution of cyanuric acid; both solutions prepared at 90 °C) and subsequent heat treatment of the assemblies (550 °C/4 h; 5 $°C{\cdot}min^{-1}$). This material (sample MNCA-75; 75 ppm of 1,5-naphthalene disulfonic acid) degraded SMX (10 ppm), with 400 ppm of photocatalyst, to 100% in 120 min (unmodified gCN only reached 15% in 140 min) under visible light, 300 W Xe lamp (Table 2), attributing the improved photocatalytic behavior to the increased lifetime of the charges, the reduced gap, and a higher specific area (Figure 9a,b). The removal efficiency of the SMX was still above 82.5% in the fifth cycle.

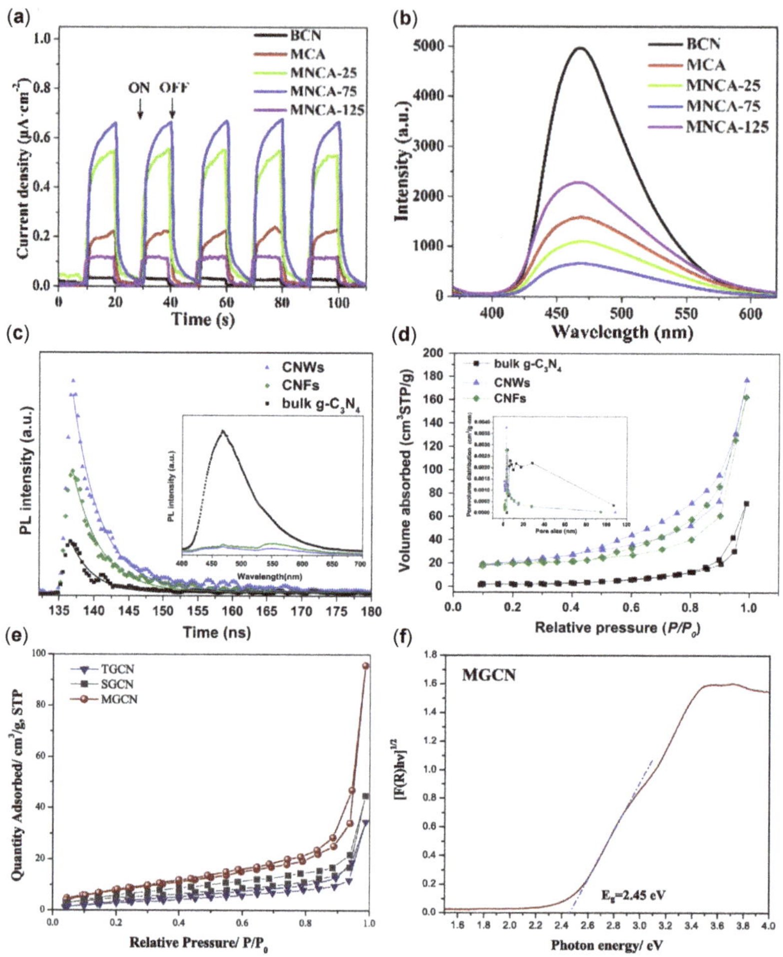

Figure 9. (**a**,**b**) Transient photocurrent and photoluminescence spectra (PL) of gCNNTs [81], (**c**,**d**) PL (time and space resolved) and adsorption–desorption isotherms of gCNNWs/gCNNFs [82], and (**e**,**f**) adsorption–desorption isotherms of gCNNFs prepared by thermal (TGCN), thermal (TGCN), solvothermal (SGCN), and microwave (MGCN) methods and Kubelka–Munk plot of MGCN [83].

Another example of 1D morphologies without templates (by molecular assembly) could be the work of Xie et al. [82], regarding gCN nanowires and nanofibers (gCNNWs and gCNNFs, respectively) evaluated in methylene blue (MB) degradation. These 1D nanostructures were synthesized by molecular assembly of cyanuric acid and melamine (in acetonitrile with stirring 12 h, washed also with acetonitrile, and subsequent drying at 80 °C 12 h), followed by two-step treatment: solvothermal (180 °C for 24 and 48 h) and heat treatment at 500 °C for 1 h (10 °C·min^{-1}). Two materials were obtained: gCNNWs with the 24 h solvothermal treatment, and gCNNFs with the 48 h solvothermal treatment. In this way, an MB degradation degree (10 ppm) of 98.5% was achieved with gCNNWs, and 90.9% with gCNNFs (g-CN, 71.1%) in 120 min, with 1000 ppm of photocatalyst under visible light (Table 2). These gCN-derived 1D structures showed improved degradation efficiency compared to pristine material mainly due to a considerable reduction in band gap, an increase in specific area (also higher microporosity), and longer photogenerated charge lifetime (Figure 9c,d).

As an example of template-free synthesis, also by molecular assembly, but by microwave treatment, the work of Mohamed et al. [83] could be mentioned, in which they prepared gCN nanofibers by this method and evaluated their photocatalytic performance in the degradation of methyl orange (MO) and phenol under simulated sunlight. The nanofibers were synthesized by mixing melamine and cyanuric acid in acetonitrile, which after magnetic stirring was poured into a pressurized vial and introduced into a microwave reactor (CEM; Discover 2.0) at 300 W (270 °C and 16.55 bar) with stirring for 60 min. The material thus prepared (1000 ppm), MGCN, showed a gap of 2.45 eV (a thermally treated sample a gap of 2.8 eV for TGCN, and of 2.4 eV for the solvothermal-prepared material, SGCN) and was able to degrade phenol (20 ppm) by 85% in 180 min and MO (20 ppm) by 92% in 150 min (SGCN sample 60% phenol and 65% MO; TGCN sample 50% phenol and 62% MO, within the same time) under simulated sunlight (SLB-300A, 300 W) (Table 2). The authors attribute the efficiency improvements over the thermal- and solvothermal-treated materials to an improvement in surface area (31.84 $m^2 \cdot g^{-1}$; 1.5 times the area of SGCN, and 3.7 times the area of TGCN), and a smaller gap than the TGCN sample (Figure 9e,f).

2.3. *Two-Dimensional Structures (2D)*

One of the most important morphology modifications to increase low surface area and improve mobility and charge separation in gCN is to prepare 2D structures by exfoliation. This flat, flexible, two-dimensional structure can improve compatibility with various modification strategies, such as heterojunction construction, doping, 0D and 1D materials, and the introduction of vacancies [13]. Among the different types of possible 2D structures are nanosheets, a few layered structures, nanoplates, etc. Several methods are possible for the exfoliation of this material, e.g., exfoliation with ultrasound (in a suitable solvent such as IPA, cheap and easy to remove; NMP, quite effective; DMF; 1,3-butanediol; EtOH; MeOH; or mixtures with water such as EtOH/H_2O, IPA/H_2O, and DMF/H_2O); thermal exfoliation; microwave-assisted exfoliation; mechanical exfoliation (e.g., ball milling); or by chemical exfoliation with mineral acids such as HCl, H_2SO_4, or HNO_3 [1,6,13]. Chemical exfoliation has great advantages over physical or thermal exfoliation, such as better performance, better exfoliation rate, and the possibility of functionalization [1]. Thus, treatments with different acids impart distinct properties to the photocatalytic material; however, chemical exfoliation has the disadvantage of not being a very environmentally friendly method [70]. With thermal oxidation, which is cheap, sustainable, and easily scalable, surface areas of up to [300–390] $m^2 \cdot g^{-1}$ can be achieved, and in addition, for example, heat treatment in an H_2 atmosphere has led to improvements not only in the degree of exfoliation but also in the π-conjugated structure [6]. Microwave exfoliation leads to improvements in both the surface area and separation—and therefore lifetime—of photogenerated charges and is a more sustainable method than chemical exfoliation [92].

In a comparison between thermal, chemical, and ultrasonic exfoliation methods for the degradation of CIP and RhB, as carried out by Wang et al. [84], gCN (550 CN) was exfoliated in different ways (ultrasonic; thermal; acidic; thermal/ultrasonic; acidic/ultrasonic; acidic/thermal; acidic/thermal/ultrasonic). The smallest gap was obtained by the heat-treated exfoliated sample (475 CN). The sample with the largest specific area was the one obtained by acid exfoliation with HNO_3 (550 H-CN), and the best charge separation was achieved with thermal treatment + ultrasound (ul) + acid (475 ul-H-CN), also obtaining the best photocatalytic behavior, both in the photodegradation of CIP (5 ppm), with a degree of degradation of 57.2% in 60 min (for the unmodified gCN 8.3%), and in the degradation of RhB (5 ppm), with 91.52% in 40 min (gCN only 57.6%), with 1000 ppm of photocatalyst, under visible light (Table 2). After five cycles, the degradation rates of 475 ul-H-CN and 550 CN to RhB were 70% and 63%. The improvements achieved with the 475 ul-H-CN sample, with respect to gCN, and the other samples are mainly due to better photogenerated charge separation (Figure 10a).

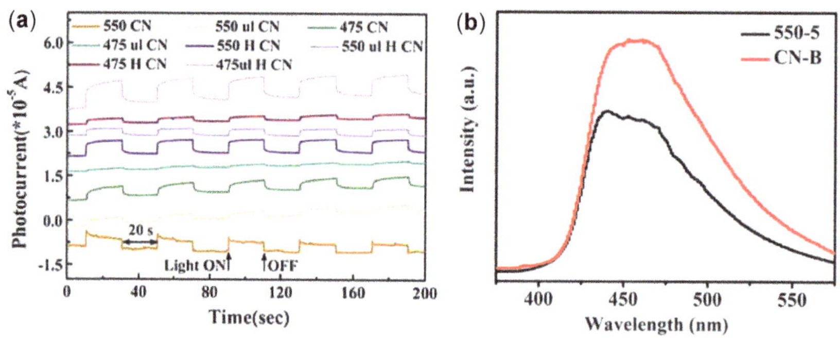

Figure 10. (**a**) Photoconductivity of different exfoliation methods (ultrasounds; ul, acid; H) at different T [84]; (**b**) PL of thermally exfoliated gCN at 550 °C in 5 h (550-5) and bulk (CN-B) [85].

Another comparison between chemical and thermal exfoliation can be found in the work of Papailias et al. [93], where it is concluded that chemical exfoliation produces a wider band gap than thermal exfoliation. It was also found that the specific surface area, porosity, and gap increased with the duration of the heat treatment, although not significantly.

It is worth mentioning that heat treatment is the cheapest and simplest treatment, as Li et al. [85] proposes in a work about the exfoliation of gCN, and it is also an effective method for obtaining ultra-thin gCN nanosheets. These authors started by thermally condensing (inert atmosphere) melamine at 600 °C (5 $°C \cdot min^{-1}$) for 2 h. Thermal exfoliation was then carried out at different temperatures and different times. The highest surface area was obtained for the sample exfoliated at 550 °C 6 h (295 $m^2 \cdot g^{-1}$ compared to 18 $m^2 \cdot g^{-1}$ for the untreated gCN, CN-B). The band gap of this sample was 2.89 eV, showing that as the degree of exfoliation increases with T, the gap increases, but more so with condensation time. In addition, an increase in process temperature results in more specific area (more exfoliation) and more nitrogen vacancies (N_V), while an increase in process time leads to a higher degree of exfoliation, more carbon vacancies (C_V), and larger pore volume. Both the degree of exfoliation and the photocatalytic activity of Rh6G (5 ppm) improve with increasing heat treatment time and temperature under visible light with 500 ppm photocatalyst (Table 2). The CN 550-5 sample also shows a longer lifetime of photogenerated charges (compared to CN-B), mainly due to a reduction in migration distances (Figure 10b).

Zhang et al. [94] prepared, by thermal polymerization of a urea solution in water (in crucible at 550 °C for 3 h; 5 $°C \cdot min^{-1}$), exfoliated gCN in one step, and documented, by comparison with other exfoliation methods (chemical and ultrasonic), that the one-step synthesis produces a material with higher photocatalytic efficiency (200 ppm photocatalyst) than the other methods studied in the degradation of BPA (20 ppm) under visible light. After five repeated cycles, the BPA degradation rate remains at more than 92% with this material.

A 'green' method for the synthesis of exfoliated gCN was documented by Pattnaik et al. [86] in the degradation of CIP under sunlight, by a so-called 'bi-thermal' method, in aqueous solution, starting from gCN powder (obtained from urea by thermal polymerization), by refluxing for 6 h. It was then allowed to cool, followed by freezing for another 6 h, and thus eight reflux–freezing steps were repeated. They achieved an improved specific area with respect to bulk, and up to 2.5 times more photocatalytic activity, than the unmodified gCN (despite a wider band gap, due to the quantum confinement effect). A CIP degradation rate (20 ppm) of 78% in 60 min was achieved with 1000 ppm photocatalyst under solar irradiation (untreated gCN 45% at the same time) (Table 2). The improvement in efficiency was attributed to an improvement in the surface area, as well as a slight improvement in the lifetime of the photogenerated loads.

2.4. Three-Dimensional Porous Structures (3D)

To conclude this subsection, the discussion focuses on the preparation of porous 3D materials. These porous materials have a larger surface area (a higher number of active sites) and better charge separation. The three main types of pores present in this type of structure are initially described: Micropores (<2 nm) allow for a greater specific surface area with an increased number of active sites. Mesopores (2–50 nm) are more conducive to the transport of reagents or solvents than the former due to a smaller distance between pores; in addition, the multiple reflection of light within mesopores improves the efficiency of light utilization. Lastly, macropores (>50 nm) have better adsorption capacity for organic macromolecules. Thus, good porous structure design achieves a good balance between light absorption and charge transport and separation, for timely optimization of quantum efficiency [95]. Among the most referenced synthesis methods are hard template with carbon nanotubes (CNNTs), soft template with ionic liquids or polymers, but also template-free methods by molecular assembly, acid treatment, or ammonium carbonate treatment, to name a few examples [1,6,45] (Figure 11).

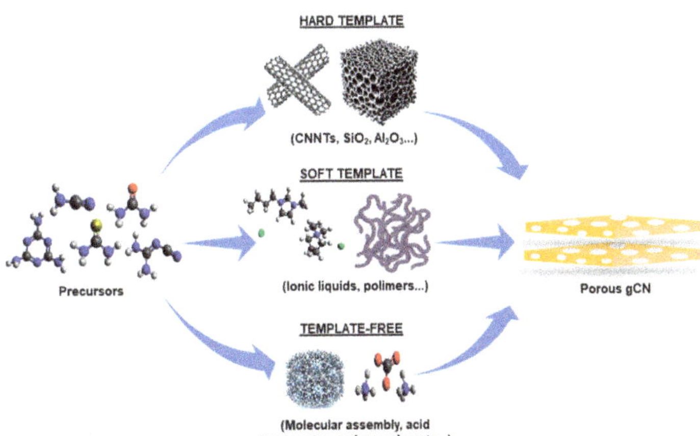

Figure 11. Different strategies to synthesize 3D porous gCN structures; hard template in the top of the figure, soft template in the middle, and template-free at the bottom of the figure.

2.4.1. Hard Template

For hard template methods, it can be mentioned that, in general, three synthesis steps are necessary:

i. Coating the chosen hard template with the gCN precursor;
ii. Treatment for conversion to gCN on the template;
iii. Removal of the template.

Silica or alumina derivatives (SiO_2 or Al_2O_3, respectively) are the most used hard templates. This method provides a structure that will favor optimal migration of the photogenerated charges. The lengthy template removal process, which often requires toxic acids such as HF or NH_4HF_2, makes this method unattractive [6,13]. A more ecofriendly alternative could be the use of salts as a hard template, e.g., ClNa [13,95,96].

Another example of the use of hard templates in the synthesis of porous gCN, which are sustainable and easily removed and recovered, is proposed by Chen et al. [97] in a paper on the use of CNTs as an ecofriendly and recyclable hard template for the photodegradation of organic pollutants. These authors prepared the photocatalyst by mixing dicyandiamide and CNTs by grinding and heat treating at 550 °C 2 h in N_2 atmosphere (condensation). The obtained product was dispersed in deionized water (DW) and treated with ultrasound to separate the CNTs. The authors achieved a surface area in the material of 103.3 $m^2 \cdot g^{-1}$, with a pore volume of 0.61 $cm^3 \cdot g^{-1}$ (compared to 10.5 $m^2 \cdot g^{-1}$ and 0.091 $cm^3 \cdot g^{-1}$, respectively, for the unmodified gCN). The band gap of the material thus obtained (2.78 eV) hardly changed with respect to the unmodified gCN (2.79 eV). A degradation of RhB (10 ppm) of 63% in 90 min (25% for gCN within the same time) was achieved with 1000 ppm of photocatalyst under visible light (300 W; Xe lamp) (Table 3). The performance of porous $g\text{-}C_3N_4$ before the fourth cycle was still 92%. The better efficiency was attributed to the improved surface area and pore volume in addition to the increase in photogenerated charge lifetime (Figure 12a,b).

Table 3. Parameters of photodegradation of different pollutants in water with porous 3D morphologies of gCN.

Morphology	Template	Photocatalyst	Area ($m^2 \cdot g^{-1}$)	Pore ($cm^3 \cdot g^{-1}$)	Gap (eV)	Pollutant (ppm)	Radiation	Efficiency (%)	Time (min)	k (min^{-1})	Refs.
3D	HARD	gCN (1000 ppm)	10.5	0.091	2.79	RhB: 10	300 W Xe lamp (Vis.)	25	90	0.004	[97]
		Porous 3D gCN (1000 ppm)	103.3	0.61	2.78			63		0.018	
	SOFT	gCN (500 ppm)	9.75	0.061	2.7	RhB: 10	Visible	25	40	0.007	[98]
		Porous gCN (P123-6) (500 ppm)	73.29	0.27	2.75			98.7		0.1	
	WITHOUT	gCN (1000 ppm)	8.36	0.02	2.6	RhB: 10	500 W Xe lamp (Vis.)	40	80	0.004	[99]
		Porous hexagonal gCN (1000 ppm)	67.3	0.32	2.43			100		0.053	
High crystalline gCN		BCN (1000 ppm)	-	-	2.67	NPX: 8	350 W Xe lamp (Vis.)	20.9	70	0.013	[100]
		CCN (1000 ppm)	-	-	2.72			98.4		0.092	

2.4.2. Soft Template

Soft template methods include the use of surfactants as template, for example, the use of ionic liquids such as 1-butyl-3-methylimidazolium dicyanamide (BmimDCN), 1-butyl-3-methylimidazolium chloride (BmimCl), or 1-butyl-3-methylimidazolium hexafluorophosphate (BmimPF6). However, the removal of this type of template can also result in gCN pores being resealed and a high amount of excess carbon being produced, thus decreasing photocatalytic activity [6,95].

An example of the use of soft templates to obtain porous gCN was proposed by Yan et al. [98], where they synthesized porous gCN using Pluronic-P123 (triblastic copolymer based on polypropylene oxide (PPO) and polyethylene oxide (PEO)) as a soft template in two steps: (i) first mixing melamine and Pluronic-P123 (in different mass ratios with respect to P123) in acidic medium and mixed with stirring for 24 h, and then solvothermal treatment in an autoclave at 130 °C/24 h; (ii) removing the P123 polymeric template by heat treatment at 550 °C for 6 h in an aerobic environment.

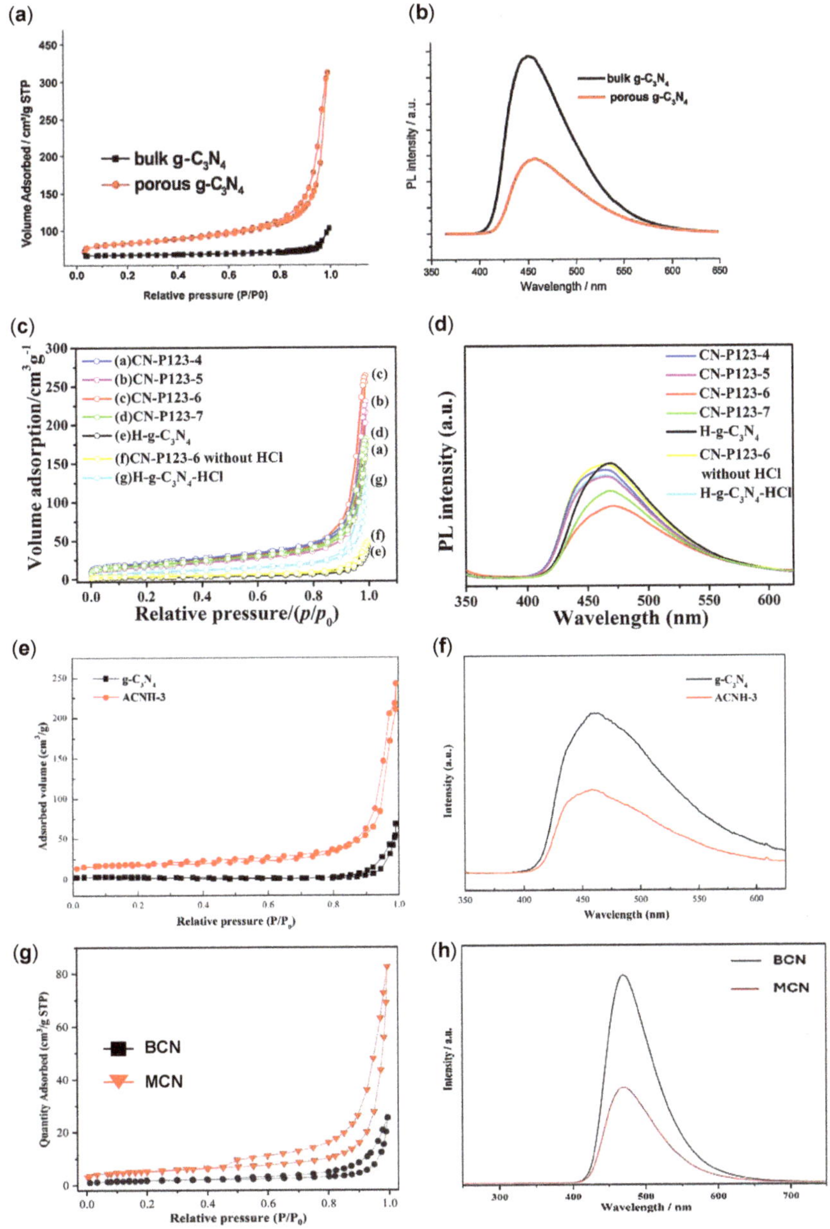

Figure 12. (**a**,**b**) adsorption isotherms (**left**) and PL spectra (**right**) of porous gCN with CNNTs as hard template [97]; (**c**,**d**) adsorption isotherms (**left**) and PL spectra (**right**) of porous gCN with Pluronic-P123 as a soft template [98]; (**e**,**f**) adsorption isotherms (**left**) and PL spectra (**right**) of the porous gCN without template by supramolecular assembly [99]; and (**g**,**h**) adsorption isotherms (**left**) and PL spectra (**right**) of porous gCN, also without template, from melamine hydrochloride [101].

The photocatalytic efficiency of this material (gCN-P123-x; 500 ppm) was evaluated with RhB (10 ppm) under visible light. The gCN-P123-6 sample presented a surface area of 73.3 $m^2 \cdot g^{-1}$ with a specific pore volume of 0.27 $cm^3 \cdot g^{-1}$ (the solvothermal synthesized gCN, without template and without acid presence, presented a surface area of 9.75 $m^2 \cdot g^{-1}$ with a specific pore volume of 0.061 $cm^3 \cdot g^{-1}$). The band gap of the sample gCN-P123-6 was 2.75 eV (gCN without template, 2.70 eV). With the porous material (gCN-P123-6), they achieved a photocatalytic degradation efficiency of RhB of 98.7% in 40 min (gCN 25% within the same time) (Table 3). The efficiency of this material before five cycles was 90%. The authors attributed the increase in surface area and pore volume, together with improved exciton separation, to improvements in RhB degradation efficiency under visible light (Figure 12c,d).

2.4.3. Template Free

As template free methods for porous 3D structures, it could be said that the molecular assembly (or supramolecular pre-organization) is the most widely used method, e.g., the use of melamine as a precursor together with triazine derivatives, which gives rise to hydrogen-bonded assemblies, such as melamine–cyanuric acid or melamine–trithiocyanuric acid, with subsequent heat treatment (by changing the solvents used, different morphologies are possible) [1,6,96,101,102].

An example of template-free porous gCN synthesis can be seen in the work of Chen et al. [99], about the synthesis of porous hexagonal prisms of gCN, wherein they were evaluated in terms of RhB degradation. The authors prepared the photocatalyst by supramolecular assembly of melamine and cyanuric acid in a solution of water and acetic acid (in different volumetric ratios of HAc:H_2O—5:1, 3:1, 1:1, 1:3, and 1:5), which, after being stirred, was first autoclaved at 180 °C/12 h, and then, after being freeze-dried, was heat-treated at 520 °C for 2 h (5 $°C \cdot min^{-1}$). They obtained a photocatalytic material (the ACNH-3 sample) with a specific area de 67.3 $m^2 \cdot g^{-1}$ and a specific pore volume of 0.37 $cm^3 \cdot g^{-1}$ (the bulk gCN 8.36 $m^2 \cdot g^{-1}$ and 0.02 $cm^3 \cdot g^{-1}$, respectively). The band gap of this material was 2.43 eV (gCN 2.6 eV). They achieved a degradation efficiency of RhB (10 ppm) of 100% in 80 min (gCN 40% at the same time) with the thus-prepared photocatalyst (1000 ppm) under visible light (Table 3). After four cycles, this sample maintains nearly 90% of its original degradation capacity. The efficiency improvements over the gCN bulk were attributed to a larger specific pore area and volume, a reduction in band gap, and an improvement in photogenerated charge separation (Figure 12e,f).

To conclude this subsection, another example of porous gCN without template could be the work of Dou et al. [101] in the degradation of amoxicillin (AMX) and cefotaxime (CFX) under visible light. The authors prepared the photocatalyst by first mixing an aqueous solution of melamine and HCl with stirring, thus obtaining melamine hydrochloride after drying. It was then heat treated at 500 °C 2 h (20 $°C \cdot min^{-1}$) in N_2 atmosphere and treated a second time at 520 °C 2 h (4 $°C \cdot min^{-1}$). The material thus prepared (MCN) showed a surface area of 18.06 $m^2 g^{-1}$ (5.64 $m^2 \cdot g^{-1}$ of the untreated gCN, BCN), with a gap almost the same as the untreated material (2.6 eV vs. 2.63 eV for the untreated material). They thus achieved a degree of degradation of 90% for AMX and 99% for CFX (for the unmodified gCN, 40% for AMX, and 80% for CFX, within the same time), both 2 ppm, in 60 min under visible light (300 W; Xe lamp) with 1000 ppm of photocatalyst (Table 3). After five cycles, the degradation of AMX still remains at 75%, also for the CFX 90% with this gCN derivative. The improvements in efficiency are mainly due to a slight improvement in surface area, and an improvement in exciton lifetime (Figure 12g,h).

2.5. High Crystalline gCN Structures

Another aspect that can be considered in this section is the crystallinity of the photocatalyst. A higher degree of crystallinity generally results in a faster migration rate of charge carriers and a reduced recombination rate of electron–hole pairs, though often at the cost of decreased surface area. The main methods to synthesize crystalline gCN derivatives are the molten salt method (the most widely used method), template-based method, solvothermal method, pure O_2-assisted method, and thermionic synthesis method [103,104]. One way to prepare highly crystalline gCN would be by post-thermal treatment of gCN with a mixture of KCl and LiCl (or NaCl/KCl) in a nitrogen atmosphere at 550 °C [1,49].

Few studies have reported the degradation of pollutants using highly crystalline gCN. One notable example is the work carried out by Wang et al., 2020 [100], which examined the degradation of pharmaceuticals and personal care products (eight different PPCPs; each at 8 ppm) using the thus-modified gCN. The authors prepared crystalline carbon nitride (CCN) in two steps: (i) by calcining 5 g of dicyandiamide at 550 °C for 3 h with a heating rate of 2.8 $°C{\cdot}min^{-1}$ to obtain the bulk material (BCN); (ii) by grinding the BCN with KCl and LiCl, followed by thermal treatment under an inert atmosphere at 550 °C for 4 h using the same heating rate. The band gaps of BCN and CCN were measured to be 2.67 eV and 2.72 eV, respectively. Using this modified material, they achieved 98.4% degradation of naproxen (NPX) in 70 min, k = 0.092 min^{-1} (the BCN, k = 0.013 min^{-1}) under visible light (350 W Xe lamp with a 420 nm cut-off filter) with 1000 ppm of CCN. The authors attributed the enhanced degradation efficiency compared to BCN to improved separation of photogenerated charges and increased O_2 adsorption on the photocatalyst surface, which leads to higher H_2O_2 production. The H_2O_2 subsequently decomposes into HO• radicals that effectively oxidize the PPCPs.

It is important to note, however, that the main application of this type of gCN modification is hydrogen (H_2) production via water splitting, or the reduction of CO_2, and less the degradation of EPs [104–108].

In gCN, the synthesis of 0D structures (quantum dots) improve charge carrier migration, enlarge the active surface area, and enhance light absorption, although their quantum confinement effect increases the band gap; their use is mainly restricted to their integration in heterojunctions (type II, Z, Schottky) or homojunctions mitigating charge recombination, optimizing applications in decontamination. One-dimensional morphologies (nanotubes, nanofibers) reduce charge diffusion distances, increase the surface area, and decrease the band gap, favoring efficient exciton separation and higher photocatalytic activity of the gCN structure. Two-dimensional structures (exfoliated nanosheets) maximize surface area, improve charge separation, and adapt the band gap by sustainable methods (thermal, ultrasound), while three-dimensional porous materials balance high surface area, reagent transport, and light exploitation by hierarchical design of micro/meso/macropores, achieving synergy between active sites and charge mobility. High crystalline gCN, mainly obtained by heat treatment with salts, optimizes carrier mobility, but may slightly reduce surface area. It is also worth mentioning that although not all the referenced studies report on the stability of these gCN derivatives after several cycles of use, those that do generally demonstrate good stability and maintain high efficiency after four or five cycles.

3. Surface Functionalization

These surface modifications related to the introduction of functional groups which, apart from enlarging the active surface (more active sites) can enhance charge separation and carrier mobility (the introduction of surface defects, such as vacancies, was considered more appropriate to deal with in the next section, dedicated to vacancies and defects). Thus, on the surface of the gCN itself, there are different groups of interest, including Brønsted and Lewis basic sites, a hydrogen bonding group, cyano groups, hydroxyl groups, etc. The nitrogen of the ring (imino) acts as Lewis basic site, while the amino groups at the edges act as Brönsted bases [15].

As an example, basic sites can be used to improve the anchoring of metal oxides during doping (as well as to improve their dispersion on the gCN surface), the adsorption of weak acids such as phenols, or cationic dyes. Such sites are favored by incomplete condensations, or by increasing the porosity. Protonation with inorganic acids not only improves charge separation but—as has been previously seen—also facilitate exfoliation. Thus, given the reactivity of the amino and imino groups in gCN, their covalent functionalization is also possible, which lowers the energy barrier for intramolecular charge transfer [4,42].

The main functional groups inserted into the gCN are as follows (Figure 13):

- Hydroxyls (introduced by hydrothermal treatment, for example). They can enhance adsorption of organic matter, and act as h^+ trapping centers; however, they can slightly widen the band gap (they are electro-donators) [41,42,109].
- Cyano groups (by treatment of gCN with $NaBH_4$ and subsequent treatment at 150–350 °C, or with potassium thiocyanate (KSCN) and subsequent heat treatment at 500 °C, among other methods) [42,110]. These functional groups can considerably reduce the band gap (they are electron-accepting groups) [109]. The introduction of the cyano group into gCN structures increases O_2 adsorption and introduces lone pair electrons into the structure that participate in photoexcitation [111].
- Carboxyl groups (by oxidation with HNO_3, for example) improve the mobility and separation of the photogenerated charges as well as improve the surface hydrophilicity by being electron acceptor groups [42,48,112,113].
- Ureido groups (introduced in the same way as cyano groups, with the addition of a subsequent treatment with HCL and stirring for a long time) also improve charge separation [42].
- Aromatic groups (introduced by thermal copolymerization with aromatic precursors such as thiophene, aniline, benzonitrile, etc.) provide more active areas as well as extend the π-conjugated system and decrease the band gap [42,114].

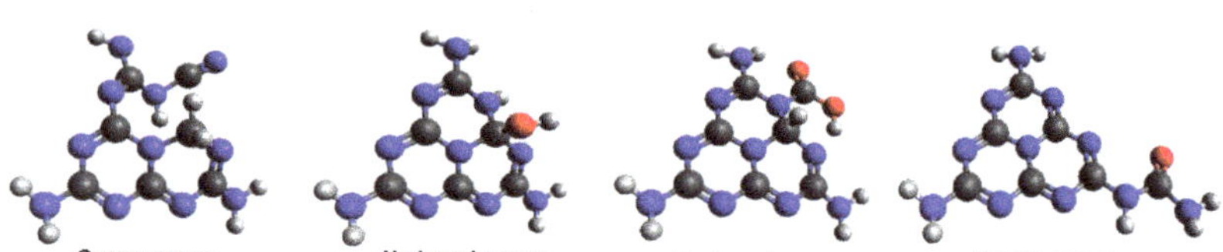

Figure 13. Four examples of functional groups in heptazine-based gCN. Carbon atoms: grey sphere, Nitrogen atoms: blue spheres, Oxigen atom: red spheres, and Hydrogen atoms: white spheres.

These functional groups can enhance the absorption of visible light, improve charge separation, increase the specific surface area, and thereby increase the number of active sites on the gCN. Such surface modifications are mainly effective in CO_2 reduction, H_2 production, ammonia synthesis, H_2O_2 production, and disinfection [42,47].

An example of modifications of surface functional groups is proposed by Zhu et al. [48] in a paper on the surface carboxylation of gCN and its impact on the adsorption and photodegradation of MB and RhB (both 15 ppm). These authors modified gCN by first mixing melamine and urea, with subsequent heat treatment at 550 °C for 4 h. The synthesized gCN was then treated with HNO_3 (10 M) for 24 h at 80 °C. The product thus obtained was separated, washed, and freeze-dried. This material had a surface area of 88.5 $m^2 \cdot g^{-1}$ and a pore volume of 0.48 $cm^3 \cdot g^{-1}$ (38.17 $m^2 \cdot g^{-1}$ and pore volume of 0.24 $cm^3 \cdot g^{-1}$ for the untreated gCN). This material achieved a gap of 2.65 eV, a degree of degradation for MB of 79%, and 62% degradation for RhB, with the thus-prepared photocatalyst (gCN-HNO_3; 200 ppm) in 180 min under visible light (300 W Xe lamp) (Table 4). The authors attributed the improvement in efficiency to a higher surface area and pore volume, and to a better separation of the photogenerated charges (Figure 14a,b).

Table 4. Parameters of photodegradation of pollutants in water with gCN with -COOH, -C≡N, and thiophene groups.

Group	Photocatalyst	Area ($m^2 \cdot g^{-1}$)	Gap (eV)	Pollutant (ppm)	Radiation Xe Lamp (Vis.)	Efficiency (%)	Time (min)	k (min^{-1})	Refs.
-COOH	gCN (200 ppm)	38.2	-	MB and RhB: 15	300 W	-	180	-	[48]
	gCN-HNO_3 (200 ppm)	88.6	2.65			MB: 79 RhB: 62		RhB: 0.026; MB: 0.033	
-C≡N	gCN (1000 ppm)	21.49	2.72	RhB and TC: 15	300 W	100	30	RhB: 0.025; TC: 0.029	[115]
	Cyano-gCN (1000 ppm)	51.34	2.63			100		RhB and TC: 0.099	
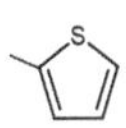	gCN (500 ppm)	16.6	2.61	RhB: 5	500 W	45	90	0.0072	[116]
	Thiophen-gCN (500 ppm)	78.4	2.64			96		0.036	

Another example in which functional groups were introduced into gCN was the work of Razavi-Esfal M. et al. [115], in which they prepared porous gCN rich in cyano groups and evaluated the degradation of RhB and tetracycline (TC) under visible light. The authors prepared the photocatalyst by mixing melamine with SiO_2 as a hard template and then heat-treated it at 550 °C/3 h (2.3 °C·min^{-1}). After eliminating the SiO_2 template with HF, the material was washed and dried. The modified photocatalyst presented a specific surface area of 51.34 $m^2 \cdot g^{-1}$ and a pore volume of 0.23 $cm^3 \cdot g^{-1}$, much higher than those of gCN (21.49 $m^2 \cdot g^{-1}$ and 0.126 $cm^3 \cdot g^{-1}$, respectively). The effective band gap of the group-modified photocatalyst was 2.63 eV (2.72 eV for unmodified gCN). With this material, they were able to degrade RhB (15 ppm) by 100% in 30 min (gCN by 40% in 90 min) and TC (15 ppm) by 100% in 30 min (gCN by 27% in 90 min) with 1000 ppm of photocatalyst under visible light (300 W Xe lamp) (Table 4). According to the authors, the improvement in efficiency over the unmodified material is mainly due to improvements in specific area and porosity, reduction in band gap, and separation of photogenerated charges (Figure 14c,d).

A final example of gCN derivatives with functional groups is given by Xu et al. [116], with work on functionalization with aromatic groups, specifically thiophene, which evaluated the degradation of RhB, MO, MB, and CV (5 ppm) under visible light (500 W Xe lamp). Thiophene grafted in a gCN porous nanosheet photocatalyst was prepared employing a one-step method by mixing cyanuric acid and melamine in DW for 12 h, which after being separated and dried were mixed with different amounts of 3-thiophenecarboxylic acid and heat-treated at 550 °C/1 h (3 °C·min^{-1}). This material presented a surface area of 74.8 $m^2 \cdot g^{-1}$ and a gap of 2.64 eV (gCN had a surface area of 16.6 $m^2 \cdot g^{-1}$ and a gap of 2.61 eV). Thus, the sample with 30 mg of 3-thiophenecarboxylic acid and a heating ramp of 3 °C (3-CM-CN-Th A30) degraded RhB 96.1% in 90 min (gCN only 45% at the same time) under visible radiation with 500 ppm of photocatalyst. They also achieved a degradation efficiency of 53.2% of MO, 89.4% of MB, and 88.8% of CV (Table 4). The RhB removal efficiency remained at 89.22% after five cycles of use. This improvement in efficiency is attributed to the increase in specific pore area and volume, as well as to the enhanced separation of the photogenerated charges (Figure 14e,f).

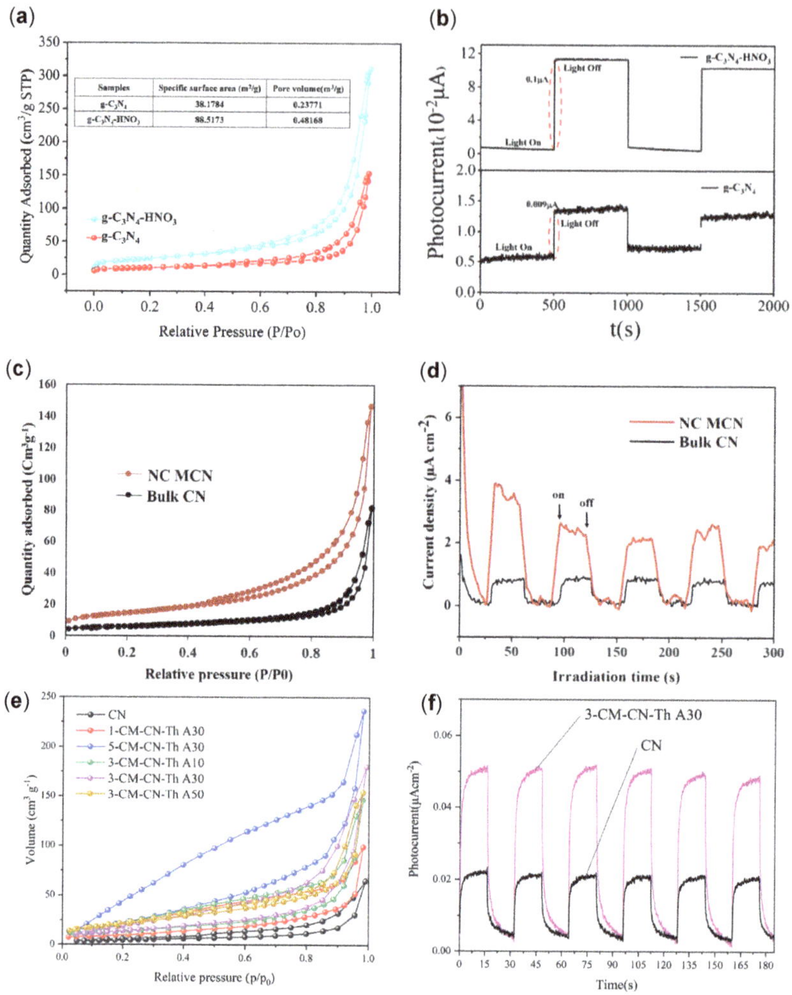

Figure 14. (**a**,**b**) BET area (**left**) and transient photocurrent (**right**) of bulk and gCN treated with HNO_3 for introduction of -COOH groups [48]; (**c**,**d**) BET area (**left**) and transient photocurrent (**right**) for bulk and gCN with cyano groups [115]; and (**e**,**f**) BET area (**left**) and transient photocurrent (**right**) for bulk and gCN functionalized with thiophene groups synthesized at different ramp-up rates (x) and with different mounts of thiophene (y) in mg (x-CM-CN-Th Ay) [116].

Thus, the introduction of hydroxyl groups (-OH) improves the adsorption of organic matter and acts as a hole trap (h^+), although it can slightly increase the band gap as they are electro-donating groups. The cyano groups (-C≡N) reduce the band gap (electron acceptor effect), favor the adsorption of O_2, and introduce free electron pairs, optimizing photoexcitation. Carboxyl (-COOH) increases mobility and charge separation, as well as improve surface hydrophilicity. Ureido groups enhance charge separation, while aromatic groups (e.g., thiophene) extend the π-conjugated system, reduce the band gap, and increase the number of active sites. These modifications increase surface area, tailor electronic properties, and improve pollutant degradation efficiency, as demonstrated by studies with RhB, MB, and TC, where functionalization with -COOH, -C≡N, and aryls achieved >90% degradations under visible light, outperforming unmodified gCN. Among the three references provided in this section, only one reports a stability study after several treatment cycles (RhB degradation using thiophene-functionalized gCN), showing good material stability after five cycles.

4. Electronic Structure Optimization

Modifications aimed at modifying the band structure of gCN are not only aimed at improving absorption in the visible spectrum, but also at optimizing the redox potentials to improve photocatalytic activity. Thus, several possibilities have been considered for this purpose, such as, for example, the design of vacancies, or the different types of doping (metallic, non-metallic), which are examined in the next section.

4.1. Defect Engineering

Vacancies refer to the absences in lattice positions, in this case of gCN. It can be noted that there are two main types of vacancies in gCN: nitrogen vacancies, N_v (anionic), and carbon vacancies, C_v (cationic).

4.1.1. Nitrogen Vacancies (N_v)

The formation of nitrogen vacancies (N_v) in gCN plays an important role both in charge separation (increase of exciton lifetime) and in influencing the band gap. The insertion of nitrogen vacancies can be carried out by controlling the condensation temperature, by reduction in H_2 atmosphere, by hydrothermal treatment in the presence of $(NH_4)_2S_2O_3$, or by heat treatment in alkaline medium (e.g., with KOH, NaOH, or $Ba(OH)_2$). By thermal synthesis in an alkaline medium, controlling the base/precursor ratio, the gap reduction can be modulated. The gap decreases with increasing base/precursor ratio, mainly due to the reduction in conduction band potential towards less negative values [4,6].

Similar conclusions were reached by Katsumata et al. [51] in a study on the degradation of BPA with N_v-gCN. These authors introduced the N_v, mixing in a first step KOH and urea (in different mass ratios) in distilled water with stirring, and then drying. The powder thus obtained was mixed with OA in a solution of H_2O/ethanol. The solvent was then evaporated, and the resulting solid was calcined at 600 °C for 2 h (2.5 $°C{\cdot}min^{-1}$). This material (KOH-OA-gCN) presented a band gap of 2.6 eV, with a surface area of only 29 $m^2{\cdot}g^{-1}$ (the untreated gCN had a gap of 2.85 eV and a specific area of 85 $m^2{\cdot}g^{-1}$). This material (600 ppm) achieved 90% degradation of 10 ppm BPA in 150 min under visible radiation (300 W Xe lamp), while gCN only achieved 25% (Table 5). After five usage cycles, the photocatalytic activity dropped significantly to 80%, which the authors attributed primarily to the loss of ultrafine particles during the recycling process. The authors also concluded that the better efficiency was due to the reduction in band gap and a better separation of generated charges (Figure 15a,b) and inferred that the addition of KOH leads to a substantial loss of specific area due to a decrease in mesoporosity.

Table 5. Parameters of photodegradation of pollutants in water with gCN with vacancies.

Vacancy	Photocatalyst	Area ($m^2{\cdot}g^{-1}$)	Gap (eV)	Pollutant (ppm)	Radiation	Efficiency (%)	Time (min)	k (min^{-1})	Refs.
N_v	gCN (600 ppm)	2.85	85	BPA:10	300 W Xe lamp (Vis.)	25	150	0.0021	[51]
	KOH-OA-gCN (600 ppm)	2.6	29			90		0.0147	
	gCN (600 ppm)	75.7	2.79	OTC:20	24 W; LED (Vis.)	45.8	135	0.0046	[117]
	N_vrCN (600 ppm)	64.3	2.73			92.5		0.018	
	gCN (250 ppm)	85.4	2.75	TC: 15 SMX: 5	300 W Xe lamp (Vis.)	TC: 40 SMX: 52	TC: 60 SMX: 120	0.009 0.025	[118]
	ODH-CN-2 (250 ppm)	108.2	2.61			TC: 79.9 SMX: 91.5		0.007 0.020	

Table 5. *Cont.*

Vacancy	Photocatalyst	Area ($m^2 \cdot g^{-1}$)	Gap (eV)	Pollutant (ppm)	Radiation	Efficiency (%)	Time (min)	k (min^{-1})	Refs.
C_v	gCN (300 ppm)	30.1	2.76	BPA: 10	350 W Xe lamp (Vis.)	78	120	0.003	[43]
	C_v-gCN (300 ppm)	14.7	2.65			90		0.006	
	gCN (250 ppm)	272	2.94	TC; 16	300 W halogen lamp (Vis.) + ul (600 W/40 kHz)	30	60	0.0003	[44]
	C_v-gCN-20 (250 ppm)	331	2.9			96		0.0010	
	gCN (1000 ppm)	79.7	2.73	4-clorophenol: 10	300 W Xe lamp (Vis.)	33.8	120	0.003	[119]
	C_v-gCN-575 (1000 ppm)	64.2	2.71			60.1		0.008	

Likewise, Lee et al. [117] carried out the synthesis of N_v-gCN by alkaline solvothermal method. They used a suspension of Mg-gCN in EG together with KOH, heated at 160 °C for 12 h, in a work on the degradation of OTC. The modified material had a band gap of 2.73 eV (gCN 2.79 eV and MgCN 2.89 eV), and a surface area of 64.3 $m^2 \cdot g^{-1}$, with a pore volume of 0.16 $cm^3 \cdot g^{-1}$ (75.7 $m^2 \cdot g^{-1}$, 0.38 $cm^3 \cdot g^{-1}$, and 87.3 $m^2 \cdot g^{-1}$, 0.43 $cm^3 \cdot g^{-1}$, for gCN and Mg-CN, respectively). This photocatalyst thus prepared achieved an OTC degradation (20 ppm) of 92.5% in 135 min (gCN 45.8% within the same time), under visible light with 600 ppm of photocatalyst (Table 5). After three cycles, the removal efficiency of OTC was still 82%. The better efficiency was attributed to the reduction in band gap and a better separation of generated charges (Figure 15c,d); also, the authors concluded that the reduction in surface area is mainly due to the decrease in pore volume in N_v-gCN.

And as a final example of this type of modification with N_v in the gCN structure, it is worth mentioning the research by Wang Y. et al. [118] on the degradation of tetracycline hydrochloride (TC-HCl) and sulfamethoxazole (SMX) with the thus-modified gCN. The authors synthesized the photocatalytic material by mixing urea and oxalyl dihydrazide (ODH) in different amounts and heat-treating them (polymerization in an aerobic environment) at 550 °C/4 h (5 °C·min^{-1}). The ODH-CN2 (Urea:ODH, 100:1) sample achieved a surface area of 108.2 $m^2 \cdot g^{-1}$ (the unmodified gCN 85.4 $m^2 \cdot g^{-1}$) and a band gap of 2.61 eV (the gCN 2.75 eV). This photocatalyst thus modified can degrade to 79.9% of TC-HCl (15 ppm) in 60 min and 91.5% of SMX (5 ppm) in 120 min under visible radiation with 250 ppm of photocatalyst (Table 5). The photocatalytic degradation efficiencies of TC-HCl and SMZ remained at 70.85% and 79.33%, respectively, after five cycles. The improvement in efficiency over the unmodified material is due (according to the authors)—as in the previous examples—to a higher surface area, to a narrowed band gap, and an improvement in photogenerated charge separation (Figure 15e,f).

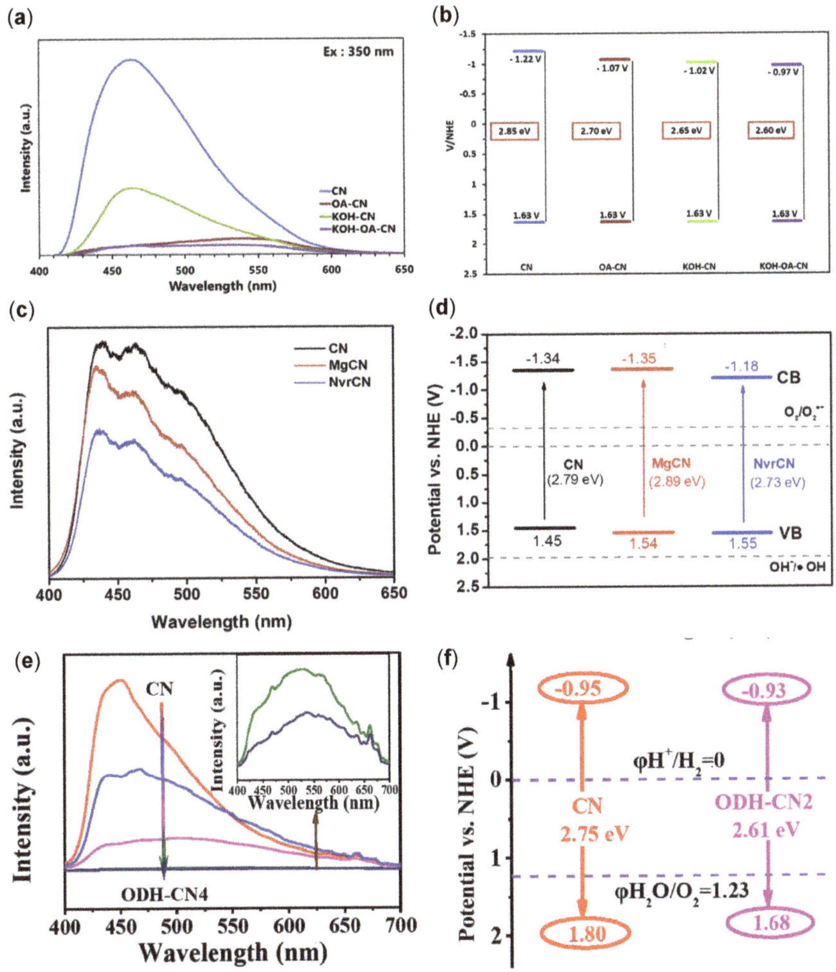

Figure 15. (**a**,**b**) PL spectra (**left**) and band gap diagrams (**right**) of N_V-gCN introduced by mixing urea and KOH and with oxamide (OA) [51]; (**c**,**d**) PL spectra (**left**) and band gap diagrams (**right**) of bulk (CN), Mg-doped gCN (MgCN), and N vacancy rich gCN (N_VrCN) by solvothermal method [117]; and (**e**,**f**) PL spectra (**left**) and band gap diagrams (**right**) for gCN (CN) and N_V-gCN by mixing urea and oxalyl dihydrazide (ODH-CN-2) [118].

4.1.2. Carbon Vacancies (C_V)

For C_V-modified gCN, it can be mentioned that these vacancies can serve as a reservoir of photogenerated electrons, thereby inhibiting the recombination of h^+ and e^-. These vacancies can also serve as electron transfer centers to the adsorbed molecular oxygen, favoring the production of superoxide radicals ($O_2^{\bullet-}$), thus improving photocatalytic activity [43,44].

As proposed by Liang et al. [43], these vacancies may be introduced by thermal treatment of melamine in two steps:

i. Calcination in an air muffle furnace at 550 °C 2 h;
ii. Pyrolysis at 520 °C in a tube furnace in Ar atmosphere (2 h).

Under these conditions, the obtained material exhibited a surface area of 14.7 $m^2 \cdot g^{-1}$ and a gap of 2.65 eV (the untreated gCN 30.1 $m^2 \cdot g^{-1}$ and a band gap of 2.76 eV), and the degradation of BPA (10 ppm) was 1.65 times better with the C_V-gCN (90% degradation in 120 min) than with the untreated gCN (78% in the same time) using 300 ppm photocatalyst under visible light (350 W Xe lamp) (Table 5). After five cyclic experiments, the photocatalytic ability of C_V-gCN was almost unchanged. The authors concluded that, despite the decrease in surface area due to the Ar atmosphere treatment, after the introduction of the carbon vacancy, the recombination rate of photogenerated h^+ and e^- is greatly decreased. In addition, the carbon vacancies, which serve as conversion centers, transferred most of the photogenerated electrons, trapped there, to the absorbed O_2 facilitating the generation of radicals (Figure 16a,b).

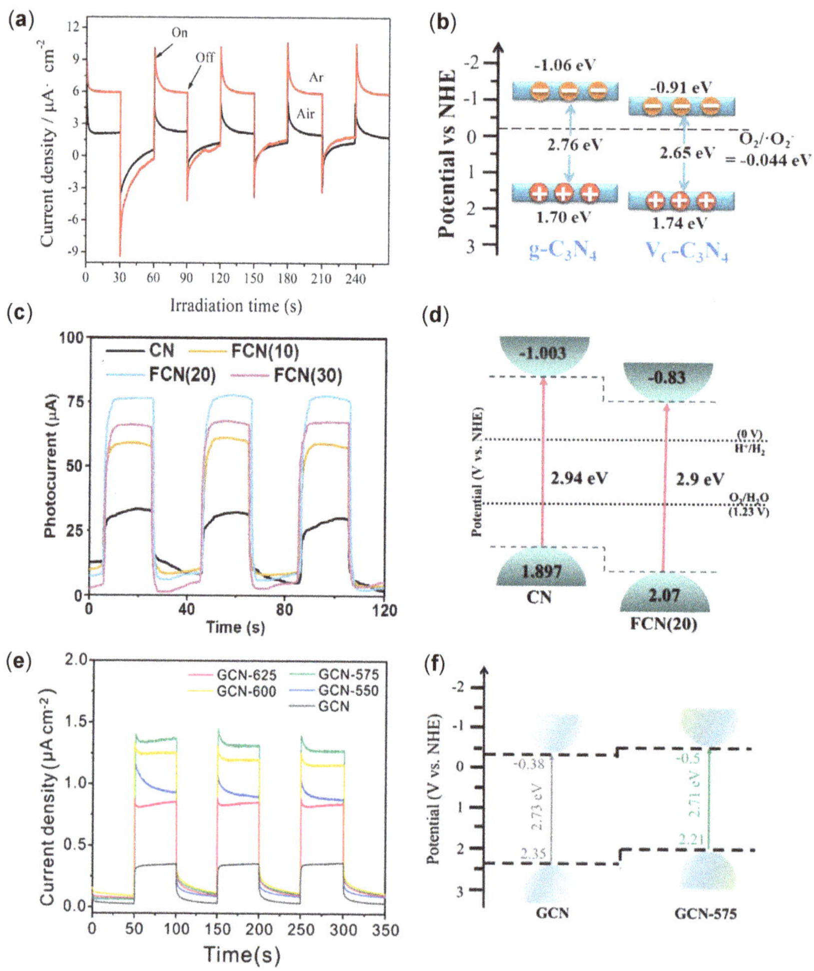

Figure 16. (**a**,**b**) transient photocurrent (**left**) and band gap diagrams (**right**) of gCN, and C_V-gCN introduced by thermal treatment of melamine [43]; (**c**,**d**) transient photocurrent (**left**) and band gap diagrams (**right**) of gCN, and C_V-gCN from urea and formaldehyde in different amounts (FCN) [44]; and (**e**,**f**) transient photocurrent (**left**) and band gap diagrams (**right**) of gCN, and C_V-gCN from gCN treated with powdered Mg at different T [119].

Another example of C_V-gCN would be, for example, the one proposed by Preeyanghaa et al. [44], in the TC degradation by sono-photocatalysis with C_V-gCN nanosheets. They prepared the photocatalyst in two steps: (i) they first melted the urea at 140 °C and added formaldehyde (different amounts) to the molten mixture, and then (ii) heat-treated both reagents at 550 °C for 3 h (5 °C·min^{-1}). Thus, they obtained material with a band gap of 2.9 eV for the C_V-gCN-20 sample (2.94 eV for gCN). They also achieved a specific area of 331 m^2·g^{-1} (272 m^2·g^{-1} for the gCN without formaldehyde with the same treatment), reaching a degree of TC degradation (16 ppm) of 96% in 60 min, under visible light and ultrasound, with 250 ppm of photocatalyst (Table 5). The sono-photocatalytic degradation efficiency of TC remains almost unchanged even after five consecutive cycles. The better efficiency of this photocatalyst thus modified was mainly due to an increased surface area and for an improvement in the separation of excitons (Figure 16c,d).

A final example of this type of vacancy in the gCN structure could be the work of Huang Z. et al. [119], on the introduction of C_v in gCN for the degradation of 4-chlorophenol under visible light. The authors synthesized the thus-modified gCN in three steps: (i) calcining urea at 550 °C/4 h (2.5 °C min^{-1}) in a flask; (ii) a second heat treatment at 500 °C/2 h (5 °C min^{-1}) for thermal exfoliation; and (iii) for the introduction of C_v, they mixed the obtained gCN (bulk and exfoliated) with powdered Mg and treated it in a quartz tube furnace in an inert environment at temperatures between 550 and 625 °C (10 °C min^{-1}). They obtained a sample (gCN-575) with a band gap of 2.71 eV (gCN a band gap of 2.73) and a surface area of 64.2 m^2·g^{-1} (gCN 79.7 m^2·g^{-1}). The same sample achieved a 4-chlorophenol degradation efficiency (10 ppm) of 60.1% in 120 min (gCN 33.8% within the same time), with 1000 ppm photocatalyst under visible radiation (Table 5). After four consecutive cycles, GCN-575 exhibited a slight decrease in performance. The authors concluded that the improvement in efficiency, despite the decrease in surface area, is due to an improvement in the separation of photogenerated charges (Figure 16e,f).

In general, both defects (N_v and C_v vacancies) improve charge separation and introduce additional energy levels in the band gap that improve the absorption of the visible spectrum. Both vacancies promote the presence of unsaturated centers such as cyano groups (carbon vacancies) and carbonyl groups (nitrogen vacancies), to give just a couple of examples, which expand the number of active sites in the material. Optimization of the electronic structure by defect engineering allows for tuning both the visible light absorption and the redox potentials of gCN. The introduction of nitrogen vacancies helps to reduce the energy gap and improve charge separation. On the other hand, carbon vacancies act as electron reservoirs, decreasing recombination and favoring electron transfer to adsorbed oxygen to form radicals. Together, these defects generate additional energy levels that optimize carrier mobility and enhance photocatalytic efficiency. gCN with Nv and Cv vacancies remains stable after several degradation cycles, indicating that these modifications are stable in the reaction medium. These strategies are key to tailoring gCN for applications in pollutant degradation.

4.2. Element Doping Strategies

Generically, doping in gCN is defined as the process of replacing atoms in the lattice positions (non-metallic doping; anionic) or introducing them into lattice interstices (metallic doping; cationic). Both influence the electronic structure of the gCN, improving not only the absorption range (due to the introduction of intermediate energy levels in the band gap, as well as vacancies) but also enhancing the lifetime of the charge carriers and improving their separation. This doping can also compensate for the increased gap produced in nanostructures (e.g., nanosheets) by the quantum confinement effect [1,13].

4.2.1. Non-Metal Doping

Non-metal doping, as well as the introduction of vacancies, is a promising strategy to prepare metal-free photocatalysts. This type of doping (usually with B, P, S, C, N, O, and halogens, such as F, Cl, Br, and I) can modify the structure, improve conductivity and charge separation, and reduce the band gap. For electron-poor dopants, such as B, the intermediate energy levels in the band gap are closer to the valence band, increasing the p-type conductivity; for electron-rich dopants, such as O or S, increasing the n-type conductivity, the introduced levels are closer to the conduction band [2,13,40]. In addition, due to the high ionization energies and electronegativity of non-metals, they can quickly form covalent bonds with other compounds, facilitating the functionalization of the photocatalyst [13].

Doping with O

This subsection begins with a description of oxygen (O) doping. There are several possibilities for doping gCN with O, for example, Deng et al. [120] refer to their preparation in one step by thermal copolymerization (calcination at 550 °C 3 h with a ramp of 5 °C·min^{-1}) starting from urea (10 g) and different amounts of oxalic acid, $H_2C_2O_4$. The photocatalyst thus obtained has a considerably reduced band gap with respect to untreated gCN (1.93 eV for the doped material and 2.65 eV for the untreated gCN) (Figure 17a), which, according to the authors, ostensibly improves the degradation capacity of the antibiotic lincomycin (100 ppm), 99% in 180 min (with the gCN at 50% within the same time), with 300 ppm of photocatalyst under visible light (Table 6). This O-doped gCN derivative also maintains its efficiency in lincomycin degradation even after three consecutive degradation cycles.

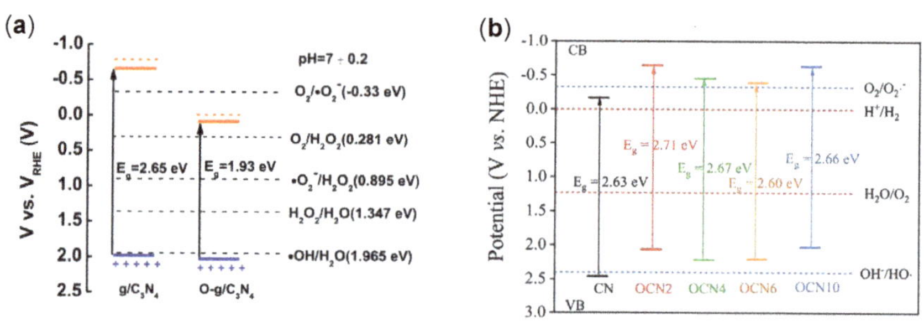

Figure 17. (**a**) Potentials and band gaps of bulk gCN and O-gCN by thermal copolymerization of urea and oxalic acid [120]; (**b**) potentials and band gaps of bulk gCN and O-gCN samples from urea at different times of polymerization [121].

Table 6. Parameters of photodegradation of pollutants in water with non-metal-doped gCN.

Dopant	Photocatalyst	Area ($m^2 \cdot g^{-1}$)	Gap (eV)	Pollutant (ppm)	Radiation	Efficiency (%)	Time (min)	k (min^{-1})	Refs.
O	gCN (300 ppm)	-	2.65	Lincomycin: 100	90 W LED (Vis.)	45	180	0.005	[120]
	O-gCN (300 ppm)	-	1.93			99		0.034	
	gCN (200 ppm)	62.50	2.63	BPA: 10	300 W Xe lamp (Vis.)	14	120	0.001	[121]
	O-gCN (200 ppm)	70.32	2.60			99		0.032	
P	gCN (1000 ppm)	26.86	2.69	RhB: 10	300 W, Xe lamp (Vis)	100	30	-	[122]
	P-gCN (1000 ppm)	40.5	2.84				10	-	
	gCN (100 ppm)	73.8	-	RhB: 20	300 W Xe lamp (Vis.)	64.2	70	0.039	[123]
	P-gCN (100 ppm)	202.9	-			99.5		0.120	

Table 6. *Cont.*

Dopant	Photocatalyst	Area ($m^2 \cdot g^{-1}$)	Gap (eV)	Pollutant (ppm)	Radiation	Efficiency (%)	Time (min)	k (min^{-1})	Refs.
S	gCN (100 ppm for MB and 200 ppm for TC)	11	2.55	MB and TC: 20	300 W Xe lamp (Vis.)	MB: 4; TC: 10	MB: 300 TC: 240	MB: 0.00014 TC: 0.0003	[124]
	S-gCN (100 ppm for MB and 200 ppm for TC)	15	1.83			MB: 60; TC: 89		MB: 0.0014 TC: 0.037	
	gCN (1000 ppm)	-	-	OTC: 10	300 W Xe lamp (Vis.)	57.1	40	-	[125]
	S-gCN (1000 ppm)	31.2	2.83			93.3		0.133	
B	gCN (68 mg gCN/62 mg CFs)	-	2.71	RhB: 5	8 W LED lamp (Vis.)	82	120	0.015	[126]
	B-gCN (68 mg B-gCN/62 mg CFs)	-	2.69			95		0.024	
	gCN (500 ppm)	17.5	2.73	RhB: 2	500 W Xe lamp (Vis.)	-	30	0.026	[127]
	B-gCN (500 ppm)	105.1	2.70			97		0.086	
Halogen	gCN (500 ppm)	42.3	2.75	TC: 10	300 W Xe lamp (Vis.)	32	120	0.004	[128]
	Cl-gCN (500 ppm)	114.4	2.7			92		0.02	
	gCN (1000 ppm)	-	2.78	OTC: 10	35 W LED lamp (Vis.)	30	150	0.004	[129]
	Br-gCN (1000 ppm)	-	2.75			75		0.018	
	Cl-gCN (1000 ppm)	-	2.73			75		0.017	
C	gCN (400 ppm)	-	2.77	TC: 30	35 W LED lamp (Vis.)	40	60	0.01	[130]
	C-gCN (400 ppm)	-	2.71			77		0.03	
	gCN (1000 ppm)	43	2.88	BPA: 10	300 W Xe lamp (Vis.)	25	60	0.005	[131]
	C-gCN (1000 ppm)	85	2.19			96		0.053	
N	gCN (500 ppm)	18.4	2.51	TC: 10	300 W Xe lamp (Vis.)	52.2	60	0.013	[132]
	N-gCN (500 ppm)	74.79	2.47			81.7		0.026	
	gCN (1000 ppm)	76.69	2.51	PhOH: 10	300 W Xe lamp (Vis.)	70.1	180	0.002	[133]
	N-gCN (1000 ppm)	72.26	1.82			37.6		0.006	

Long et al. [121], in a work on the degradation of BPA with O-doped gCN, point out the synthesis by thermal polymerization (at 550 °C/4 h, 5 °C·min^{-1}) of this material without adding any precursor other than urea, by removing the gCN at room temperature after 300 °C (on cooling) until reaching room temperature. With heat treatment (at 550 °C) for different times (2, 4, 6, and 10 h), different degrees of doping were achieved by air cooling to room temperature. The sample treated at 550 °C/6 h, OCN-6, reached a surface area of 70.32 $m^2 \cdot g^{-1}$ (gCN 62.5 $m^2 \cdot g^{-1}$) and a band gap of 2.6 eV (gCN 2.63 eV). The best result in BPA degradation (10 ppm) was obtained with OCN6 (200 ppm), 99% in 120 min (gCN 14% within the same time) under visible light (Table 6). The degradation efficiency of BPA remained above 96% after four cycles. The authors also concluded that the improved efficiency of the O-doped material with respect to bulk is mainly due to an improvement in surface area, but above all to a substantial improvement in the mobility and separation of the photogenerated charges (Figure 17b).

Doping with P

Doping gCN with P improves the delocalization of the π cloud, improving the transfer of photogenerated electrons [40], improving visible light absorption, the separation of charge carriers, and overall photocatalytic activity [1].

A good way to introduce P into the gCN structure is by thermal copolymerization (in air atmosphere) at different temperatures (550–650 °C) with guanidinium hydrochloride (CH_5N_3-HCl; GndHCl) as a precursor of gCN, and hexachlorocyclotriphosphazene (HCCP), in different percentages with respect to gCN, as a cheap and sustainable source of P, with excellent photocatalytic activity in the degradation of organic dyes (RhB), as Zhou et al. [122] concluded in a paper on the degradation of this dye with P-gCN under visible light. These authors controlled the degree of doping by adjusting the HCCP amounts and the calcination temperature. This material shows a specific area for the P10-600 sample of 40.5 $m^2 \cdot g^{-1}$ (26.86 $m^2 \cdot g^{-1}$ for gCN) and a band gap of 2.84 eV (2.69 eV for gCN). The photocatalyst thus modified achieved 100% degradation of RhB (10 ppm) in 10 min with the same sample (gCN 100% in 40 min) at a concentration of 1000 ppm under visible light (Table 6). After four cycles of degradation, this photocatalyst still maintains 75% degradation efficiency. Overall, the authors attributed this improvement in efficiency over gCN to the effective suppression of electron and hole recombination photogenerated and a better redox capacity (Figure 18a).

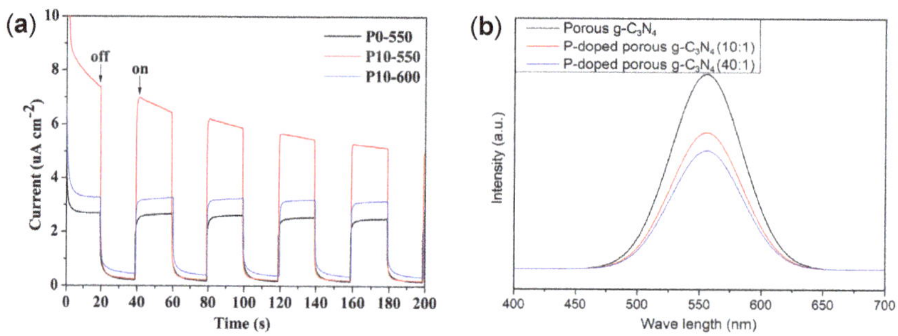

Figure 18. (**a**) Transient photocurrent of P-gCN doped with HCCP at different T [122]; (**b**) PL spectra of the P-gCN doped with sodium hypophosphite (SHP) [123].

In other study, Li et al. [123] prepared P-doped gCN in three steps, such that (i) synthesis of gCN by thermal polymerization of urea at 500 °C/4 h (heating rate: 2 °C$\cdot$min^{-1}); (ii) porous gCN was synthetized with a new thermal treatment at 500 °C/2 h (5 °C$\cdot$min^{-1}); (iii) and the subsequent doping of the gCN was made with sodium hypophosphite, SHP, prepared with different mass ratio for SHP to gCN, in a tube furnace at 350 °C/1 h. The resulting material exhibited a surface area of 202.9 $m^2 \cdot g^{-1}$ (compared for gCN 73.8 $m^2 \cdot g^{-1}$). The photocatalyst thus doped degraded 99.5% of the RhB (20 ppm) in 70 min (gCN 64.2% within the same time) with 100 ppm of photocatalyst under visible light (Table 6). The improvement in efficiency of the sample P-gCN (40:1) over undoped porous gCN is mainly due to its higher surface area and improved charge separation (Figure 18b).

P-doping can also be achieved by thermal condensation of gCN precursors with adenosine phosphates, sodium hypophosphite (NaH_2PO_2), etc. [54] Co-doping with other elements such as O or S is also interesting and gives good results in the degradation of pollutants in water bodies [134,135].

Doping with S

Another widely used non-metallic dopant is sulfur, S, which, like the non-metallic dopants mentioned above, enhances light absorption and improves charge mobility and separation [52–54]. The insertion of S into the gCN structure can be carried out by different techniques, such as by heat treatment of the gCN precursor in an SH_2 atmosphere, thermal polymerization (in air atmosphere) with S-containing precursors (thiourea, trithiocyanuric acid, thioacetamide, etc.), or by copolymerization of the precursors with S-containing compounds (benzyl disulfide, H_2SO_4, etc.) [54].

A representative example of S-gCN synthesis was reported by Guan et al. [124], who used dicyandiamide as the main precursor and trithiocyanuric acid (and thiourea) as S sources. The thermal treatment was carried out in a molten salts mixture (LiBr/KCl) at 500 °C for 3 h. By controlling the percentage ratio of trithiocyanuric acid to dicyandiamide, they adjusted the degree of doping. Thus, for a trithiocyanuric acid/dithiodiamide ratio equal to 5%, the material shows a gap in the material of 1.83 eV (gCN a gap of 2.55 eV) and a surface area of 15 $m^2 \cdot g^{-1}$ (gCN 11 $m^2 \cdot g^{-1}$). The material thus prepared achieved an improvement in photocatalytic activity in the degradation of MB and TC of 10 and 20 times, respectively; the effectiveness of the undoped gCN, i.e., 60% in 300 min for MB and 89% in 240 min for TC (gCN without any additive 4% for MB and 10% for TC, within the same time); with 100 ppm photocatalyst for MB (20 ppm) and 200 ppm photocatalyst for TC solution (20 ppm) under visible light (Table 6). The improvements in MB and TC degradation efficiency are mainly due to a slight improvement in surface area (at the expense of an improvement in microporosity), an improvement in charge separation, and an effective reduction in band gap (Figure 19a,b).

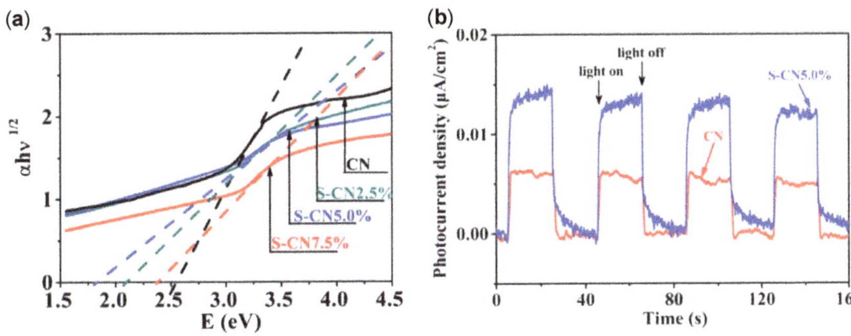

Figure 19. (**a**) Kubelka–Munk plot of bulk (black line) and gCN doped with different trithiocyanuric acid/dicyandiamide ratios (green 2.5%, blue 5%, and red 7.5%) and (**b**) transient photocurrent of bulk (red line) and S-gCN synthesized with a trithiocyanuric acid/dicyandiamide ratio of 5% [124].

A 'green' option to prepare S-gCN was proposed by Dou et al. [125] in a study on the degradation of OTC under visible light with the nanosheets of the thus-doped gCN. These authors proposed a synthesis by thermal polymerization of thiourea in three steps: (i) the precursor was heated from room temperature to 259 °C and held for 30 min; (ii) then to 426 °C and held for 30 min; and (iii) finally to 550 °C and held for 4 h. The material thus prepared had a gap of 2.83 eV and a surface area of 31.17 $m^2 \cdot g^{-1}$. The authors succeeded in 40 min to degrade almost completely (93.3%) the OTC (10 ppm) with 1000 ppm photocatalyst under visible light (Table 6). This modified gCN derivative retained 79% of the OTC degradation efficiency after four consecutive degradation cycles. The improvement in degradation compared to the unmodified material is mainly due to enhanced charge separation.

Doping with B

Doping with boron (B), like its non-metallic predecessors, improves both the surface area and the band gap. The doped gCN changes from an n-type semiconductor to a p-type semiconductor. As before, it is the introduction of intermediate energy states that reduces the gap [1]. Thermal condensation is the most common means used for this purpose. Thus, among the most frequent reagents as a source of boron are boron oxide, B_2O_3, borazane (BNH_6), boric acid ($B(OH)_3$), or sodium or potassium borohydrides ($NaBH_4$ or KBH_4) [1,2].

Among the most reported syntheses, in the different works consulted on the subject, mention should be made of doping by thermal synthesis in a muffle furnace, starting from melamine and $B(OH)_3$, as carried out by Lei et al. [126]. They prepared B-gCN supported on textile carbon fibers on RhB degradation, starting from melamine (5 g) and different quantities of boric acid in ethanol (EtOH), and coating with this solution, by immersion, carbon textile fibers (CFs) with subsequent thermal polymerization (calcination) at 570 °C for 2 h. The band gap for this sample (B-gCN-0.4) only reached a value of 2.69 eV (2.71 eV for the undoped and equally supported material). With the photocatalyst thus doped (B-gCN-0.4), a degradation of RhB (5 ppm) of 95% was achieved in 120 min (the undoped gCN 82%) with 68 mg of B-gCN on 62 mg of CFs under visible light (8 W LED lamp) (Table 6). This material not only achieves good degradability but also maintains the degree of degradation after 10 cycles of use. The improvement in activity is mainly due, according to the authors, to an improvement in charge mobility, charge separation, and porosity (Figure 20).

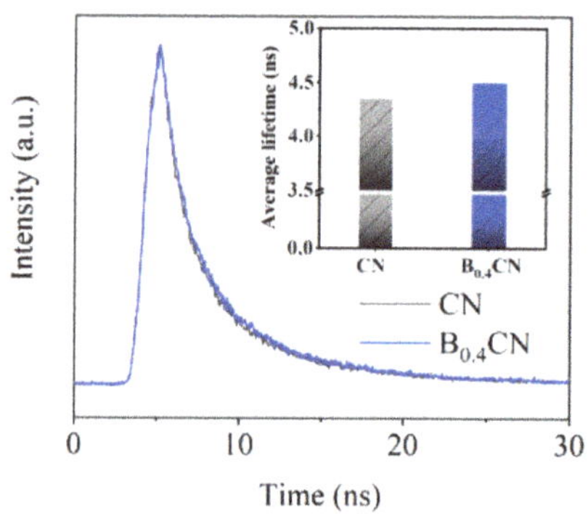

Figure 20. PL in time for the composite boron-doped and carbon fibers (B-gCN/CFs) [126].

A final example of B-doping may be the work of Zou, Jingye et al. [127], in a paper on the degradation of RhB under visible light with the gCN thus doped. These authors synthesized the B-doped gCN by first dissolving melamine and boric acid ($B(OH)_3$) in EtOH. They then removed the EtOH by heating at 80 °C, followed by drying at 100 °C/6 h. The material was then treated in a vacuum microwave oven (4.0 kW) at 560 °C/10 min. They obtained one material (BCNNs) with a surface area and a pore volume of 105.06 $m^2 \cdot g^{-1}$ and 0.229 $cm^3 \cdot g^{-1}$, respectively (PCN, 17.45 $m^2 \cdot g^{-1}$, 0.08 $cm^3 \cdot g^{-1}$), and a band gap of 2.73 eV (2.7 eV for the PCN sample). This thus-doped material reached an RhB degradation percentage (2 ppm) of 97% (50% for the undoped material) in 30 min under visible light with 500 ppm of photocatalyst (Table 6). The modified gCN showed no loss in RhB degradation efficiency after four degradation cycles. The authors attributed the improvement in photodegradation to a longer lifetime of the photogenerated charges and an increase in the surface area and pore volume due to the formation of nanosheets (despite the increase in the band gap due to the quantum confinement effect).

Doping with Halogens

It is now time to mention halogen doping (F, Cl, Br, and I). Like their preceding non-metallic partners, they can enhance absorption and improve separation and charge mobility. Of the halogens, Cl seems to improve semiconductor properties the most. Possible syntheses are the thermal (pyrolytic) condensation of melamine (and/or urea) and ammonium chloride (NH_4Cl). These materials thus prepared have good efficiency in both CO_2 reduction and H_2 production [52,53].

A similar synthesis was carried out by Guo et al. [128], in the degradation of TC. The authors prepared the photocatalyst by mixing melamine with various amounts of NH_4Cl in deionized water. This solution was dried and then calcined at 550 °C for 3 h (0.5 °C·min^{-1}). The thus-modified gCN (CN-Cl-2 sample) showed a gap of 2.69 eV and the CN-Cl-1 sample of 2.7 eV, where the pristine gCN showed a gap of 2.75 eV (the gap decreases with Cl content), with a surface area of 114.4 m^2·g^{-1} and a pore volume of 0.34 cm^3·g^{-1} (42.3 m^2·g^{-1} and 0.11 cm^3·g^{-1} for the undoped gCN). The CN-Cl-1 sample achieved a TC degradation (10 ppm) of 92% in 120 min, k = 0.02 min^{-1} (32%, k = 0.004 min^{-1}, for pristine gCN), under visible light (300 W; Xe lamp) with 500 ppm of photocatalyst (Table 6). After three consecutive cycles, this photocatalyst showed a scarce decrease in TC degradation. According to the authors, the improvement in degradation is mainly due to the improvement in separation and mobility of the photogenerated charges, reduction in the gap, and increase in specific area (Figure 21a).

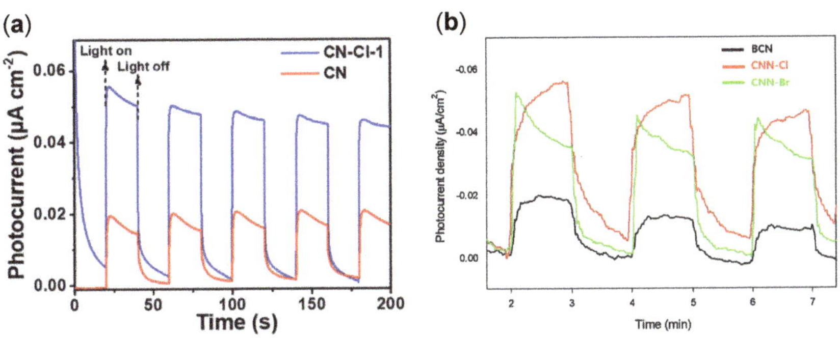

Figure 21. (**a**) Transient photocurrent of bulk (red, CN) and chlorine-doped gCN with mass ratio of melamine and ammonium chloride of 1:1 (blue, CN-Cl-1) [128]; (**b**) chlorine doped (red, CNN-Cl) and bromine doped from melamine and the corresponding ammonium halide (green, CNN-Br) [129].

A similar strategy (thermal copolymerization of melamine and the corresponding ammonium halide (NH_4Cl and NH_4Br), at 550 °C, 3h (15 °C min^{-1})) for Br and Cl doping is provided by Hong et al. [129] in a paper on the degradation of OTC with Br-gCN and Cl-gCN nanosheets under visible light. They obtained a band gap of 2.78 eV for gCN, 2.73 eV for Cl-gCN, and 2.75 eV for Br-gCN. The authors concluded that with less Cl doping (5 g NH_4Cl/5 g melamine; CNN-Cl5, 1000 ppm), it is possible achieve the same percentage of OTC degradation (10 ppm) within the same time (150 min) as with more Br doping (15 g NH_4Br/5 g melamine; CCN-Br15, 1000 ppm), i.e., 75% degradation (gCN 30%) under visible light (35 W; LED), k = 0.004, 0.017, and 0.0179 min^{-1} for the BCN, CNN-Cl, and CNN-Br samples, respectively (Table 6). In addition, they achieved a facile exfoliation route for this doped material compared to the conventional two-step techniques. During the calcination (polymerization) of melamine, these ammonium halides decompose into their respective hydrogen chloride, or fluoride, plus ammonia, which act as exfoliating agents. These small gaseous and polar molecules can easily intercalate in the gCN, facilitating its exfoliation, obtaining a yield for the exfoliated material of close to 50%. The photodegradation efficiency of the CNN-Br sample remained unchanged after the three cycles. The better efficiency of this type of doping was due to improvement in exciton lifetime (mainly) and a reduction in the band gap (Figure 21b).

Self-Doping

This list of non-metallic doping will end with self-doping; carbon doping, C-gCN; and nitrogen doping, N-gCN. This is a convenient method to improve photocatalytic activity without the introduction of defects, and/or impurities.

Carbon doping increases the number of delocalized π-bonds, which improves conductivity and thus charge transfer and separation, as well as adsorption capacity, and greatly broadens the absorption spectrum in the visible range [53,54].

Thus, regarding the doping of gCN with C, the work of Shi et al. [130] about the degradation of TC under visible light with C-gCN nanosheets can be mentioned. These authors prepared the photocatalyst by mixing 10 g of urea with different amounts of n-octanol (0.1; 0.5; 3; 5; 10 mL). It was stirred for 30 min and then heat-treated in a muffle furnace at 550 °C for 2 h (with a heating ramp of 5 $°C \cdot min^{-1}$), achieving a slight improvement in the band gap of the material (from 2.77 eV for gCN to 2.71 eV for C-gCN-5), and also reaching with the sample treated with 5 mL of n-octanol (C-CN-5) a 77% degradation of tetracycline (TC; 30 ppm), at pH = 7; gCN was 40% at the same pH, in 60 min under visible light (30 W; LED) with 400 ppm of photocatalyst (Table 6). After three cycles, the degradation efficiency of TC was reduced by only 5%, indicating the good stability of this material. The authors point out that the degradation of TC in a basic medium is improved; at higher pH, the production of hydroxyl radicals, $HO^{\bullet}$, is favored. The better efficiency was mainly due to the improvement in the charge separation (Figure 22a).

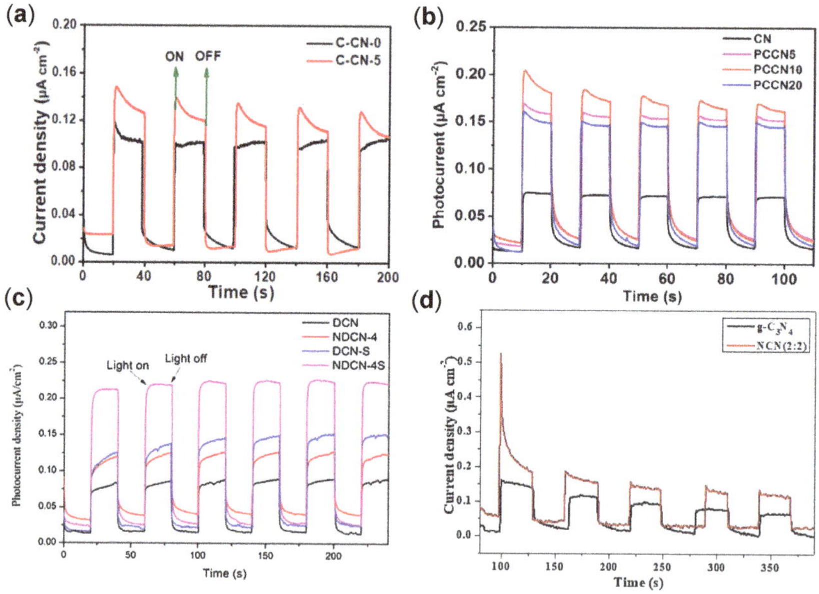

Figure 22. (**a**) Transient photocurrent of bulk and C-doped gCN (C-CN) from urea and octanol [130]; (**b**) transient photocurrent of bulk and carbon-doped gCN and different amounts of dead leaves (PCCNx) from urea [131]; (**c**) transient photocurrent of bulk (DCN) and N-doped gCN from dicyandiamide dissolved in DMF (NDCN-4) and N-doped gCN nanosheets (NDCN-4S) [132]; (**d**) transient photocurrent of bulk gCN and N-doped gCN from urea and citric acid [133].

The synthesis proposed by Li et al. [131] can also be added, regarding the synthesis of porous C-gCN, in a work on the production of H_2O_2 with this photocatalyst and the subsequent degradation of Bisphenol A (BPA). They used dried and ground dead leaves as a carbon source and morphology regulator. Different materials were obtained by mixing different amounts of dead leaf powder with urea and then calcined at 550 °C for 2 h. The sample with 10 mg of leaves (PCCN10) had an effective band gap of 2.19 eV (2.88 gCN) and a specific area of 85 $m^2 \cdot g^{-1}$ (43 $m^2 \cdot g^{-1}$ gCN). This sample achieved a 96% degradation of BPA (25% the gCN), 10 ppm, in 60 min under visible light with 1000 ppm of photocatalyst (Table 6). The improvement in BPA degradation efficiency over undoped gCN was mainly due to an improvement in surface area, a reduction in the band gap, and an improvement in the separation of photogenerated charges. (Figure 22b).

Other reagents as a carbon source are DMF, agar gel and melamine, sucrose [54], and ammonium citrate [136].

As for nitrogen doping (N-gCN), it can be said to provide similar improvements to C. Thus, the transfer and separation of charge carriers are improved, apart from extending the absorption range. Nitrogen doping is less widely used than its non-metallic precedents, perhaps due to synthesis processes with not very 'ecofriendly' reagents such as hydroxylammonium chloride (NH_3OHCl) or hydrazine (N_2H_4). Other friendlier syntheses could consist of using trichloromelamine ($C_3H_3Cl_3N_6$) as the only precursor together with heat treatment at 600 °C, but also with precursors such as 3,6-di(azido)-1,2,4,5-tetrazine, 2,4,6-triazido-1,3,5-triazine, etc., precursors with which excellent results have been achieved both in the production of H_2 and in the degradation of pollutants [52,53].

An example of N-gCN synthesis is that proposed by Jiang et al. [132] on the degradation of TC with N-gCN nanosheets, with dicyandiamide dissolved in DMF, stirred (12 h), dried (100 °C), thermally condensed at 550 °C for 4 h, and with subsequent heat treatment at 500 °C (4 h) for exfoliation. The authors thus achieved a material with a specific surface area of 74.79 $m^2 \cdot g^{-1}$ (18.4 $m^2 \cdot g^{-1}$ for unmodified gCN); they reduced the band gap from 2.51 eV of the pristine material (DCN) to 2.47 eV of the N-doped material (NDCN-4), but due to the quantum confinement effect of exfoliation, the band gap increased to 2.54 eV for the N-gCNNSs (NDCN-4S). With the NDCN-4S sample (the doped and exfoliated material), they achieved TC degradation (10 ppm) of 81.7% in 60 min (52.2% for the unmodified gCN) with 500 ppm of photocatalyst under visible light (Table 6). After the fifth consecutive degradation cycle, this material showed a slight decrease in TC degradation efficiency (79.55% compared to 81.7% in the first cycle). The authors were thus able to verify improvements in the BET area, light absorption, and separation of the photoexcited charges, and thus in efficiency (Figure 22c).

A final example of N-doped gCN could be the work carried out by Zhu et al. [133], wherein they evaluated in the degradation of phenol (PhOH). The authors prepared the photocatalytic material by mixing urea and citric acid monohydrate ($HOC(COOH)(CH_2COOH)_2$-H_2O) and heat treating them (polymerization in a non-inert environment) at 550 °C/4 h (2 $°C \cdot min^{-1}$). Different degrees of doping were achieved by changing the urea/citric acid ratio. They obtained an N-gCN sample (NCN-(2:2); 2 g urea/2 mg citric acid) with a slightly lower surface area than the undoped material, i.e., 72.26 $m^2 \cdot g^{-1}$ versus 76.69 $m^2 \cdot g^{-1}$ for undoped material. The band gap was reduced from 2.51 eV for the pristine material to 1.82 eV for the doped material (NCN-(2:2)). This sample showed a 70.1% PhOH degradation efficiency (37.6% degradation efficiency of pristine gCN), 10 ppm, within 180 min with 1000 ppm photocatalyst under visible light (300 W Xe lamp; Vis.) (Table 6). After three consecutive cycles of degradation, the efficiency decreases up to 50%, according to the authors, due to a significant loss of photocatalyst during the process. According to researchers, both the reduction in band gap and an improvement in the separation and lifetime of the excitons are responsible for the better efficiency compared to the undoped material (Figure 22d).

Non-metal doping in gCN allows for tuning its electronic structure by introducing intermediate levels, which improves conductivity and charge separation and reduces the gap. Oxygen doping significantly decreases the energy gap and optimizes carrier mobility, enhancing the degradation of pollutants. Phosphorus and sulfur doping enhances absorption in the visible range by favoring the delocalization of the π-system and improving charge separation, while boron doping transforms the material from an n-type to a p-type semiconductor, increasing electron mobility. Halogen doping not only reduces the gap and improves charge separation, but also facilitates exfoliation, increasing the surface area. Finally, self-doping with carbon or nitrogen extends the absorption in the visible spectrum and improves charge transfer without introducing additional defects, which together increase photocatalytic efficiency. As a final remark, it can be noted that this type of doping has minimal impact on the stability of the photocatalysts after several cycles (and if there is any decrease in efficiency, it is due to the loss of photocatalytic material during the process), indicating good stability of these derivatives in the reaction medium.

4.2.2. Metal Doping

The photocatalytic activity of gCN doped with metal nanoparticles from surface plasmon resonance (SPR)—by resonating with the incident radiation—enhances the excitation of electrons to the BC (more free electrons are introduced into the system) while introducing intermediate energy levels into the band gap that broaden the absorption spectrum and reduce the effective gap (especially with alkali metals) [1,13]. Nitrogen atoms located in the interstitial sites (6 N in heptazine- and 3 N in triazine-based structures) are responsible for interacting with metals via electrostatic interactions.

Several strategies have been adopted for this type of doping, such as deposition–precipitation techniques, thermal polymerization, the solvothermal method, photodeposition, etc., as pointed out by Thomas et al. [1] in a review article. Thus, these authors refer to the doping of gCN with Pt nanoparticles by solvothermal method (after thermal condensation of the precursor) with chloroplatinic acid hexahydrate (H_2PtCl_6-$6H_2O$), mixing the gCN obtained with this Pt source in ethanol, with ethanol being the reducing agent that will reduce the Pt^{+4} nanoparticles. Another similar example, for Ni doping, reported in the same literature review [1], is also carried out by solvothermal method with Ni acetylacetonate ($Ni(CH_3COCHCOCH_3)_2$) dispersed in DMF, which, after being mixed with the gCN, was autoclaved. Other salts used as metal sources, for the same (solvothermal) method, would be, for example, $CoCl_2$ and $HAuCl_4$ for Co and Au, respectively [1], or potassium fluoride KF and $NaBH_4$ for K and Na insertion, respectively [53]. Another way to carry out this type of doping would be by thermal condensation, for example, of H_2PdCl_4 and dicyandiamide, dispersed in deionized water and mixed at 80 °C, and heat-treated in an Ar atmosphere. Doping with alkali and alkaline earth metals by heat treatment with the corresponding halides is also possible (Figure 23a) [137]. To illustrate the diversity in metal doping strategies, the following section describes example syntheses for each type of metal dopant.

Alkaline and Alkaline Earth Metal Doping

For alkali metal doping, Zhang et al. [138] studied the optical properties of alkali metal-doped gCN (not focused on EP degradation button CO_2 reduction, though the synthesis remains of interest). They synthesized doped gCN by mixing 3 g of urea with 0.1 g of different alkali metal nitrates ($LiNO_3$, $NaNO_3$, KNO_3, and $RbNO_3$), mixed in DW, and treated with ultrasound for 5 h. The water content was then evaporated at 70 °C, followed by heat treatment (calcination) at 500 °C 2 h. (2.3 °C·min^{-1}), and then the powder thus obtained was heat treated at 350 °C 3 h. They obtained reduced gaps for all the doped materials, e.g., 2.2 eV for K-gCN, and only 2 eV for Rb-gCN.

In another synthesis, Yan et al. [137] dissolved urea in an aqueous solution of alkali (Na and K) and alkaline earth (Ca and Mg; 3% by weight) bromides, mixed them overnight, and then heat-treated them at 550 °C for 2 h (15 °C·min^{-1}). Gaps ranging from 2.46 eV for Mg-gCN to 2.29 eV for Ca-gCN, 2.41 eV for K-gCN, and 2.43 eV for Na-gCN (undoped gCN had a gap of 2.57 eV) were obtained, in addition to a considerable increase in the half-life of the charge carriers. The ENR degradation kinetic coefficient (5 ppm) was multiplied by 3.3 for Ca-gCN (50%), by 4.4 for Mg-gCN (70%), by 4.8 for Na-gCN (80%), and by 5.4 for K-gCN (95%), with respect to gCN (22%) in 120 min under visible radiation, with 400 ppm of photocatalyst (Table 7). For SMX and TC degradation, the degradation efficiency followed the same order as for ENR (K-gCN > Na-gCN > Mg-gCN > Ca-gCN > gCN). The improvements in respect to the efficiency of undoped gCN are mainly due to the reduction in band gap and an improvement in exciton lifetime (Figure 23a).

Table 7. Parameters of photodegradation of pollutants in water with metal doped gCN.

Dopant	Photocatalyst	Area ($m^2 \cdot g^{-1}$)	Gap (eV)	Pollutant (ppm)	Radiation	Efficiency (%)	Time (min)	k (min^{-1})	Refs.
Alkaline and alkaline earth metals	gCN (400 ppm)	177	2.57	ENR, SMX, and TC: 5	300 W Xe lamp (Vis.)	ENR: 22	120	0.014	[137]
						TC: 42		0.036	
						SMX: 19		0.001	
	Mg-gCN (400 ppm)	107	2.46			ENR: 78		0.062	
						TC: 79		0.067	
						SMX: 40		0.003	
	Ca-gCN (400 ppm)	100	2.29			ENR: 58		0.046	
						TC: 78		0.067	
						SMX: 35		0.003	
	K-gCN (400 ppm)	81	2.41			ENR: 82		0.075	
						TC: 80		0.072	
						SMX: 67		0.006	
	Na-gCN (400 ppm)	90	2.43			ENR: 81		0.067	
						TC: 78		0.070	
						SMX: 65		0.006	
Rare earths (Ln)	gCN	46.33	2.73	RhB, MB, and PR: 15	300 W Xe lamp (Vis.)	RhB: 54	150	0.0062	[139]
						MB: 52		0.0061	
						PR: 56		0.007	
	Yb-gCN	52.56	2.5			RhB: 88		0.017	
						MB: 91		0.019	
						PR: 88		0.018	
	Nd-gCN	46.38	2.56			RhB: 74		0.012	
						MB: 87		0.018	
						PR: 74		0.015	
	Ce-gCN	58.09	2.47			RhB: 91		0.019	
						MB: 93		0.023	
						PR: 91		0.020	
Noble metals	gCN (1000 ppm) Ag-gCN (1000 ppm)	59.5	2.66	TC, OTC, and CTC: 20	300 W Xe lamp (Vis.)	TC: 35	120	0.0037	[140]
						OTC: 50		0.0053	
						CTC: 43		0.0039	
		58.4	2.55			TC: 83		0.0142	
						OTC: 81		0.0130	
						CTC: 85		0.0139	
	gCN (1000 ppm)	-	2.7	BZF: 3	500 W Xe lamp (Vis.)	27	90	0.012	[141]
	Pd-gCN (1000 ppm)	-	2.62			100		0.036	

Table 7. *Cont.*

Dopant	Photocatalyst	Area ($m^2 \cdot g^{-1}$)	Gap (eV)	Pollutant (ppm)	Radiation	Efficiency (%)	Time (min)	k (min^{-1})	Refs.
Transition metals	gCNNSs (600 ppm)	91	2.7	RhB: 20	300 W Xe lamp (Vis.)	60	60	0.012	[142]
	Fe-gCNNs (600 ppm)	132	2.5			100	30	0.117	
	gCN (1000 ppm)	95.95	2.73	MB; 10CIP; 15	300 W Xe lamp (Vis.)	MB: 67	150	0.013	[143]
						CIP: 30		0.002	
	gCNNSs (1000 ppm)	70.17	2.76			MB: 81		0.014	
						CIP: 37		0.003	
	Ni-gCNNSs (1000 ppm)	8.23	2.68			MB: 93		0.017	
						CIP: 52		0.003	
	gCN (600 ppm)	11.4	2.7	TC: 30	300 W Xe lamp (Vis.)	47	120	-	[144]
	Cu-gCN (600 ppm)	142.8	2.45			98		-	

Rare Earths (Ln) Doping

For rare earth (lanthanide; Ln) doping, the available literature is scarce, and only the research by Yu et al. [139] is discussed. The study focused on the degradation of organic dyes (RhB, MB, and PR) using gCN doped with rare earths (Yb, Nd, and Ce). They used 10 g of urea and 2% molar $Yb(NO_3)_3{\cdot}5H_2O$, $Nd(NO_3)_3{\cdot}6H_2O$, and $Ce(NO_3)_3{\cdot}6H_2O$, dissolved in 15 mL of deionized water with stirring 30 min, then dried it, followed by heat treatment at 550 °C for 2 h (2 $°C{\cdot}min^{-1}$). They found that photocatalytic activity improved for all three metals. Ce-gCN is the material with the largest surface area of the three, with 58.09 $m^2{\cdot}g^{-1}$ (46.33 $m^2{\cdot}g^{-1}$ for undoped gCN). The calculated band gap was 2.47 eV for Ce-gCN (2.73 eV for gCN). The most efficient doping was Ce doping, reaching 91.23% for RhB, 93.45% for MB, and 91.25 for PR, all 15 ppm, in 150 min (54%, 53%, and 56% with the undoped gCN for RhB, Mb, and PR, respectively) with 300 ppm photocatalyst under visible light (Table 7). The authors studied the stability of this material only in the degradation of RhB and MB. Thus, after five degradation cycles, this sample (Ce-gCN) showed a slight decrease in the degradation efficiency of RhB, and no loss in MB degradation. The enhanced efficiency was mainly attributed to band gap narrowing and improved separation of photogenerated charges.

Noble Metals Doping

As an example of noble metal doping, the work of Ren et al. [140] on the degradation of different tetracyclines (OTC; CTC; and TC) with Ag-gCN under visible light can be highlighted. These authors prepared the semiconductor thus doped by dissolving gCN (obtained by thermal treatment, calcination, at 550 °C/4 h., 2.3 $°C{\cdot}min^{-1}$) in deionized water, and mixing this solution with another aqueous solution of $AgNO_3$ (Ag/gCN: 4%, 6%, 8%, 10%, and 12% by weight), with subsequent photodeposition (with a 300 W Xe lamp for 2 h). It was then centrifuged, washed, and dried. The material obtained (Ag-CN-8) had a surface area of 58.4 $m^2{\cdot}g^{-1}$ (gCN of 59.5 $m^2{\cdot}g^{-1}$) and a band gap of 2.55 eV (2.66 eV for gCN). They obtained a photocatalyst with an improved band gap, and with a longer half-life of the charge carriers. The best degree of degradation was obtained with the Ag-gCN-8 sample (8% Ag): 83% for TC; 81% for OTC; and 85% for CTC (all three antibiotics 20 ppm) in 120 min, and with 1000 ppm of the photocatalyst under visible light (Table 7). After four consecutive photocatalytic cycles, a reduction in Ag content from 7.47% to 6.20% was observed, indicating that the diminished catalytic efficiency could be attributed to silver leaching during the recovery process. They (the authors) concluded that better efficiency was due to the better separation of generated charges and a reduced band gap, and that by increasing the pH, the tetracyclines are negatively charged, and although this may decrease the adsorption on gCN, at the same time, the production of hydroxyl radicals is enhanced, thus improving degradation (Figure 23b).

Another example of noble metal doping is provided by Yin et al. [141] in the degradation of bezafibrate (BZF; a lipid regulator) under simulated sunlight, using Pd-gCN as a photocatalyst. These authors prepared the Pd-doped material by mixing gCN powder synthesized from melamine by thermal condensation in two steps (first at 500 °C for 2 h, at 2 °C·min^{-1}; and then, after cooling and grinding, a subsequent treatment at 550 for 3 h, also at 2 °C·min^{-1}). They then added $Pd(NO_3)_2$ (different percentage ratios) to 0.2 g gCN and treated with ultrasound 1 h. It was then stirred for 12 h at 25 °C. $NaBH_4$ (0.1 M) was added and stirred magnetically for another 2 h. The product obtained was then washed and dried overnight. The narrowest band gap was for the sample with Pd/gCN ratio of 1%, with 2.62 eV (2.7 eV for undoped g-CN). The material thus obtained (Pd-gCN-1%) had a BZF degradation efficiency (3 ppm) of 100% in 90 min (27% for gCN at the same time) with 1000 ppm photocatalyst (Table 7). After five consecutive degradation cycles, this doped $g\text{-}C_3N_4$ derivative exhibited only a 10% loss in BZF removal efficiency. In this case, the better efficiency was due to the reduced band gap and due to the improvement in charge separation (Figure 23c).

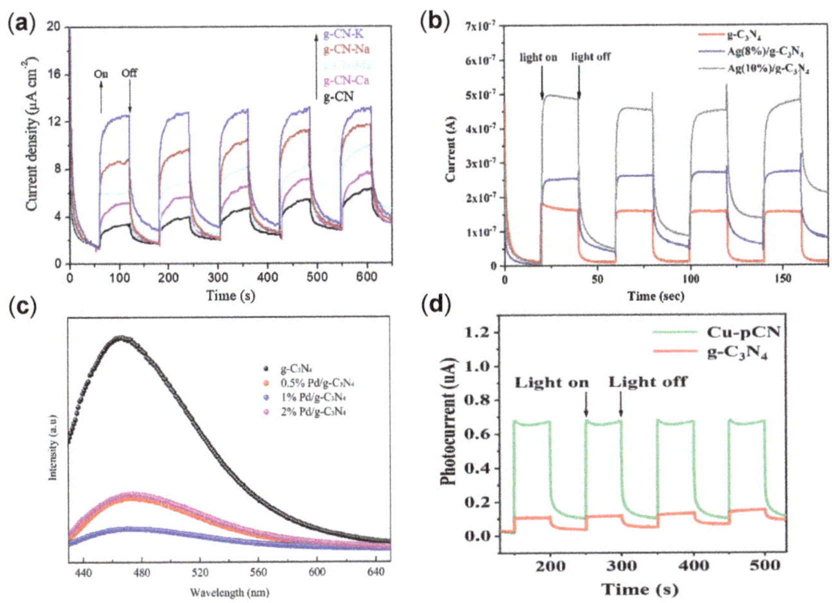

Figure 23. (**a**) Transient photocurrent of alkaline- and alkaline earth-doped gCN [137]; (**b**) transient photocurrent of gCN doped with Ag in different amounts [140]; (**c**) PL spectra of Pd-gCN [141]; (**d**) the transient photocurrent of gCN doped with Cu [144].

Finally, this section discusses the Au (Au-gCN) doping reported by Faisal et al. [145], in a study focused on the degradation of MB and gemifloxacin mesylate (an antibacterial compound) under visible light. Thus, these authors report a synthesis (like the previous method with Pd) by treating gCN (synthesized by thermal condensation of urea at 550 °C/3 h.), in aqueous solution, with different amounts of $HAuCl_4$ (0.2%, 0.5%, 1%, 2%, and 5%), agitated by ultrasound for 1 h. It was then filtered by centrifugation, washed, and dried. The band gap of these doped materials was almost the same for all of them (2.86 eV, compared to 2.85 eV for undoped gCN). These authors concluded that the sample with the best BZF degradation efficiency was the Au-gCN-1% sample with 95.13% dye degradation in 90 min (69% for gCN at the same time) with 400 ppm photocatalyst under visible light (400 W; lamp metal halide). This photocatalyst showed no loss in degradation efficiency after five consecutive degradation cycles. The authors attributed the increase in efficiency for Au-gCN-1%, with respect to gCN, to a substantial increase in the lifetime of the excited charge pairs.

Transition Metal (Non-Noble) Doping

To conclude this subsection, transition metal doping is mentioned. It was precisely because of the high cost of noble metals (and their scarcity) that transition metals became an alternative to noble metals as dopants [4]. Like their metallic partners, transition metals can also improve the absorption range by decreasing the gap, as well as improving mobility and charge separation [2,13,52].

To begin with, as an example of Fe doping, the work carried out by Nguyen Van et al. [142] on the degradation of RhB with Fe-gCN is worth mentioning. They carried out the synthesis by dissolving 0.5 g of gCN nanosheets (obtained from urea by thermal polymerization at 550 °C, 2 h) in distilled water with magnetic stirring for 30 min, followed by ultrasound treatment for 1 h. To this solution, different amounts of $FeCl_3$-$6H_2O$ (3, 5, 7, 8, and 10 molar) were added, followed by heat treatment at 90 °C for 12 h with magnetic stirring. This solution was centrifuged, and the residue was rinsed with EtOH and dried. This sample had a surface area of 132 $m^2 \cdot g^{-1}$ (91 $m^2 \cdot g^{-1}$ for gCN nanosheets) and a band gap of 2.5 eV (2.7 eV for gCN). The material thus prepared (Fe-gCN-7 sample) showed a degree of degradation of RhB (20 ppm) of 100% in 30 min, with a kinetic coefficient 10 times higher than that of gCN with 6000 ppm of photocatalyst under visible light (Table 7). The photodegradation percentage of RhB was still up to 95% after three cycles. They also concluded that, on the one hand, Fe^{+3} traps the photoexcited electrons, allowing for a longer lifetime of the excited pairs, which, together with the increase in specific area, improves the degradation of RhB.

Another doping procedure with this type of metal, namely, Ni, was carried out by Zhou et al. [143], where they managed to prepare Ni-gCN nanosheets, with good degradation of MB and CIP under visible light. The photocatalyst was prepared in several steps: (i) synthesis of gCN by calcination at 520 °C for 4 h; (ii) hydrothermal exfoliation of the gCN (after ultrasonic treatment in HCl solution for 30 min) at 150 °C for 5 h in an autoclave; (iii) additional calcination at 500 °C for 2 h; (iv) mixing of the exfoliated nanosheets with the Ni precursor ($Ni(CH_3COCHCOCH_3)_2$; 5, 10, 15, and 20 wt.% Ni) in DMF under ultrasonic stirring; (v) the resulting solution was solvothermal treated at 200 °C for 10 h. The authors were able to verify that the surface area of the nanosheets decreased with increasing Ni %; for Ni-CN-10%, this was 70.17 $m^2 \cdot g^{-1}$ (95.95 $m^2 \cdot g^{-1}$ for undoped nanosheets, CN-500). This photocatalyst thus doped showed a band gap for Ni-gCN-10% equal to 2.68 eV (nanosheets 2.76 eV and unmodified gCN 2.73 eV). The degradation of MB (10 ppm) and CIP (15 ppm) with 1000 ppm photocatalyst improved with increasing Ni content up to 10% with 93% degradation (67% untreated gCN) for MB, and 52% (30% untreated gCN) for CIP, in 150 min under visible light (Table 7). The Ni-doped gCN, after three degradation cycles, showed no loss in efficiency for either MB or CIP degradation. According to the authors, the improvement in the efficiency of this material (Ni-CN-10%) compared to gCN nanosheets (CN-500) and bulk gCN is mainly due to a reduction in the band gap and an improvement in photogenerated charge separation. The material is also magnetically recoverable.

Finally, Cu doping is added to this subsection, mentioning the work of Bao et al. [144] on the degradation of TC with Cu-doped gCN. The synthesis was carried out by calcination at 550 °C 6 h (1.5 $°C \cdot min^{-1}$) from a mixture of melamine, cyanuric acid, and copper acetate ($Cu(OAc)_2$), with 0, 4, 8, 12, and 16 wt.% copper acetate, dissolved in EtOH (after evaporation). The 3Cu-pCN sample had a surface area d 142.8 $m^2 \cdot g^{-1}$ and a cumulative pore volume of 1.15 $cm^3 \cdot g^{-1}$ (11.4 $m^2 \cdot g^{-1}$ and 0.11 $cm^3 \cdot g^{-1}$ for the gCN). The best gap was achieved with a 3Cu-pCN sample (12% Cu) equal to 2.45 eV (2.7 eV for gCN), reaching the best TC degradation (30 ppm), 98% in 2 h (47% for gCN within the same time), with 600 ppm of photocatalyst under visible light (Table 7). This photocatalyst (3Cu-pCN sample) showed a decrease in efficiency from 98% to 80.6% after seven cycles of TC degradation, attributed to catalyst loss during washing and centrifugation steps. The improved performance of degradation was mainly attributed to the narrowed band gap, enhanced visible light absorption, and increased exciton lifetime (Figure 23d).

Metal doping in $g\text{-}C_3N_4$ takes advantage of plasmon resonance to increase electronic excitation and reduce the energy gap, which broadens the absorption of visible light. This modification introduces intermediate energy levels that facilitate more efficient electron transfer and improve charge separation. Doping with alkali, alkaline earth, and rare earth metals decreases the gap and extends the lifetime of excitons, boosting the degradation of pollutants. In addition, the incorporation of noble and transition metals further optimizes carrier mobility, resulting in higher photocatalytic efficiency. Taken together, these strategies enhance adsorption, with charge transfer notably strengthening the performance of gCN in environmental applications. Finally, it should be noted that, regarding the stability of metal-doped photocatalysts, their efficiency is generally maintained over multiple reuse cycles, with observed declines typically attributable to physical losses of material during handling rather than to structural degradation.

5. Conclusions

This review provides a comprehensive and detailed overview of the different methodologies for gCN synthesis and modification and a set of reference data for photocatalyst design strategies. These contributions pursue the development of more efficient, sustainable, and economically competitive water decontamination systems.

gCN is a highly promising and sustainable alternative to conventional catalysts based on metal oxides, chalcogenides, nitrides, or carbides. Its use minimizes the environmental impact associated with the extraction and processing of traditional metal-based materials while reducing energy and operating costs, since its ability to absorb light in the visible range enables the use of renewable energy sources. Moreover, the synthesis of gCN using low-cost and environmentally benign precursors, such as urea, presents a competitive advantage that paves the way for large-scale production.

In this context, surface functionalization, defect introduction, and doping strategies (both with non-metals and metals) are of fundamental relevance, as they allow for fine tuning the electronic and structural properties of gCN, improving light absorption, charge separation, surface area, and, consequently, its performance in the degradation of a wide variety of pollutants. The extensive body of literature reviewed here demonstrates the high efficiency of these materials in the degradation of EPs in aqueous media, and their stability in reaction media without significant losses in photocatalytic efficiency, placing them at the epicenter of research in heterogeneous photocatalysis. However, several limitations have been identified in the materials and their evaluation:

- The documented leaching of metals when doping gCN and taking place during degradation cycles could be an environmental and/or public health issue, and further research into doping techniques to overcome this effect would be desirable.
- The stability of the synthesized materials is a matter of concern and there is no consensus in this respect. Not all the studies present these stability studies, whether carried out with only 2 or 3 or up to 10 consecutive degradation cycles, which makes

it difficult to have a clear perception of this. Given the operational requirements, i.e., the high flow rates to be treated (both in WWTPs and DWTPs), and the need for the photocatalysts to be economically viable, more than five cycles would be necessary to get an idea of the operational stability of these materials in real situations. It is also noteworthy that 2D and porous 3D structures frequently exhibit superior recyclability, which positions them as particularly promising for long-term applications. It would also be advisable to accompany with stability studies—under the same conditions—of unmodified gCN in order to obtain a clear idea of the effect of the modifications on the stability of these gCN derivatives. Also, accompanying these stability studies with characterizations (FTIR, XRD, etc.) of the modified photocatalyst would reinforce this perception.

- The lack of a standardized protocols for both the study and preliminary optimization of catalysts (generally with organic dyes) and for the degradation conditions of model pollutants—as exists for the evaluation of photocatalytic ceramic materials in aqueous media (ISO 10678) [146], for the evaluation of methods for air-purification performance of semiconductor photocatalytic materials (ISO 22197) [147], or for water purification performance of semiconductor photocatalytic materials by measurement of forming ability of active oxygen (ISO 10676) [148,149]—is a drawback. Standardization of these processes would be desirable, as it would facilitate comparability between materials.
- For the degradation of model EPs, experimental studies should also include assessments of the potential toxicity of both the materials and the degradation products of the target EPs. This approach would improve the positive perception of heterogeneous photocatalysis within industrial sectors.
- Finally, in the 21st century, and amid the planet's environmental crisis, there is requirement for a commitment by researchers to align their work with Green Chemistry principles (the 12 tenets established in 1998 by Paul Anastas and John Warner).

Other aspects that can be considered as future directions or opportunities identified in this review are as follows:

- Recent advances in DFT and machine learning offer a powerful means to pre-screen and predict the performance of modified gCN, allowing for researchers to optimize band structure, charge dynamics, and catalytic behavior before experimental validation. Integrating these computational tools could greatly accelerate materials discovery while aligning with the principles of green chemistry. Few studies incorporate prior in silico analysis to guide and justify the selection of gCN modification methodologies, which would enhance the sustainability of experimental designs and processes.
- The improvement of photocatalytic capabilities by the combination of photocatalysts in the form of heterojunctions or homojunctions is also an important aspect for the improvement of the photodegradation efficiency of PEs that must be assessed by photocatalysis researchers. Although this review deliberately focuses on modifications of gCN itself, it is essential to highlight that the integration of gCN into heterojunction systems (e.g., type II, Z-scheme, Schottky) offers exciting avenues for synergistic performance improvements, meriting dedicated and systematic exploration in future studies.
- Immobilization of photocatalytic materials, although addressed in numerous studies, requires solutions and more investigations that do not compromise the inherent photocatalytic efficiency of gCN. Even though it is not an aim of this review, the fine and dispersed nature of gCN nanoarchitecture complicates its recovery after the photocatalytic process, as was detected in the reuse of the materials relying mostly on the loss of catalytic material. It is a challenge shared with other nano-photocatalysts, both metallic and those based on carbonaceous compounds.

The integration of these approaches, combined with innovative support or immobilization techniques, is essential to advance the scalability and economic viability of heterogeneous photocatalysis as a large-scale decontamination solution.

Finally, while gCN represents a powerful tool for the degradation of pollutants in water bodies, it is essential to remember that the ultimate solution to water pollution lies in prevention. The integration of advanced technologies with prevention and environmental management strategies will undoubtedly be the key to achieving effective and long-lasting decontamination outcomes.

Author Contributions: Conceptualization, X.B.-X., E.R. and M.Á.S.; resources, X.B.-X.; writing—original draft preparation, X.B.-X.; writing—review and editing, X.B.-X., E.R. and M.Á.S.; visualization, X.B.-X., E.R. and M.Á.S.; supervision, M.Á.S. and E.R.; project administration, E.R. and M.Á.S.; funding acquisition, E.R. and M.Á.S. All authors have read and agreed to the published version of the manuscript.

Funding: This research was supported through Projects PCI2022-132941 (BiodivRestore Cofund 2020) and TED2021-129590A-I00 funded by MICIU/AEI/10.13039/501100011033 and European Union Next Generation EU/PRTR.

Data Availability Statement: The data used in this review will be made available from the authors upon request.

Conflicts of Interest: The authors declare no conflicts of interest.

Abbreviations

The following abbreviations are used in this manuscript:

AAO	anodic aluminum oxide
BET	Brunauer–Emmett–Teller
B-gCN	boron-doped graphitic carbon nitride
BPA	Bisphenol A
BZF	bezafibrate
CCDC	Cambridge Crystallographic Data Centre
CFO	cobalt ferrite
C-gCN	carbon-doped graphitic carbon nitride
CIP	ciprofloxacin
CN-B	untreated gCN
CQD	carbon quantum dot
CTC	chlortetracycline
CTO	cobalt titanate
C_V	carbon vacancies
CVDTLA	Chemical Vapor Deposition Three Letter Acronym
DCN	pristine carbon nitride
DFT	density functional theory
DMF	dimethylformamide
DW	distilled water
ENR	enrofloxacin
EPDOAJ	Emerging Pollutants Directory of Open Access Journals
EtOH	ethanol
FCN	formaldehyde carbon nitride
FTIR	Fourier transform infrared spectroscopy
GCC	GCNQDs-$CoTiO_3/CoFe_2O_4$
gCN	graphitic carbon nitride
gCNNF	graphitic carbon nitride nanofiber

gCNNR	graphitic carbon nitride nanorod
gCNNT	graphitic carbon nitride nanotube
gCNNW	graphitic carbon nitride nanowire
gCNQD	graphitic carbon nitride quantum dot
IPA	isopropyl alcohol
MB	methylene blue
MCN	melamine-derived gCN
MeOH	Methanol
MGCN	microwave graphitic carbon nitride
MNCA	supramolecular aggregates melamine (MA) and cyanuric acid (CA)
MO	methyl orange
MUCN	homojunction (type II) between 0D and 1D structures of gCN
NDCN	nitrogen-doped carbon nitride
NFs	nanofibers
N-gCN	Nitrogen-doped graphitic carbon nitride
NMP	N-methyl-2-pyrrolidone
NOR	norfloxacin
NPX	naproxen
NR	nanorod
NTs	nanotubes
N_V	nitrogen vacancies
NW	nanowire
OA	oxamide
ODH	oxalyl dihydrazide
O-gCN	oxygen-doped graphitic carbon nitride
OTC	oxytetracycline
PhOH	phenol
PL	photoluminescence
PPCPs	pharmaceuticals and personal care products
PR	phenol red
PVDLD	physical vapor deposition linear dichroism
QDs	quantum dots
RhB	Rhodamine B
SGCN	solvothermal graphitic carbon nitride
SHP	sodium hypophosphite
S-gCN	sulfur-doped graphitic carbon nitride
SMX	sulfamethoxazole
SPR	surface plasmon resonance
TC	tetracycline
TGCN	thermal graphitic carbon nitride

References

1. Thomas, S.; Anas, S.; Joy, J. (Eds.) *Synthesis, Characterization, and Applications of Graphitic Carbon Nitride*; Elsevier: Amsterdam, The Netherlands, 2023; ISBN 9780128230381.
2. Li, C.; Li, J.; Huang, Y.; Liu, J.; Ma, M.; Liu, K.; Zhao, C.; Wang, Z.; Qu, S.; Zhang, L.; et al. Recent Development in Electronic Structure Tuning of Graphitic Carbon Nitride for Highly Efficient Photocatalysis. *J. Semicond.* **2022**, *43*, 021701. [CrossRef]
3. Balakrishnan, A.; Chinthala, M. Comprehensive Review on Advanced Reusability of G-C_3N_4 Based Photocatalysts for the Removal of Organic Pollutants. *Chemosphere* **2022**, *297*, 134190. [CrossRef] [PubMed]
4. Wang, L.; Wang, K.; He, T.; Zhao, Y.; Song, H.; Wang, H. Graphitic Carbon Nitride-Based Photocatalytic Materials: Preparation Strategy and Application. *ACS Sustain. Chem. Eng.* **2020**, *8*, 16048–16085. [CrossRef]
5. Veerakumar, P.; Sangili, A.; Lin, K.C. *Nanoscale Graphitic Carbon Nitride*; Elsevier: Amsterdam, The Netherlands, 2022; ISBN 9780128230343.
6. Liu, Y.; Mao, B.; Shi, W. *Novel Carbon Materials and Composites*; Jiang, X., Kang, Z., Guo, X., Zhuang, H., Eds.; Wiley: Hoboken, NJ, USA, 2019; ISBN 9781119313397.

7. Qi, K.; Liu, S.; Zada, A. Graphitic Carbon Nitride, a Polymer Photocatalyst. *J. Taiwan Inst. Chem. Eng.* **2020**, *109*, 111–123. [CrossRef]
8. Antonietti, M.; Savateev, A.; Lotsch, B.V. Preface. In *Carbon Nitrides*; De Gruyter: Berlin, Germany, 2023; pp. V–VI.
9. Kathiresan, M. Graphitic Carbon Nitrides: Synthesis and Properties. In *Nanoscale Graphitic Carbon Nitride*; Elsevier: Amsterdam, The Netherlands, 2022; pp. 1–16.
10. Teter, D.M.; Hemley, R.J. Low-Compressibility Carbon Nitrides. *Science* **1996**, *271*, 53–55. [CrossRef]
11. Wang, X.; Maeda, K.; Thomas, A.; Takanabe, K.; Xin, G.; Carlsson, J.M.; Domen, K.; Antonietti, M. A Metal-Free Polymeric Photocatalyst for Hydrogen Production from Water under Visible Light. *Nat. Mater.* **2009**, *8*, 76–80. [CrossRef]
12. Fakhrul Ridhwan Samsudin, M.; Bacho, N.; Sufian, S. Recent Development of Graphitic Carbon Nitride-Based Photocatalyst for Environmental Pollution Remediation. In *Nanocatalysts*; IntechOpen: London, UK, 2019.
13. Wang, Q.; Li, Y.; Huang, F.; Song, S.; Ai, G.; Xin, X.; Zhao, B.; Zheng, Y.; Zhang, Z. Recent Advances in G-C_3N_4-Based Materials and Their Application in Energy and Environmental Sustainability. *Molecules* **2023**, *28*, 432. [CrossRef]
14. Wang, A.; Wang, C.; Fu, L.; Wong-Ng, W.; Lan, Y. Recent Advances of Graphitic Carbon Nitride-Based Structures and Applications in Catalyst, Sensing, Imaging, and LEDs. *Nano-Micro Lett.* **2017**, *9*, 47. [CrossRef]
15. Thomas, A.; Fischer, A.; Goettmann, F.; Antonietti, M.; Müller, J.-O.; Schlögl, R.; Carlsson, J.M. Graphitic Carbon Nitride Materials: Variation of Structure and Morphology and Their Use as Metal-Free Catalysts. *J. Mater. Chem.* **2008**, *18*, 4893. [CrossRef]
16. Kroke, E.; Schwarz, M.; Horath-Bordon, E.; Kroll, P.; Noll, B.; Norman, A.D. Tri-s-Triazine Derivatives. Part I. From Trichloro-Tri-s-Triazine to Graphitic C_3N_4 StructuresPart II: Alkalicyamelurates M3[$C_6N_7O_3$], M = Li, Na, K, Rb, Cs, Manuscript in Preparation. *New J. Chem.* **2002**, *26*, 508–512. [CrossRef]
17. Dong, Q.; Mohamad Latiff, N.; Mazánek, V.; Rosli, N.F.; Chia, H.L.; Sofer, Z.; Pumera, M. Triazine- and Heptazine-Based Carbon Nitrides: Toxicity. *ACS Appl. Nano Mater.* **2018**, *1*, 4442–4449. [CrossRef]
18. Zhu, W.; Song, H.; Lv, Y. Triazine-Based Graphitic Carbon Nitride: Controllable Synthesis and Enhanced Cataluminescent Sensing for Formic Acid. *Anal. Bioanal. Chem.* **2018**, *410*, 7499–7509. [CrossRef] [PubMed]
19. Algara-Siller, G.; Severin, N.; Chong, S.Y.; Björkman, T.; Palgrave, R.G.; Laybourn, A.; Antonietti, M.; Khimyak, Y.Z.; Krasheninnikov, A.V.; Rabe, J.P.; et al. Triazine-Based Graphitic Carbon Nitride: A Two-Dimensional Semiconductor. *Angew. Chem. Int. Ed.* **2014**, *53*, 7450–7455. [CrossRef]
20. Liu, N.; Li, T.; Zhao, Z.; Liu, J.; Luo, X.; Yuan, X.; Luo, K.; He, J.; Yu, D.; Zhao, Y. From Triazine to Heptazine: Origin of Graphitic Carbon Nitride as a Photocatalyst. *ACS Omega* **2020**, *5*, 12557–12567. [CrossRef] [PubMed]
21. Sun, S.P.; Gu, S.; Sun, J.H.; Xia, F.F.; Chen, G.H. First Principles Investigation of the Electronic Properties of Graphitic Carbon Nitride with Different Building Block and Sheet Staggered Arrangement. *J. Alloys Compd.* **2018**, *735*, 131–139. [CrossRef]
22. Wang, F.; Ye, Y.; Cao, Y.; Zhou, Y. The Favorable Surface Properties of Heptazine Based G-C_3N_4 (001) in Promoting the Catalytic Performance towards CO_2 Conversion. *Appl. Surf. Sci.* **2019**, *481*, 604–610. [CrossRef]
23. Melissen, S.; Le Bahers, T.; Steinmann, S.N.; Sautet, P. Relationship between Carbon Nitride Structure and Exciton Binding Energies: A DFT Perspective. *J. Phys. Chem. C* **2015**, *119*, 25188–25196. [CrossRef]
24. Kaewsud, K.; Vchirawongkwin, V.; Ruangpornvisuti, V. Hydrogen Molecule Adsorption on Heptazine and Triazine Carbon Nitride Nanosheets and Their Selected-Elements (Periods 3 and 4)-Decorated Derivatives as Hydrogen Storage Materials. *J. Energy Storage* **2024**, *84*, 110926. [CrossRef]
25. Tyborski, T.; Merschjann, C.; Orthmann, S.; Yang, F.; Lux-Steiner, M.-C.; Schedel-Niedrig, T. Crystal Structure of Polymeric Carbon Nitride and the Determination of Its Process-Temperature-Induced Modifications. *J. Phys. Condens. Matter* **2013**, *25*, 395402. [CrossRef]
26. Fina, F.; Callear, S.K.; Carins, G.M.; Irvine, J.T.S. Structural Investigation of Graphitic Carbon Nitride via XRD and Neutron Diffraction. *Chem. Mater.* **2015**, *27*, 2612–2618. [CrossRef]
27. Devthade, V.; Kulhari, D.; Umare, S.S. Role of Precursors on Photocatalytic Behavior of Graphitic Carbon Nitride. *Mater. Today Proc.* **2018**, *5*, 9203–9210. [CrossRef]
28. Jiménez-Calvo, P.; Marchal, C.; Cottineau, T.; Caps, V.; Keller, V. Influence of the Gas Atmosphere during the Synthesis of G-C_3N_4 for Enhanced Photocatalytic H_2 Production from Water on Au/g-C_3N_4 Composites. *J. Mater. Chem. A Mater.* **2019**, *7*, 14849–14863. [CrossRef]
29. Florentino-Madiedo, L.; Díaz-Faes, E.; Barriocanal, C. Relationship between GCN Structure and Photocatalytic Water Splitting Efficiency. *Carbon* **2022**, *187*, 462–476. [CrossRef]
30. Zhao, Z.; Ma, Y.; Fan, J.; Xue, Y.; Chang, H.; Masubuchi, Y.; Yin, S. Synthesis of Graphitic Carbon Nitride from Different Precursors by Fractional Thermal Polymerization Method and Their Visible Light Induced Photocatalytic Activities. *J. Alloys Compd.* **2018**, *735*, 1297–1305. [CrossRef]
31. Alwin, E.; Kočí, K.; Wojcieszak, R.; Zieliński, M.; Edelmannová, M.; Pietrowski, M. Influence of High Temperature Synthesis on the Structure of Graphitic Carbon Nitride and Its Hydrogen Generation Ability. *Materials* **2020**, *13*, 2756. [CrossRef]

32. Yang, W.; Jia, L.; Wu, P.; Zhai, H.; He, J.; Liu, C.; Jiang, W. Effect of Thermal Program on Structure–Activity Relationship of g-C_3N_4 Prepared by Urea Pyrolysis and Its Application for Controllable Production of g-C_3N_4. *J. Solid State Chem.* **2021**, *304*, 122545. [CrossRef]
33. Zheng, Y.; Zhang, Z.; Li, C. A Comparison of Graphitic Carbon Nitrides Synthesized from Different Precursors through Pyrolysis. *J. Photochem. Photobiol. A Chem.* **2017**, *332*, 32–44. [CrossRef]
34. Mukhopadhyay, T.K.; Leherte, L.; Datta, A. Molecular Mechanism for the Self-Supported Synthesis of Graphitic Carbon Nitride from Urea Pyrolysis. *J. Phys. Chem. Lett.* **2021**, *12*, 1396–1406. [CrossRef]
35. Wang, Y.; Wang, X.; Antonietti, M. Polymeric Graphitic Carbon Nitride as a Heterogeneous Organocatalyst: From Photochemistry to Multipurpose Catalysis to Sustainable Chemistry. *Angew. Chem. Int. Ed.* **2012**, *51*, 68–89. [CrossRef]
36. Saman, F.; Se Ling, C.H.; Ayub, A.; Rafeny, N.H.B.; Mahadi, A.H.; Subagyo, R.; Nugraha, R.E.; Prasetyoko, D.; Bahruji, H. Review on Synthesis and Modification of G-C_3N_4 for Photocatalytic H2 Production. *Int. J. Hydrogen Energy* **2024**, *77*, 1090–1116. [CrossRef]
37. Guru, S.; Rao, G.R. Review—Strategic Design of Layered Double Hydroxides and Graphitic Carbon Nitride Heterostructures for Photoelectrocatalytic Water Splitting Applications. *J. Electrochem. Soc.* **2022**, *169*, 046515. [CrossRef]
38. Kessler, F.K.; Zheng, Y.; Schwarz, D.; Merschjann, C.; Schnick, W.; Wang, X.; Bojdys, M.J. Functional Carbon Nitride Materials—Design Strategies for Electrochemical Devices. *Nat. Rev. Mater.* **2017**, *2*, 17030. [CrossRef]
39. Alaghmandfard, A.; Ghandi, K. A Comprehensive Review of Graphitic Carbon Nitride (g-C_3N_4)–Metal Oxide-Based Nanocomposites: Potential for Photocatalysis and Sensing. *Nanomaterials* **2022**, *12*, 294. [CrossRef]
40. Lin, J.; Tian, W.; Guan, Z.; Zhang, H.; Duan, X.; Wang, H.; Sun, H.; Fang, Y.; Huang, Y.; Wang, S. Functional Carbon Nitride Materials in Photo-Fenton-Like Catalysis for Environmental Remediation. *Adv. Funct. Mater.* **2022**, *32*, 2201743. [CrossRef]
41. Yurdakal, S.; Garlisi, C.; Özcan, L.; Bellardita, M.; Palmisano, G. (Photo)Catalyst Characterization Techniques. In *Heterogeneous Photocatalysis*; Elsevier: Amsterdam, The Netherlands, 2019; pp. 87–152.
42. Wang, N.; Cheng, L.; Liao, Y.; Xiang, Q. Effect of Functional Group Modifications on the Photocatalytic Performance of G-C_3N_4. *Small* **2023**, *19*, e2300109. [CrossRef]
43. Liang, X.; Wang, G.; Dong, X.; Wang, G.; Ma, H.; Zhang, X. Graphitic Carbon Nitride with Carbon Vacancies for Photocatalytic Degradation of Bisphenol A. *ACS Appl. Nano Mater.* **2019**, *2*, 517–524. [CrossRef]
44. Preeyanghaa, M.; Erakulan, E.S.; Thapa, R.; Ashokkumar, M.; Neppolian, B. Scrutinizing the Role of Tunable Carbon Vacancies in G-C_3N_4 Nanosheets for Efficient Sonophotocatalytic Degradation of Tetracycline in Diverse Water Matrices: Experimental Study and Theoretical Calculation. *Chem. Eng. J.* **2023**, *452*, 139437. [CrossRef]
45. Molaei, M.J. Graphitic Carbon Nitride (g-C_3N_4) Synthesis and Heterostructures, Principles, Mechanisms, and Recent Advances: A Critical Review. *Int. J. Hydrogen Energy* **2023**, *48*, 32708–32728. [CrossRef]
46. Hao, X.-Q.; Yang, H.; Jin, Z.-L.; Xu, J.; Min, S.-X.; Lü, G.-X. Quantum Confinement Effect of Graphene-Like C_3N_4 Nanosheets for Efficient Photocatalytic Hydrogen Production From Water Splitting. *Acta Phys. Chim. Sin.* **2016**, *32*, 2581–2592. [CrossRef]
47. Wu, C.; Han, Q.; Qu, L. Functional Group Defect Design in Polymeric Carbon Nitride for Photocatalytic Application. *APL Mater.* **2020**, *8*, 120703. [CrossRef]
48. Zhu, X.; Duan, C.; Wang, W.; Xin, G.; Song, J. Fabrication of Carboxylated G-C_3N_4 with Excellent Adsorption and Photocatalytic Properties. *Mater. Lett.* **2022**, *317*, 132045. [CrossRef]
49. Wang, S.; Zhang, J.; Li, B.; Sun, H.; Wang, S. Engineered Graphitic Carbon Nitride-Based Photocatalysts for Visible-Light-Driven Water Splitting: A Review. *Energy Fuels* **2021**, *35*, 6504–6526. [CrossRef]
50. Iqbal, O.; Ali, H.; Li, N.; Al-Sulami, A.I.; Alshammari, K.F.; Abd-Rabboh, H.S.M.; Al-Hadeethi, Y.; Din, I.U.; Alharthi, A.I.; Altamimi, R.; et al. A Review on the Synthesis, Properties, and Characterizations of Graphitic Carbon Nitride (g-C_3N_4) for Energy Conversion and Storage Applications. *Mater. Today Phys.* **2023**, *34*, 101080. [CrossRef]
51. Katsumata, H.; Sakakibara, K.; Tateishi, I.; Furukawa, M.; Kaneco, S. Structurally Modified Graphitic Carbon Nitride with Highly Photocatalytic Activity in the Presence of Visible Light. *Catal. Today* **2020**, *352*, 47–53. [CrossRef]
52. Jiang, H.; Li, Y.; Wang, D.; Hong, X.; Liang, B. Recent Advances in Heteroatom Doped Graphitic Carbon Nitride (g-C_3N_4) and g-C_3N_4/Metal Oxide Composite Photocatalysts. *Curr. Org. Chem.* **2020**, *24*, 673–693. [CrossRef]
53. Huang, X.; Gu, W.; Ma, Y.; Liu, D.; Ding, N.; Zhou, L.; Lei, J.; Wang, L.; Zhang, J. Recent Advances of Doped Graphite Carbon Nitride for Photocatalytic Reduction of CO_2: A Review. *Res. Chem. Intermed.* **2020**, *46*, 5133–5164. [CrossRef]
54. Starukh, H.; Praus, P. Doping of Graphitic Carbon Nitride with Non-Metal Elements and Its Applications in Photocatalysis. *Catalysts* **2020**, *10*, 1119. [CrossRef]
55. Blinder, S.M. Density Functional Theory. In *Introduction to Quantum Mechanics*; Elsevier: Amsterdam, The Netherlands, 2021; pp. 235–244.
56. Du, S.; Zhang, F. General Applications of Density Functional Theory in Photocatalysis. *Chin. J. Catal.* **2024**, *61*, 1–36. [CrossRef]
57. Gusarov, S. Advances in Computational Methods for Modeling Photocatalytic Reactions: A Review of Recent Developments. *Materials* **2024**, *17*, 2119. [CrossRef]

58. Masood, H.; Toe, C.Y.; Teoh, W.Y.; Sethu, V.; Amal, R. Machine Learning for Accelerated Discovery of Solar Photocatalysts. *ACS Catal.* **2019**, *9*, 11774–11787. [CrossRef]
59. Silva, A.M.; Rojas, M.I. Electric and Structural Properties of Polymeric Graphite Carbon Nitride (g-C_3N_4): A Density Functional Theory Study. *Comput. Theor. Chem.* **2016**, *1098*, 41–49. [CrossRef]
60. Wang, J.; Hao, D.; Ye, J.; Umezawa, N. Determination of Crystal Structure of Graphitic Carbon Nitride: Ab Initio Evolutionary Search and Experimental Validation. *Chem. Mater.* **2017**, *29*, 2694–2707. [CrossRef]
61. Negro, P.; Cesano, F.; Casassa, S.; Scarano, D. Combined DFT-D3 Computational and Experimental Studies on g-C_3N_4: New Insight into Structure, Optical, and Vibrational Properties. *Materials* **2023**, *16*, 3644. [CrossRef] [PubMed]
62. Agrawal, S.; Lin, W.; Prezhdo, O.V.; Trivedi, D.J. Ab Initio Quantum Dynamics of Charge Carriers in Graphitic Carbon Nitride Nanosheets. *J. Chem. Phys.* **2020**, *153*, 054701. [CrossRef]
63. Jing, B.; Ao, Z.; Teng, Z.; Wang, C.; Yi, J.; An, T. Density Functional Theory Study on the Effects of Oxygen Groups on Band Gap Tuning of Graphitic Carbon Nitrides for Possible Photocatalytic Applications. *Sustain. Mater. Technol.* **2018**, *16*, 12–22. [CrossRef]
64. Ri, M.-H.; Ri, H.-M.; Ri, U.-S.; Kim, K.-I.; Kim, N.-H.; Sin, Y.-S. Ab Initio Study of Photocatalytic Characteristics of Graphitic Carbon Nitride Assisted by Oxalic Acid. *J. Mol. Model.* **2021**, *27*, 258. [CrossRef]
65. Srinivasu, K.; Modak, B.; Ghosh, S.K. Porous Graphitic Carbon Nitride: A Possible Metal-Free Photocatalyst for Water Splitting. *J. Phys. Chem. C* **2014**, *118*, 26479–26484. [CrossRef]
66. Chen, X.; Hu, R. DFT-Based Study of Single Transition Metal Atom Doped g-C_3N_4 as Alternative Oxygen Reduction Reaction Catalysts. *Int. J. Hydrogen Energy* **2019**, *44*, 15409–15416. [CrossRef]
67. Zhang, S.; Liu, Y.; Gu, P.; Ma, R.; Wen, T.; Zhao, G.; Li, L.; Ai, Y.; Hu, C.; Wang, X. Enhanced Photodegradation of Toxic Organic Pollutants Using Dual-Oxygen-Doped Porous g-C_3N_4: Mechanism Exploration from Both Experimental and DFT Studies. *Appl. Catal. B* **2019**, *248*, 1–10. [CrossRef]
68. Nguyen, T.T.H.; Le, M.C.; Ha, N.N. Understanding the Influence of Single Metal (Li, Mg, Al, Fe, Ag) Doping on the Electronic and Optical Properties of g-C_3N_4: A Theoretical Study. *Mol. Simul.* **2021**, *47*, 10–17. [CrossRef]
69. Lu, S.; Li, C.; Li, H.H.; Zhao, Y.F.; Gong, Y.Y.; Niu, L.Y.; Liu, X.J.; Wang, T. The Effects of Nonmetal Dopants on the Electronic, Optical and Chemical Performances of Monolayer g–C_3N_4 by First-Principles Study. *Appl. Surf. Sci.* **2017**, *392*, 966–974. [CrossRef]
70. Anastas, P.T.; Warner, J.C. *Green Chemistry*; Oxford University Press: Oxford, UK, 2000; ISBN 9780198506980.
71. Coronado, J.M. *Design of Advanced Photocatalytic Materials for Energy and Environmental Applications*; Green Energy and Technology; Coronado, J.M., Fresno, F., Hernández-Alonso, M.D., Portela, R., Eds.; Springer: London, UK, 2013; ISBN 978-1-4471-5060-2.
72. Naushad, M. (Ed.) *A New Generation Material Graphene: Applications in Water Technology*; Springer International Publishing: Cham, Switzerland, 2019; ISBN 978-3-319-75483-3.
73. Durgalakshmi, D.; Ajay Rakkesh, R.; Rajendran, S.; Naushad, M. *Green Photocatalysts*; Environmental Chemistry for a Sustainable World; Naushad, M., Rajendran, S., Lichtfouse, E., Eds.; Springer International Publishing: Cham, Switzerland, 2020; Volume 34, ISBN 978-3-030-15607-7.
74. Majdoub, M.; Sengottuvelu, D.; Nouranian, S.; Al-Ostaz, A. Graphitic Carbon Nitride Quantum Dots (G-C_3N_4 QDs): From Chemistry to Applications. *ChemSusChem* **2024**, *17*, e202301462. [CrossRef] [PubMed]
75. Wang, T.; Nie, C.; Ao, Z.; Wang, S.; An, T. Recent Progress in G-C_3N_4 Quantum Dots: Synthesis, Properties and Applications in Photocatalytic Degradation of Organic Pollutants. *J. Mater. Chem. A Mater.* **2020**, *8*, 485–502. [CrossRef]
76. Goren, A.Y.; Recepoglu, Y.K.; Vatanpour, V.; Yoon, Y.; Khataee, A. Insights into Engineered Graphitic Carbon Nitride Quantum Dots for Hazardous Contaminants Degradation in Wastewater. *Environ. Res.* **2023**, *223*, 115408. [CrossRef]
77. Li, Y.; Lv, K.; Ho, W.; Dong, F.; Wu, X.; Xia, Y. Hybridization of Rutile TiO_2 ($RTiO_2$) with g-C_3N_4 Quantum Dots (CN QDs): An Efficient Visible-Light-Driven Z-Scheme Hybridized Photocatalyst. *Appl. Catal. B* **2017**, *202*, 611–619. [CrossRef]
78. Feng, C.; Lu, Z.; Zhang, Y.; Liang, Q.; Zhou, M.; Li, X.; Yao, C.; Li, Z.; Xu, S. A Magnetically Recyclable Dual Z-Scheme GCNQDs-$CoTiO_3$/$CoFe_2O_4$ Composite Photocatalyst for Efficient Photocatalytic Degradation of Oxytetracycline. *Chem. Eng. J.* **2022**, *435*, 134833. [CrossRef]
79. Feng, C.; Wang, Y.; Lu, Z.; Liang, Q.; Zhang, Y.; Li, Z.; Xu, S. Nanoflower Ni_5P_4 Coupled with GCNQDs as Schottky Junction Photocatalyst for the Efficient Degradation of Norfloxacin. *Sep. Purif. Technol.* **2022**, *282*, 120107. [CrossRef]
80. Zheng, Y.; Liu, Y.; Guo, X.; Zhang, W.; Wang, Y.; Zhang, M.; Li, R.; Peng, Z.; Xie, H.; Huang, Y. In-Situ Construction of Morphology-Controllable 0D/1D g-C_3N_4 Homojunction with Enhanced Photocatalytic Activity. *Appl. Surf. Sci.* **2021**, *563*, 150317. [CrossRef]
81. Zhang, H.; Li, W.; Yan, Y.; Wang, W.; Ren, Y.; Li, X. Synthesis of Highly Porous G-C_3N_4 Nanotubes for Efficient Photocatalytic Degradation of Sulfamethoxazole. *Mater. Today Commun.* **2021**, *27*, 102288. [CrossRef]
82. Xie, M.; Wei, W.; Jiang, Z.; Xu, Y.; Xie, J. Carbon Nitride Nanowires/Nanofibers: A Novel Template-Free Synthesis from a Cyanuric Chloride–Melamine Precursor towards Enhanced Adsorption and Visible-Light Photocatalytic Performance. *Ceram. Int.* **2016**, *42*, 4158–4170. [CrossRef]

83. Mohamed, H.H.; Aziz, M.; Youssef, T.E.; Alomair, N.A. Microwave Synthesized G-C_3N_4 Nanofibers with Modified Properties for Enhanced Solar Light Photocatalytic Performance. *Inorg. Chem. Commun.* **2024**, *168*, 112975. [CrossRef]
84. Wang, Y.; Tan, G.; Dang, M.; Dong, S.; Liu, Y.; Liu, T.; Ren, H.; Xia, A.; Lv, L. Study on Surface Modification of G-C_3N_4 Photocatalyst. *J. Alloys Compd.* **2022**, *908*, 164507. [CrossRef]
85. Li, Y.; Wang, M.-Q.; Bao, S.-J.; Lu, S.; Xu, M.; Long, D.; Pu, S. Tuning and Thermal Exfoliation Graphene-like Carbon Nitride Nanosheets for Superior Photocatalytic Activity. *Ceram. Int.* **2016**, *42*, 18521–18528. [CrossRef]
86. Pattnaik, S.P.; Behera, A.; Martha, S.; Acharya, R.; Parida, K. Facile Synthesis of Exfoliated Graphitic Carbon Nitride for Photocatalytic Degradation of Ciprofloxacin Under Solar Irradiation. *J. Mater. Sci.* **2019**, *54*, 5726–5742. [CrossRef]
87. Wang, Y.; Liu, M.; Fan, F.; Li, G.; Duan, J.; Li, Y.; Jiang, G.; Yao, W. Enhanced Full-Spectrum Photocatalytic Activity of 3D Carbon-Coated C_3N_4 Nanowires via Giant Interfacial Electric Field. *Appl. Catal. B* **2022**, *318*, 121829. [CrossRef]
88. Wang, Y.; Yang, X.; Tian, T.; Liu, Y.; Chen, Y.; Xu, G.; Gu, L.; Li, H.; Yuan, Y. Chitosan-Assisted Synthesis of 1D g-C_3N_4 Nanorods for Enhanced Photocatalysis. *Chem. Commun.* **2023**, *59*, 10528–10531. [CrossRef]
89. Zhang, W.; Xu, D.; Wang, F.; Liu, H.; Chen, M. Enhanced Photocatalytic Performance of S/Cd Co-Doped g-C_3N_4 Nanorods for Degradation of Dyes. *Colloids Surf. A Physicochem. Eng. Asp.* **2022**, *653*, 130079. [CrossRef]
90. Jiang, Z.; Zhang, X.; Chen, H.; Hu, X.; Yang, P. Formation of G-C_3N_4 Nanotubes towards Superior Photocatalysis Performance. *ChemCatChem* **2019**, *11*, 4558–4567. [CrossRef]
91. Joy, J.; Anas, S.; Thomas, S. Graphitic Carbon Nitride: An Uprising Carbonaceous Material. In *Synthesis, Characterization, and Applications of Graphitic Carbon Nitride*; Elsevier: Amsterdam, The Netherlands, 2023; pp. 1–14.
92. Torres-Pinto, A.; Silva, C.G.; Faria, J.L.; Silva, A.M.T. The Effect of Precursor Selection on the Microwave-Assisted Synthesis of Graphitic Carbon Nitride. *Catal. Today* **2023**, *424*, 113868. [CrossRef]
93. Papailias, I.; Todorova, N.; Giannakopoulou, T.; Ioannidis, N.; Boukos, N.; Athanasekou, C.P.; Dimotikali, D.; Trapalis, C. Chemical vs Thermal Exfoliation of G-C_3N_4 for NOx Removal under Visible Light Irradiation. *Appl. Catal. B* **2018**, *239*, 16–26. [CrossRef]
94. Zhang, M.; Yang, Y.; An, X.; Zhao, J.; Bao, Y.; Hou, L. Exfoliation Method Matters: The Microstructure-Dependent Photoactivity of g-C_3N_4 Nanosheets for Water Purification. *J. Hazard. Mater.* **2022**, *424*, 127424. [CrossRef] [PubMed]
95. Dong, J.; Zhang, Y.; Hussain, M.I.; Zhou, W.; Chen, Y.; Wang, L.-N. G-C_3N_4: Properties, Pore Modifications, and Photocatalytic Applications. *Nanomaterials* **2021**, *12*, 121. [CrossRef]
96. Qian, X.; Meng, X.; Sun, J.; Jiang, L.; Wang, Y.; Zhang, J.; Hu, X.; Shalom, M.; Zhu, J. Salt-Assisted Synthesis of 3D Porous g-C_3N_4 as a Bifunctional Photo- and Electrocatalyst. *ACS Appl. Mater. Interfaces* **2019**, *11*, 27226–27232. [CrossRef]
97. Chen, X.; Wang, H.; Meng, R.; Chen, M. Porous Graphitic Carbon Nitride Synthesized via Using Carbon Nanotube as a Novel Recyclable Hard Template for Efficient Visible Light Photocatalytic Organic Pollutant Degradation. *ChemistrySelect* **2019**, *4*, 6123–6129. [CrossRef]
98. Yan, Q.; Zhao, C.; Zhang, L.; Hou, Y.; Wang, S.; Dong, P.; Lin, F.; Wang, Y. Facile Two-Step Synthesis of Porous Carbon Nitride with Enhanced Photocatalytic Activity Using a Soft Template. *ACS Sustain. Chem. Eng.* **2019**, *7*, 3866–3874. [CrossRef]
99. Chen, Y.; Ding, F.; Khaing, A.; Yang, D.; Jiang, Z. Acetic Acid-Assisted Supramolecular Assembly Synthesis of Porous g-C_3N_4 Hexagonal Prism with Excellent Photocatalytic Activity. *Appl. Surf. Sci.* **2019**, *479*, 757–764. [CrossRef]
100. Wang, Y.; Jing, B.; Wang, F.; Wang, S.; Liu, X.; Ao, Z.; Li, C. Mechanism Insight into Enhanced Photodegradation of Pharmaceuticals and Personal Care Products in Natural Water Matrix over Crystalline Graphitic Carbon Nitrides. *Water Res.* **2020**, *180*, 115925. [CrossRef]
101. Dou, M.; Wang, J.; Gao, B.; Xu, C.; Yang, F. Photocatalytic Difference of Amoxicillin and Cefotaxime under Visible Light by Mesoporous G-C_3N_4: Mechanism, Degradation Pathway and DFT Calculation. *Chem. Eng. J.* **2020**, *383*, 123134. [CrossRef]
102. Zhao, S.; Fang, J.; Wang, Y.; Zhang, Y.; Zhou, Y.; Zhuo, S. Construction of Three-Dimensional Mesoporous Carbon Nitride with High Surface Area for Efficient Visible-Light-Driven Hydrogen Evolution. *J. Colloid Interface Sci.* **2020**, *561*, 601–608. [CrossRef]
103. Liu, J.; Fu, W.; Liao, Y.; Fan, J.; Xiang, Q. Recent Advances in Crystalline Carbon Nitride for Photocatalysis. *J. Mater. Sci. Technol.* **2021**, *91*, 224–240. [CrossRef]
104. Li, Z.; Ma, L.; Yu, M.; Chang, S.; Huang, Z.; Cheng, Z.; Li, Y.; Carabineiro, S.A.C.; Lv, K. Crystalline Graphitic Carbon Nitride in Photocatalysis. *Surf. Interfaces* **2024**, *51*, 104492. [CrossRef]
105. Xu, T.; Hur, J.; Niu, P.; Wang, S.; Lee, S.; Chun, S.-E.; Li, L. Synthesis of Crystalline G-C_3N_4 with Rock/Molten Salts for Efficient Photocatalysis and Piezocatalysis. *Green Energy Environ.* **2024**, *9*, 890–898. [CrossRef]
106. Tan, Y.; Zhang, Z.; Pu, Y.; Sun, Z.; Huang, L. Progress of Crystalline Carbon Nitride in Synthesis, Atomic Structure Characterization and Photocatalysis. *Nano Res.* **2025**, *18*, 94907047. [CrossRef]
107. Pu, W.; Zhou, Y.; Yang, L.; Gong, H.; Li, Y.; Yang, Q.; Zhang, D. High-Efficiency Crystalline Carbon Nitride Photocatalysts: Status and Perspectives. *Nano Res.* **2024**, *17*, 7840–7863. [CrossRef]
108. Zhai, B.; Li, H.; Gao, G.; Wang, Y.; Niu, P.; Wang, S.; Li, L. A Crystalline Carbon Nitride Based Near-Infrared Active Photocatalyst. *Adv. Funct. Mater.* **2022**, *32*, 2207375. [CrossRef]

109. Li, H.; Zhang, Z.; Liu, Y.; Cen, W.; Luo, X. Functional Group Effects on the HOMO–LUMO Gap of g-C_3N_4. *Nanomaterials* **2018**, *8*, 589. [CrossRef]
110. Jing, L.; Xu, Y.; Xie, M.; Li, Z.; Wu, C.; Zhao, H.; Zhong, N.; Wang, J.; Wang, H.; Yan, Y.; et al. Cyano-Rich G-C_3N_4 in Photochemistry: Design, Applications, and Prospects. *Small* **2024**, *20*, e2304404. [CrossRef]
111. Li, Y.; Li, X.; Zhang, H.; Fan, J.; Xiang, Q. Design and Application of Active Sites in G-C_3N_4-Based Photocatalysts. *J. Mater. Sci. Technol.* **2020**, *56*, 69–88. [CrossRef]
112. Shiravand, G.; Badiei, A.; Mohammadi Ziarani, G. Carboxyl-Rich g-C_3N_4 Nanoparticles: Synthesis, Characterization and Their Application for Selective Fluorescence Sensing of Hg^{2+} and Fe^{3+} in Aqueous Media. *Sens. Actuators B Chem.* **2017**, *242*, 244–252. [CrossRef]
113. Teng, Z.; Yang, N.; Lv, H.; Wang, S.; Hu, M.; Wang, C.; Wang, D.; Wang, G. Edge-Functionalized g-C_3N_4 Nanosheets as a Highly Efficient Metal-Free Photocatalyst for Safe Drinking Water. *Chem* **2019**, *5*, 664–680. [CrossRef]
114. Chuang, P.-K.; Wu, K.-H.; Yeh, T.-F.; Teng, H. Extending the π-Conjugation of g-C_3N_4 by Incorporating Aromatic Carbon for Photocatalytic H_2 Evolution from Aqueous Solution. *ACS Sustain. Chem. Eng.* **2016**, *4*, 5989–5997. [CrossRef]
115. Razavi-Esfali, M.; Mahvelati-Shamsabadi, T.; Fattahimoghaddam, H.; Lee, B.-K. Highly Efficient Photocatalytic Degradation of Organic Pollutants by Mesoporous Graphitic Carbon Nitride Bonded with Cyano Groups. *Chem. Eng. J.* **2021**, *419*, 129503. [CrossRef]
116. Xu, R.; Li, J.; Sui, G.; Zhuang, Y.; Guo, D.; Luo, Z.; Liang, S.; Yao, H.; Wang, C.; Chen, S. Constructing Supramolecular Self-Assembled Porous g-C_3N_4 Nanosheets Containing Thiophene-Groups for Excellent Photocatalytic Performance Under Visible Light. *Appl. Surf. Sci.* **2022**, *578*, 152064. [CrossRef]
117. Lee, Y.-J.; Jeong, Y.J.; Cho, I.S.; Park, S.-J.; Lee, C.-G.; Alvarez, P.J.J. Facile Synthesis of N Vacancy G-C_3N_4 Using Mg-Induced Defect on the Amine Groups for Enhanced Photocatalytic •OH Generation. *J. Hazard. Mater.* **2023**, *449*, 131046. [CrossRef]
118. Wang, Y.; Rao, L.; Wang, P.; Guo, Y.; Shi, Z.; Guo, X.; Zhang, L. Synthesis of Nitrogen Vacancies G-C_3N_4 with Increased Crystallinity under the Controlling of Oxalyl Dihydrazide: Visible-Light-Driven Photocatalytic Activity. *Appl. Surf. Sci.* **2020**, *505*, 144576. [CrossRef]
119. Huang, Z.; Chen, Z.; Qayum, A.; Zhao, X.; Xia, H.; Lu, F.; Hu, L. Enhanced Photocatalytic Degradation of 4-Chlorophenol under Visible Light over Carbon Nitride Nanosheets with Carbon Vacancies. *Nanotechnology* **2021**, *32*, 415704. [CrossRef]
120. Deng, Y.; Liu, J.; Huang, Y.; Ma, M.; Liu, K.; Dou, X.; Wang, Z.; Qu, S.; Wang, Z. Engineering the Photocatalytic Behaviors of G-C_3N_4-Based Metal-Free Materials for Degradation of a Representative Antibiotic. *Adv. Funct. Mater.* **2020**, *30*, 2002353. [CrossRef]
121. Long, X.; Feng, C.; Yang, S.; Ding, D.; Feng, J.; Liu, M.; Chen, Y.; Tan, J.; Peng, X.; Shi, J.; et al. Oxygen Doped Graphitic Carbon Nitride with Regulatable Local Electron Density and Band Structure for Improved Photocatalytic Degradation of Bisphenol A. *Chem. Eng. J.* **2022**, *435*, 134835. [CrossRef]
122. Zhou, Y.; Zhang, L.; Liu, J.; Fan, X.; Wang, B.; Wang, M.; Ren, W.; Wang, J.; Li, M.; Shi, J. Brand New P-Doped g-C_3N_4: Enhanced Photocatalytic Activity for H_2 Evolution and Rhodamine B Degradation under Visible Light. *J. Mater. Chem. A Mater.* **2015**, *3*, 3862–3867. [CrossRef]
123. Li, Z.; Chen, Q.; Lin, Q.; Chen, Y.; Liao, X.; Yu, H.; Yu, C. Three-Dimensional P-Doped Porous g-C_3N_4 Nanosheets as an Efficient Metal-Free Photocatalyst for Visible-Light Photocatalytic Degradation of Rhodamine B Model Pollutant. *J. Taiwan Inst. Chem. Eng.* **2020**, *114*, 249–262. [CrossRef]
124. Guan, K.; Li, J.; Lei, W.; Wang, H.; Tong, Z.; Jia, Q.; Zhang, H.; Zhang, S. Synthesis of Sulfur Doped G-C_3N_4 with Enhanced Photocatalytic Activity in Molten Salt. *J. Mater.* **2021**, *7*, 1131–1142. [CrossRef]
125. Dou, Y.; Shen, X.; Zou, J.; Shi, R.; Yan, T.; Sun, Q.; Wang, L. Green Synthesis of Sulfur-doped G-C_3N_4 Nanosheets for Enhanced Removal of Oxytetracycline under Visible-light Irradiation and Reduction of Its N-Nitrosodimethylamine Formation Potential. *J. Chem. Technol. Biotechnol.* **2021**, *96*, 1580–1592. [CrossRef]
126. Lei, L.; Wang, W.; Wang, C.; Zhang, M.; Zhong, Q.; Fan, H. In Situ Growth of Boron Doped G-C_3N_4 on Carbon Fiber Cloth as a Recycled Flexible Film-Photocatalyst. *Ceram. Int.* **2021**, *47*, 1258–1267. [CrossRef]
127. Zou, J.; Yu, Y.; Yan, W.; Meng, J.; Zhang, S.; Wang, J. A Facile Route to Synthesize Boron-Doped g-C_3N_4 Nanosheets with Enhanced Visible-Light Photocatalytic Activity. *J. Mater. Sci.* **2019**, *54*, 6867–6881. [CrossRef]
128. Guo, F.; Li, M.; Ren, H.; Huang, X.; Shu, K.; Shi, W.; Lu, C. Facile Bottom-up Preparation of Cl-Doped Porous g-C_3N_4 Nanosheets for Enhanced Photocatalytic Degradation of Tetracycline under Visible Light. *Sep. Purif. Technol.* **2019**, *228*, 115770. [CrossRef]
129. Hong, J.; Hwang, D.K.; Selvaraj, R.; Kim, Y. Facile Synthesis of Br-Doped g-C_3N_4 Nanosheets via One-Step Exfoliation Using Ammonium Bromide for Photodegradation of Oxytetracycline Antibiotics. *J. Ind. Eng. Chem.* **2019**, *79*, 473–481. [CrossRef]
130. Shi, Y.; Li, L.; Xu, Z.; Sun, H.; Guo, F.; Shi, W. One-step Simple Green Method to Prepare Carbon-doped Graphitic Carbon Nitride Nanosheets for Boosting Visible-light Photocatalytic Degradation of Tetracycline. *J. Chem. Technol. Biotechnol.* **2021**, *96*, 3122–3133. [CrossRef]

131. Li, T.; Zhang, X.; Hu, C.; Li, X.; Zhang, P.; Chen, Z. Porous Carbon-Doped g-C_3N_4 with Tunable Band Structure for Boosting Photocatalytic H_2O_2 Production with Simultaneous Pollutants Removal. *J. Environ. Chem. Eng.* **2022**, *10*, 107116. [CrossRef]
132. Jiang, L.; Yuan, X.; Zeng, G.; Liang, J.; Wu, Z.; Yu, H.; Mo, D.; Wang, H.; Xiao, Z.; Zhou, C. Nitrogen Self-Doped g-C_3N_4 Nanosheets with Tunable Band Structures for Enhanced Photocatalytic Tetracycline Degradation. *J. Colloid. Interface Sci.* **2019**, *536*, 17–29. [CrossRef]
133. Zhu, D.; Zhou, Q. Nitrogen Doped G-C_3N_4 with the Extremely Narrow Band Gap for Excellent Photocatalytic Activities under Visible Light. *Appl. Catal. B* **2021**, *281*, 119474. [CrossRef]
134. Huang, J.; Li, D.; Li, R.; Zhang, Q.; Chen, T.; Liu, H.; Liu, Y.; Lv, W.; Liu, G. An Efficient Metal-Free Phosphorus and Oxygen Co-Doped g-C_3N_4 Photocatalyst with Enhanced Visible Light Photocatalytic Activity for the Degradation of Fluoroquinolone Antibiotics. *Chem. Eng. J.* **2019**, *374*, 242–253. [CrossRef]
135. Jiang, L.; Yuan, X.; Zeng, G.; Chen, X.; Wu, Z.; Liang, J.; Zhang, J.; Wang, H.; Wang, H. Phosphorus- and Sulfur-Codoped g-C_3N_4: Facile Preparation, Mechanism Insight, and Application as Efficient Photocatalyst for Tetracycline and Methyl Orange Degradation under Visible Light Irradiation. *ACS Sustain. Chem. Eng.* **2017**, *5*, 5831–5841. [CrossRef]
136. Lei, L.; He, X.; Lin, X.; Zhao, Y.; Yang, C.; Cui, L.; Wu, G. Preparation of Carbon Self-Doped g-C_3N_4 for Efficient Degradation of Bisphenol A under Visible Light Irradiation. *Environ. Sci. Pollut. Res.* **2023**, *30*, 65328–65337. [CrossRef]
137. Yan, W.; Yan, L.; Jing, C. Impact of Doped Metals on Urea-Derived g-C_3N_4 for Photocatalytic Degradation of Antibiotics: Structure, Photoactivity and Degradation Mechanisms. *Appl. Catal. B* **2019**, *244*, 475–485. [CrossRef]
138. Zhang, H.; Tang, Y.; Liu, Z.; Zhu, Z.; Tang, X.; Wang, Y. Study on Optical Properties of Alkali Metal Doped G-C_3N_4 and Their Photocatalytic Activity for Reduction of CO_2. *Chem. Phys. Lett.* **2020**, *751*, 137467. [CrossRef]
139. Yu, X.; Song, Z.; Dong, X.; Li, H.; Liu, H.; Zhao, B.; Ye, T.; Jiang, Y.; Li, X.; Duan, L.; et al. Enhanced Photocatalytic Activity of Rare Earth (Yb, Nd and Ce)-Doped g-C_3N_4 Nanosheets for the Degradation of Organic Dyes Under Visible Light. *J. Mater. Sci. Mater. Electron.* **2022**, *33*, 13271–13289. [CrossRef]
140. Ren, Z.; Chen, F.; Wen, K.; Lu, J. Enhanced Photocatalytic Activity for Tetracyclines Degradation with Ag Modified G-C_3N_4 Composite under Visible Light. *J. Photochem. Photobiol. A Chem.* **2020**, *389*, 112217. [CrossRef]
141. Yin, Z.; Tian, Y.; Gao, P.; Feng, L.; Liu, Y.; Du, Z.; Zhang, L. Photodegradation Mechanism and Genetic Toxicity of Bezafibrate by Pd/g-C_3N_4 Catalysts under Simulated Solar Light Irradiation: The Role of Active Species. *Chem. Eng. J.* **2020**, *379*, 122294. [CrossRef]
142. Nguyen Van, M.; Mai, O.; Pham Do, C.; Lam Thi, H.; Pham Manh, C.; Nguyen Manh, H.; Pham Thi, D.; Do Danh, B. Fe-Doped g-C_3N_4: High-Performance Photocatalysts in Rhodamine B Decomposition. *Polymers* **2020**, *12*, 1963. [CrossRef]
143. Zhou, F.Y.; Mao, J.N.; Peng, X.L.; Hong, B.; Xu, J.C.; Zeng, Y.X.; Han, Y.B.; Ge, H.L.; Wang, X.Q. Magnetically Separable Ni/g-C_3N_4 Nanocomposites for Enhanced Visible-Light Photocatalytic Degradation of Methylene Blue and Ciprofloxacin. *Diam. Relat. Mater.* **2022**, *126*, 109070. [CrossRef]
144. Bao, J.; Bai, W.; Wu, M.; Gong, W.; Yu, Y.; Zheng, K.; Liu, L. Template-Mediated Copper Doped Porous g-C_3N_4 for Efficient Photodegradation of Antibiotic Contaminants. *Chemosphere* **2022**, *293*, 133607. [CrossRef] [PubMed]
145. Faisal, M.; Jalalah, M.; Harraz, F.A.; El-Toni, A.M.; Khan, A.; Al-Assiri, M.S. Au Nanoparticles-Doped g-C_3N_4 Nanocomposites for Enhanced Photocatalytic Performance under Visible Light Illumination. *Ceram. Int.* **2020**, *46*, 22090–22101. [CrossRef]
146. *ISO 10678*; Fine Ceramics (Advanced Ceramics, Advanced Technical Ceramics)—Determination of Photocatalytic Activity of Surfaces in an Aqueous Medium by Degradation of Methylene Blue. International Organization for Standardization: Geneva, Switzerland, 2024.
147. *ISO 22197*; Fine Ceramics (Advanced Ceramics, Advanced Technical Ceramics)—Test Method for Air-Purification Performance of Semiconducting Photocatalytic Materials—Part 1: Removal of Nitric Oxide. International Organization for Standardization: Geneva, Switzerland, 2016.
148. *ISO 10676*; Fine Ceramics (Advanced Ceramics, Advanced Technical Ceramics)—Test Method for Water Purification Performance of Semiconducting Photocatalytic Materials by Measurement of Forming Ability of Active Oxygen. International Organization for Standardization: Geneva, Switzerland, 2010.
149. Mills, A.; Hill, C.; Robertson, P.K.J. Overview of the Current ISO Tests for Photocatalytic Materials. *J. Photochem. Photobiol. A Chem.* **2012**, *237*, 7–23. [CrossRef]

catalysts

MDPI

Article

The Direct Formation of an Iron Citrate Complex Using a Metallurgical Slag as an Iron Source for Micropollutant Removal via the Photo-Fenton Process

Sandra Yazmin Arzate Salgado *, Ana Yañez-Aulestia and Rosa-María Ramírez-Zamora *

Coordinación de Ingeniería Ambiental, Instituto de Ingeniería, Universidad Nacional Autónoma de México, Circuito Escolar s/n, Ciudad Universitaria, Coyoacán, Ciudad de Mexico 04510, Mexico

* Correspondence: sarzates@iingen.unam.mx (S.Y.A.S.); rramirezz@iingen.unam.mx (R.-M.R.-Z.)

Abstract: Following the goals of the circular economy, this work demonstrates that an industrial by-product can be used in environmental remediation. Metallurgical slag and citric acid were used to form an Fe:Cit complex by simultaneously carrying out the lixiviation of the iron and the chelating stages with an 87% iron recovery. This complex was evaluated in the photo-Fenton process to produce $HO^{\bullet}$ through salicylic acid dosimetry or salicylic acid hydroxylation, producing 0.13 ± 0.1 mM $HO^{\bullet}$ after 30 min of operation; such a value is three orders of magnitude higher than the one reported for the metallurgical slag (as a heterogeneous catalyst, 22 μM) in the photo-Fenton-like process. The system was tested for its ability to degrade a mixture of drugs, including dexamethasone (DEX), naproxen (NAP), and ketorolac (KTR), which are often used to treat the symptoms of COVID-19. The drug degradation tests were performed in two stages. In the first stage, the Fe:Cit complex from the metallurgical slag was compared to the one formed by analytical-grade reactants; the drug degradation was faster for the former, with the major difference being observed at 5 cm and 500 W/m^2. Here, 85–90% of the drugs was degraded in 5 min using Fe:Cit from slag, while at least 20 min was necessary to achieve such degradation with the analytical reagent, conceivably because of the trace compounds being lixiviated from the slag. Then, the effects of the liquid depth (5, 10, and 15 cm) and irradiance (250, 500, and 750 W/m^2) were tested; the pseudo-first-order kinetic degradation constants for the three model pollutants were in the range of $0.009 > k_D > 0.09$ min^{-1}, showing that degradation is more feasible for DEX than for NAP and KRT because the radical attack feasibility is related to the molecular structures.

Keywords: contaminants of emerging concern; waste valorisation; iron chelate; advanced oxidation process; organic acid

1. Introduction

Water pollution is a global issue that should be addressed to promote health, reduce environmental damage, and allow for sustainable development [1]. Ryder [2] reported that, in 2015, the percentage of untreated wastewater was 30% for high-income countries, and this value increased to 92% for low-income countries. Further, in arid and semi-arid areas, river water contaminated with unacceptable levels of pollution and raw sewage—and, in a few cases, treated wastewater—is used for irrigation near large cities (i.e., peri-urban agriculture).

Additionally, the installed Wastewater Treatment Plants (WWTP) are mainly designed to remove carbon, nitrogen, and phosphorus from wastewater. Consequently, micropollutants (MPs) and contaminants of emerging concern (CECs) are poorly removed in secondary treatments like conventional activated sludge (CAS), membrane bioreactors (MBRs), moving-bed biofilm reactors (MBBRs), and constructed wetlands (CWs) [3], making it evident that the installation of tertiary treatments is necessary [4–6]. In this way, physicochemical processes and advanced oxidation processes (AOPs) have been widely studied to determine their MP removal capacity and the feasibility of their implementation on a large scale under the specific needs of the project [7–10].

AOPs, specifically Fenton and photo-Fenton processes, have been extensively studied for their ability to degrade recalcitrant organic contaminants in wastewater effluents [11,12]. One of the advantages of this technology is its capability of using solar light, diminishing energy consumption [13,14]. Several engineering-scale demonstration plants have been erected, exhibiting the potentiality of installing such technology on a large scale [15,16].

The photo-Fenton's operation conditions have also been explored in order to enhance the treatment capacity, with continuous operation being a way to avoid the load and discharge of the reactor and, consequently, make better use of the radiation available during sunshine hours [17]. Another important advantage is the use of chelating agents to maintain the iron in dissolution at a neutral pH [18–20], which leads to fast kinetic degradation constants and increases the treatment capacity to values similar to those during operation at an acidic pH; however, the operation costs for using chelating agents under neutral pH conditions remain high [21].

Metallurgical slags are by-products generated by the pyrometallurgical processing of ferrous and non-ferrous metal ores and secondary resources (recycling). Slags from copper production typically represent approximately 2 tons per ton of produced copper [22], representing an important environmental problem that needs to be addressed. However, these slags are mainly composed of iron compounds [23] that exhibit photo-activity under visible light irradiation due to their mineralogical phases and band gap values [24]. In this way, metallurgical slags have been proposed as an alternative source of catalysts, following the idea that using a by-product of one industrial activity as a raw material for another process has a positive effect on the environment and may also be cheaper than using commercial reagents. Thus, metallurgical slags (MSs) from copper production have been used as heterogeneous and homogeneous Fenton-type photocatalysts for micropollutant degradation [25,26] and pathogen inactivation [23].

The work reported by Huanosta-Gutiérrez et al. [26] shows the feasibility of lixiviating iron from MS and using it as a catalyst during the photo-Fenton (pF) reaction for phenol degradation. The lixiviation of iron is possible due to the solubility of the iron at an acidic pH; such a condition can be achieved by adding sulphuric acid.

Thus, considering what was described above, citric acid (CA) has been successfully used in photo-Fenton treatment as a chelating agent since it is biodegradable, it has a relatively low molecular weight, and the iron/citrate (Fe:Cit) complex remains stable at high pH values (up to 9.0) [27]. Therefore, a direct complexation of MS may be achieved by the addition of an organic acid.

In this work, the direct formation of the Fe:Cit complex using metallurgical slag (MS) as an iron source is tested. Then, Fe:Cit was used for $HO^{\bullet}$ radical formation by photo-Fenton, and drug degradation under different radiation and liquid depth operation conditions was also assessed. Dexamethasone (DEX), naproxen (NAP), and ketorolac (KTR) were used as model pollutants. DEX is a corticosteroid (cortisone-like medicine or steroid), while NAP and KTR are nonsteroidal anti-inflammatory drugs (NSAIDs). The three are known for being recalcitrant to biological treatments and are often found in environmental matrices [28]. Also, since they have been used for the treatment of COVID-19 [29], their environmental concentrations have increased, principally during the first years of the pandemic [30].

2. Results

2.1. Metallurgical Slag Characterisation

A sample of metallurgical slag (MS) from the copper industry was superficially and microstructurally characterised by a N_2 adsorption–desorption isotherm analysis and scanning electron microscopy, the mineralogical phases were identified by PXRD, and its elemental composition was determined by XRF.

The N_2 isotherm showed a type II profile and a very narrow H3 hysteresis loop according to the IUPAC classification [31] Its behaviour corresponds to that of a non-porous material, which agrees with a low BET surface area of 0.35 m^2/g (Figure 1a).

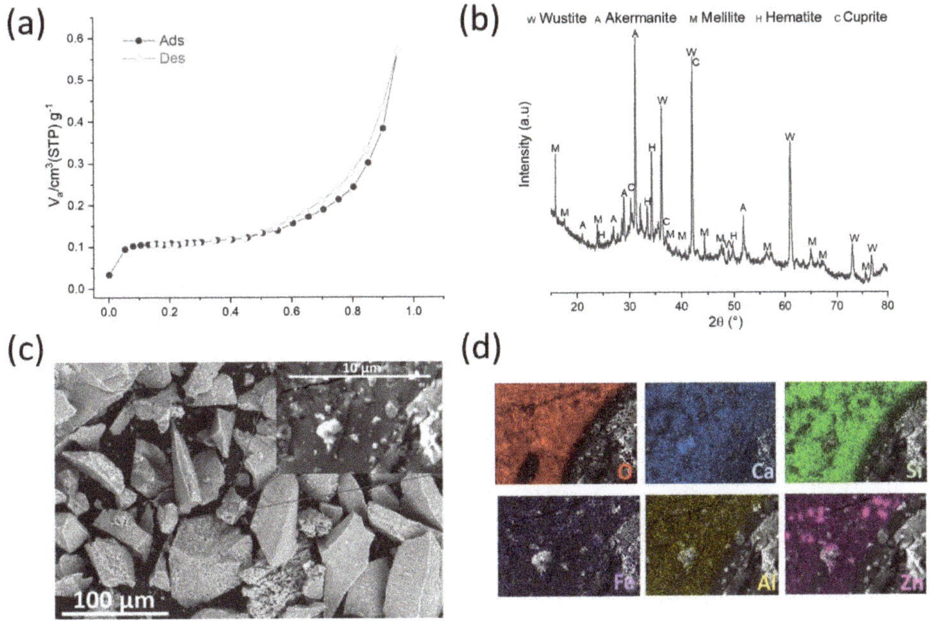

Figure 1. Metallurgical copper slag characterisation. (**a**) N_2 adsorption isotherm, (**b**) PXRD diffractogram, (**c**) SEM image using backscattering electrons, and (**d**) mapping to specific point of 10 µm on slag surface.

The MS diffractogram was analysed with different databases, with a match being found with five mineral crystalline phases: Wüstite (ICSD 98 010 876), Akermanite (ICSD 98 002 0391), Melilite (ICSD 98 017 1600), Cuprite (ICDD 04 005 4871Cuprite), and Hematite (ICDD 00 033 0664) (Figure 1b). These results are in good agreement with the XRF results, which are expressed in terms of oxide percentages as follows: Fe_2O_3, SiO_2, CaO, ZnO, and Al_2O_3 with 34.45, 22.95, 22.06, 8.07, and 4.80%, respectively. The remaining percentage (7.67%) corresponds to other elements that are present in the slag in smaller quantities.

The SEM images obtained using a backscattering detector showed irregular forms with uneven faces and particle sizes around 110 µm (Figure 1c). In addition, to further analyse the distribution of different elements on the particle surface, Figure 1d shows the results of the corresponding EDS elemental mapping analysis, which identified oxygen, calcium, silicon, iron, aluminium, and zinc. As it can be seen, the elements depicted a homogeneous distribution, and their presence correctly agreed with the results reported in the XRF analysis.

This kind of metallurgical slag was previously tested as a catalyst in the heterogeneous photo-Fenton process; the selection of this slag is the result of the characterisation of six metallurgical slags from different industries and process stages. The first selection was carried out based on the iron content. Then, two slags were compared, one from the iron and the other from the copper industry; they were tested by using the dosimetry technique with salicylic acid (SA), and the metallurgical slag was the one that produced $HO^{\bullet}$ [32].

2.2. Iron Leachate and Complexation

A volume of 7.04 ± 0.8 mL of supernatant was obtained each day after 24 h of contact between the MS and the CA, and in each of them, there was an iron recuperation around 8–20 g Fe/L, which represents a recovery of 87% from the MS's initial iron; the molar proportion between the iron and CA was variable by 0.05–0.125 CA mol for each Fe mol. Consequently, the iron recovered in some of the cases was not fully complexed because the minimum molar proportion for the complexation is 1:1 according to Silva et al. [33]. However, it guarantees the solubility of the iron at a neutral pH.

Also, the FTIR spectra (Figure 2) of the supernatant (Fe_{MS}:Cit complex, 0.1 Mm, 1:0.007) exhibited a similar spectrum to the one of the Fe:Cit complex at 0.1 mM (1:1). Because the analysis was carried out in water, the two characteristic bands of the O-H bond, 3350 and 1645 cm^{-1}, are shown in both samples [34]. The wide peak at 3350 cm^{-1} is associated with the O-H bond and the O—H produced by the hydrogen bonds that are generated when the molecules interact with each other, and the band is intense and broad. The second one, around 1645 cm^{-1}, is associated with the H-O-H bond vibration [34].

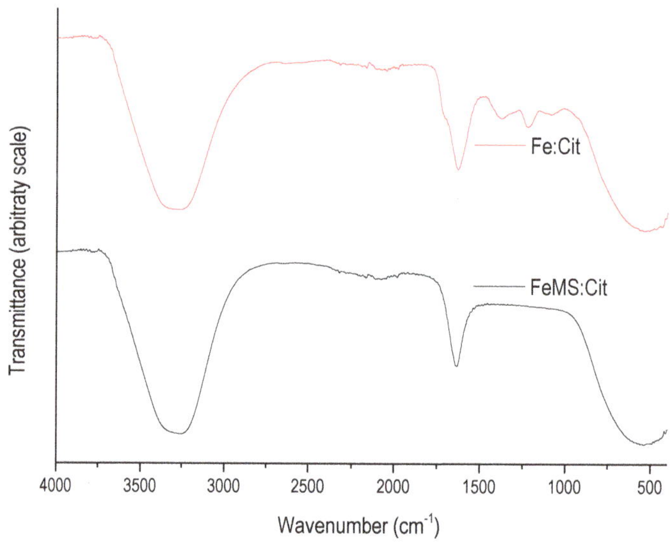

Figure 2. FTIR spectra of Fe:Cit complex 0.1 mM (1:1) and Fe_{MS}:Cit complex 0.1 Mm (1:1).

In both samples, an Fe-O bond is observed around 556 cm^{-1} [35]. This may indicate that the iron is now forming bonds with the oxygen from the citrate. Such a fact is also supported by the absence of a band at 1725 cm^{-1}, which is characteristic of the carboxylic group C=O, and that peak is absent in the complex [34].

Additionally, the bands observed at 1149 and 1112 cm^{-1} are associated with the group SO_4^{2-}. They appear in the sample of the iron citrate complex generated from the iron sulphate, but they are absent in the complex generated by the slag. The complexation of the iron provided by the cooper slag is an alternative to performing the photo-Fenton process in a homogeneous phase at a neutral pH while the MS is revalorised.

2.3. Hydroxyl Radical Formation

Hydroxyl radical formation was tested using each one of the supernatants obtained by the procedure explained in Section 3.4. From the measurement of the iron and DOC concentration, it is known that the molar proportion between the iron and the citric is not equal between the supernatants and the performance of the reaction may be affected. In this sense, Miralles et al. [13] reported an increase in the kinetic degradation constant when the molar ratio was augmented from 1:0.5 to 1:5 when they degraded the pharmaceuticals present in the concentration from the ultrafiltration of a secondary effluent of the WWTP. However, the dosimetry experiments carried out in the current study using the Fe_{MS}:Cit complex show that in the degradation of SA, the molar proportion is not a determinant factor (Figure 3); this may be related to the pH evolution throughout the experiment, which starts at pH 7 because of the previous neutralisation of the SA solution, but as the experiment continues, the pH drops to ~4.5–5.

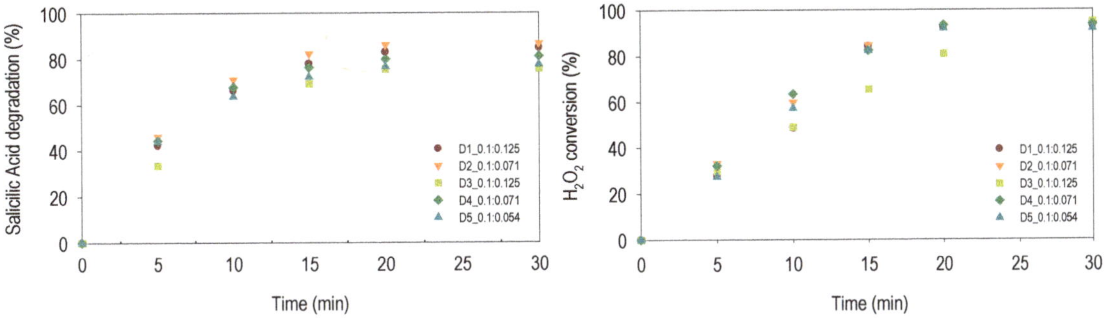

Figure 3. Dosimetry experiments. (**a**) Salicylic acid degradation; (**b**) H_2O_2 consumption for each one of the supernatants identified by the obtention day (Dx) and the molar proportions between iron and citric in the complex.

Thus, the dosimetry experiments show a SA degradation of about 81 ± 4% at 30 min, while in all of the cases, 16% of the H_2O_2 turned into $HO^•$, with a final concentration of 0.13 ± 0.1 mM of $HO^•$, which is three orders of magnitude higher than the concentration of $HO^•$ previously reported with the MS operated in a heterogeneous system [25]. The difference between the $HO^•$ formationcan be explained by the mass transfer limitations observed in heterogeneous systems when they are compared to homogeneous ones [25,26]. On the other hand, it is known that the use of iron complexing agents has been used as a strategy to enhance the reaction performance because iron complexes absorb light in the near UV and visible regions more efficiently than aqua complexes, and they undergo photoreduction through a ligand to carry out metal charge transfer, generating Fe(II) ions [36].

It is important to note that there is not a stoichiometric proportion between the H_2O_2 conversion and the $HO^•$ produced because, during the photo-Fenton process, $HO_2^•$, H_2O, O_2, and H^+ are also formed from the decomposition of H_2O_2 [37]. However, since $HO^•$ has a higher oxidant power and is the majority in comparison to $HO_2^•$, most of the reaction mechanisms have been simplified [38].

2.4. The Effects of the Iron Source and Liquid Depth on Drug Degradation

Since the Fe_{MS}:Cit complex was able to produce hydroxyl radicals in a photo-Fenton system, it was compared to the Fe:Cit complex formed from iron sulphate (analytical grade reagent). In both cases, the Fe:Cit molar ratio was adjusted to 1:1. The results show that, at the beginning of each experiment, the concentration of Fe^{2+} was below the LOQ (limit of quantification), but this parameter increased throughout the reaction (Figure 4a). It is important to note that the total dissolved iron concentration represents the sum of Fe^{2+} and Fe^{3+}, and it is the same throughout the reaction because the complex is stable and iron sedimentation does not occur.

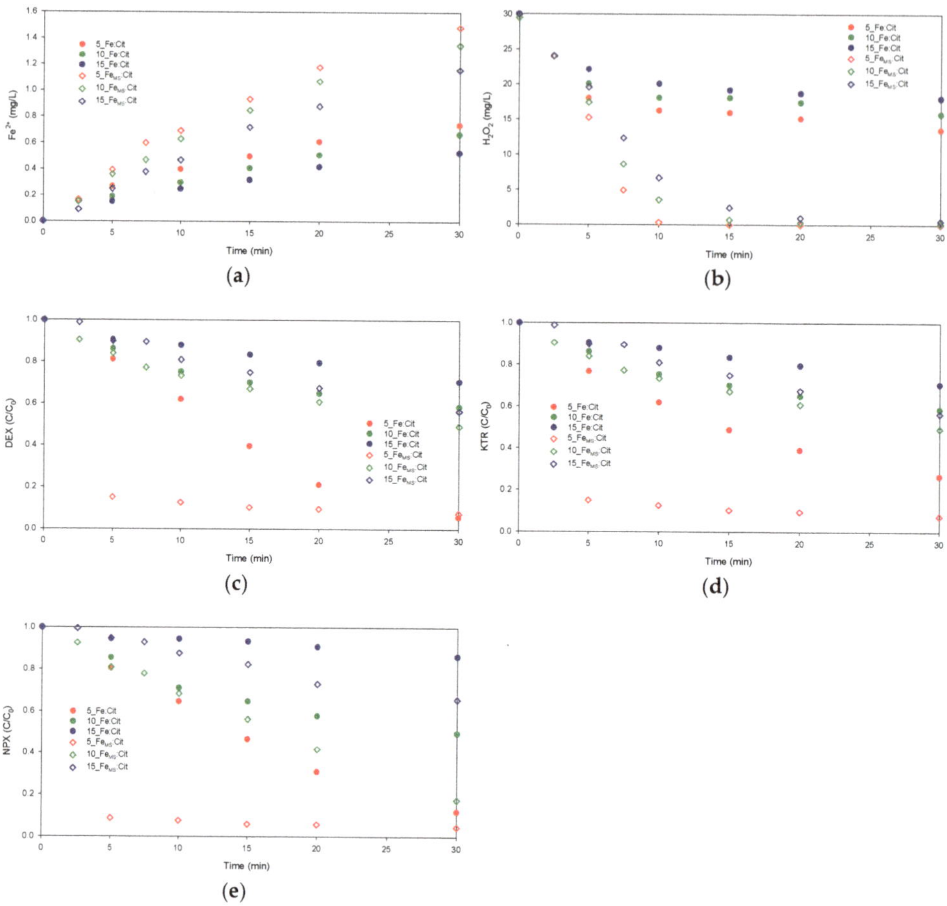

Figure 4. Oxidation experiments using Fe_{MS}:Cit complex (open symbols) or Fe:Cit complex (closed symbols) as catalyst in reactors whose depths were 5 (red), 10 (green), and 15 (blue) cm. (**a**) Fe^{2+} concentration, (**b**) H_2O_2 concentration, and (**c**–**e**) normalised concentrations of DEX, KTR, and NPX, respectively.

This behaviour can be explained by reaction 1, which describes a ligand-to-metal charge transfer, with the ligand (L, citric acid) acting as a sacrificial electron donor to Fe^{3+}, resulting in Fe^{2+}. This process is of high importance at a neutral pH where iron presents low solubility and the photo-active iron complexes can assist in maintaining an effective photo-catalytic cycle [39].

$$\left[Fe^{3+} + L\right] + hv \rightarrow \left[Fe^{3+} + L\right]^{*} \rightarrow Fe^{2+} + L_{ox}^{+} \tag{1}$$

The increase in the Fe^{2+} concentration has an inverse relationship with the liquid depth. This means that iron photoreduction is faster when the liquid depth is 5 cm than when it has higher values. Such an effect is explained by the photon availability, which is inversely proportional to the optical path length [40]. On the other hand, it is noticeable that the Fe^{2+} has a faster increment when the complex is formed from the slags; this may be related to the presence of other metals in trace concentrations that could be leached from the slag. For instance, the presence of Zn in the water acts as a reductor agent according to the following reaction: $Fe^{3+} + Zn \rightarrow Fe^{2+} + Zn^{2+}$. Zn may be present in the system by the lixiviation of the Zn by the citric acid, with a maximum calculated concentration of 2 mg/L.

Then, once the iron is photoreduced, the classic photo-Fenton equations take part (2)–(4). $HO^{\bullet}$ is formed by the catalytic decomposition of H_2O_2 in the presence of Fe^{2+} or $Fe^{3+}/h\nu$, and $HO_2^{\bullet}$ is formed when the light does not interact in the reaction between Fe^{3+} and H_2O_2 [41]. The reaction starting with Fe^{2+} has a faster kinetic constant (k = 53–76 L mol^{-1} s^{-1}) than the reaction with Fe^{3+} ($k = 3.1 \times 10^{-3}$ L mol^{-1} s^{-1}) [42]; for this reason, the H_2O_2 concentration drops faster as the reactor has a higher concentration of Fe^{2+} (Figure 4b).

$$Fe^{2+} + H_2O_2 \rightarrow Fe^{3+} + HO^{\bullet} + OH^{-} \quad (2)$$

$$Fe^{3+} + H_2O_2 \rightarrow Fe^{2+} + HO_2^{\bullet} + H^{+} \quad (3)$$

$$Fe^{3+} + H_2O_2 \xrightarrow{hv} Fe^{2+} + HO^{\bullet} + H^{+} \quad (4)$$

The drugs were degraded as a consequence of the decomposition of H_2O_2 and radical formation (Figure 4c–e). The DEX, NAP, and KTR degradation kinetics showed that at 5 min, at least 80% of the drug was degraded when the liquid depth was 5 cm, and the catalyst was the Fe_S:Cit complex, but as the liquid depth increased, the degradation rate diminished; the same behaviour was seen with the reactions that were catalysed by the Fe:Cit complex, but they presented a slower degradation.

Regarding pollutant degradation, it is important to recall that in the system, there is a scavenger effect to emulate the competition between the matrix composition (15 mg IC/L of $NaHCO_3$ and 286 mg DOC/L of MeOH) and the model pollutants (20 mg DOC/L, DEX, NPX, and KTR), whose concentrations are at least three orders of magnitude higher compared to their environmental contents; therefore, complete degradations were not expected. Instead, the observation of the effects of the liquid depth and the irradiance are the valuable results that were expected.

2.5. The Effects of the Irradiance and Liquid Depth and Radiation on Drug Degradation

Thus, the combined effect of the liquid depth and the radiation variation was evaluated (Figure 5); the results show that the Fe^{2+} (Figure 5a) and H_2O_2 (Figure 5b) concentrations are mainly grouped by irradiances. Such effect is more evident with the Fe^{2+} concentration of 750 W/m^2, with radiation promoting faster photoreduction. While the effect of the liquid depth is not evident at 250 W/m^2, it is noticeable in the experiments at 500 and 750 W/m^2; the data are scattered according to the different liquid depths: a deeper liquid depth results in less photoreduction. This behaviour is related to the photolimitation and photosaturation concepts; the first means that an increment in irradiance promotes a higher reaction rate, while the second indicates that it does not cause any effect [38].

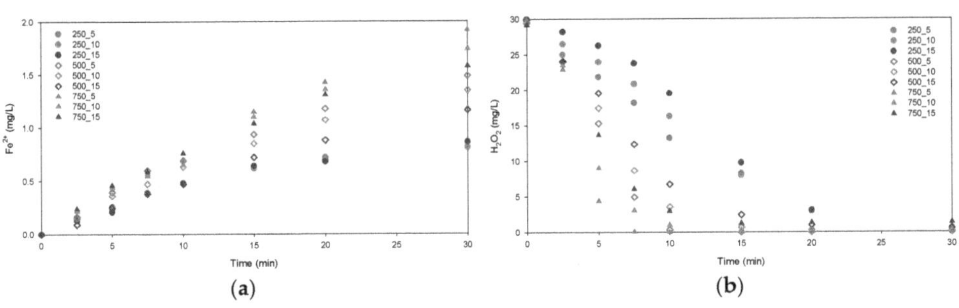

Figure 5. (**a**) Fe^{2+} and (**b**) H_2O_2 concentration profiles for oxidation experiments using Fe_{MS}:Cit complex as catalyst in reactors whose depths were 5 (red), 10 (green), and 15 (blue) cm, with Io identified as follows: 250 circles, 500 diamonds, and 750 triangles facing upward.

The H_2O_2 concentrations dropped under the LOQ (0.05 mg H_2O_2/L) at 7.5 min when the LD was 5 cm and the I_0 was 750 W/m^2, and they dropped at 30 min for the experiments carried out at 250 W/m^2; the data for these three experiments show that there was a certain level of dispersion before 15 min, but as the Fe^{2+} concentration started to maintain a constant value, the dots started to become closer. On the contrary, the dots in the experiments carried out at 500 and 750 W/m^2 exhibited major dispersion throughout the experiment, and there was an overlap between the profiles, displaying that the same H_2O_2 conversion can be achieved by modifying the LD or the I_0; for example, at 10 min, there are two points with values of 3.53 and 2.98 mg of H_2O_2/L, which means they are inside the experimental variation, and the operating conditions are I_0 = 500, LD = 10, and I_0 = 750 LD15, respectively. This is particularly important in systems that are photo-assisted by sunlight because changes in solar radiation are not controllable, but the liquid depth is controlled [21], so this information is imperative to set the operation guidelines.

In this order of ideas, the kinetic degradation constants for the model pollutants exhibit that degradation occurs faster when the liquid depth is lower and irradiance is higher (Figure 6); such behaviour was described by Rivas et al. [43], who highlighted that pollutant degradation is a consequence of H_2O_2 conversion and it is dependent on the Fe concentration and light availability, which are related to irradiance on the reactor surface and the liquid depth.

On the other hand, the kinetic constants for the three model pollutants are ranked between 0.009 and 0.09 min^{-1}. The colour maps indicate that DEX degrades faster than NAP and KTR in all of the conditions tested.

This may be related to the fact that the degradation of dexamethasone when it is attacked by HO$^\bullet$ may start in two active zones: the methylene group or the double bonds of the ring adding C-unsaturated bonds [44]. This is in contrast to the degradation of NAP, which is attacked by HO$^\bullet$ in the naphthalene ring of naproxen [45], while the KTR pyrrolizine skeleton is attacked by HO$^\bullet$ in its first degradation stages [46]; consequently, NAP and KRT are more stable compounds than DEX.

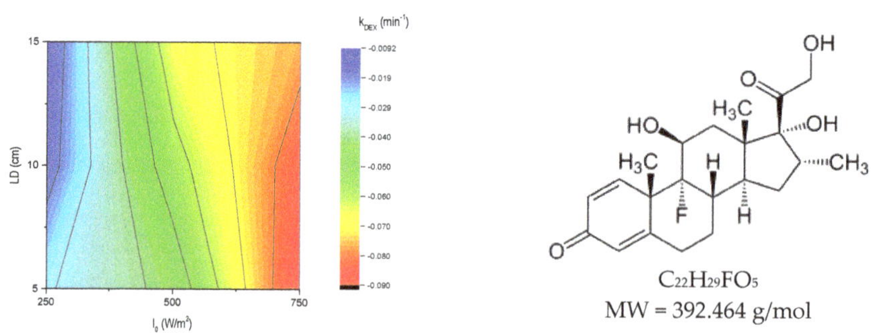

(a)

Figure 6. *Cont.*

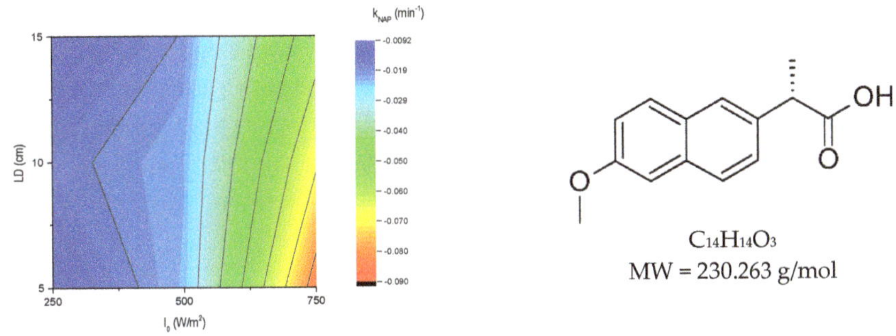

(b)

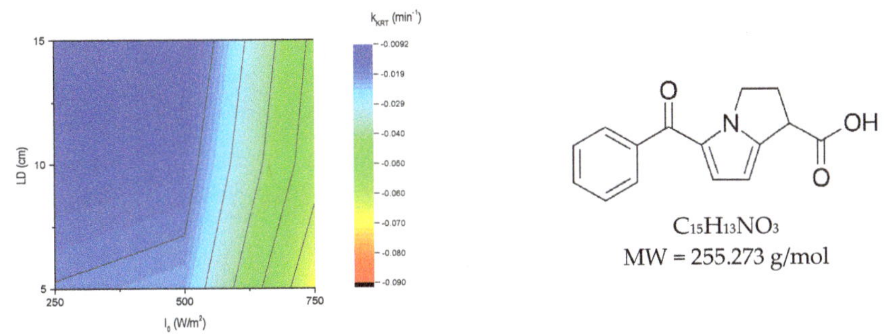

(c)

Figure 6. The variation in the kinetic constant degradation of the model pollutants according to the liquid depth and irradiance tested. (**a**) Dexamethasone, (**b**) naproxen, and (**c**) ketorolac.

3. Material and Methods

3.1. Reagents

H_2O_2 (30% *w*/*w*), sulphuric acid, and sodium hydroxide (ACS grade) were acquired from J.T. Baker (JT Baker, Phillipsburg, NJ, USA). Titanium (IV) oxysulfate solution (1.9–2.1%) was purchased from Sigma-Aldrich (Sigma–Aldrich, St. Louis, MO, USA). $Fe_2(SO_4)_3{\cdot}nH_2O$, citric acid monohydrate ($C_6H_8O_7{\cdot}H_2O$), and $NaHCO_3$ (95–100%) were provided by Meyer (Reactivos Quimica Meyer, Tláhuac, Mexico). For the HPLC analysis, phosphoric acid (85%) and HPLC-grade acetonitrile were supplied by Sigma-Aldrich. Ultra-pure water was produced using a Milli-Q System (Merck KGaA, Darmstadt, Hessen, Germany) with a specific resistance of 18.2 MΩ cm and a total organic carbon (TOC) concentration of 2 μg/L.

According to high-purity analytical standards, salicylic acid (98%) and its oxidation by-products, 2,3-dihydroxybenzoic acid (2,3-dHBA, 99%) and 2,5-dihydroxybenzoic acid (2,5-dHBA, 98%), as well as the drugs dexamethasone (DEX), naproxen (NAP), and ketorolac (KTR) (>99%), were purchased from Sigma-Aldrich. Individual stock standard solutions were prepared in methanol at a concentration of 10 mg/mL and stored in amber glass vials at 20 °C. Working solutions were prepared using the appropriate mixture and dilution of the stock solutions.

3.2. Metallurgical Slag Characterisation

A metallurgical slag (MS) from the copper industry, issued from a Northern Mexican smelter, with a particle size between 105 and 125 μm (140–120 mesh), was the iron source.

The chemical composition of the major constituents in the slag and its fractions were determined using X-ray fluorescence (XRF) with a Rigaku Fluorescent Spectrometer (Rigaku Industrial Co., Osaka, Japan). The crystalline mineral phases were identified using the powder X-ray diffraction (PXRD) method with an Empyrean XRD Diffractometer

(Malvern-PANalytical, Almelo, The Netherlands) coupled to a Cu anode X-ray tube. The PXRD pattern was used to determine the phase compositions that were identified using the ICDD and ICSD databases. Textural properties were analysed by using the N_2 adsorption–desorption isotherm in a Bel-Sorp mini II device (Microtrac BEL, Inc., Tokyo, Japan) at 77 K, and a BET analysis was used to determine the surface area. The degassing temperature of the sample was 140°C using a dynamic flow of nitrogen for 12 h. Scanning electron microscopy (SEM) was performed using a JEOL JMS-7600 F microscope (JEOL Ltd., Tokyo, Japan), obtaining backscattered electron images, and an elemental mapping analysis was carried out.

3.3. Experimental Setup

Experimentation was carried out in four stages: (1) iron leachate and complexation using citric acid and (2) an oxidation test, which involves hydroxyl radical formation to determine whether or not drug degradation was carried out by radical oxidation, followed by (3) the evaluation of the mix drug (DEX, NAP and KTR) by comparing the iron/citrate complex formed by the metallurgical slag (Fe_{MS}:Cit) and the one from the analytical-grade reactants (Fe:Cit), and (4) the effect of the irradiance and liquid depth, as operating conditions of the process, on drug degradation.

3.4. Iron Leachate and Complexation and Its Characterisation

A sample of metallurgical slag from the copper production process was put in contact with a 1 M dissolution of CA (pH = 1.80) in a hypothetical proportion that considers that all surface iron from the MS would be lixiviated; thus, it would form a complex with the CA at 2 g of MS for 10 mL. The solution was stirred in an orbital shaker (100 rpm) for 24 h, and then the supernatant was separated from the mixture and 1 M of fresh dissolution was refilled; this operation was carried out for five days.

Finally, the Fourier-Transform Infrared (FTIR) spectrum of the Fe_{MS}:Cit supernatant was compared to the Fe:Cit supernatant obtained by the complexation of $Fe_2(SO_4)_3$ and citric acid. The Fourier-Transform Infrared (FTIR) spectra were obtained using a Thermo Scientific Nicolet 6700 apparatus (Thermo Fisher Scientific™, Waltham, MA, USA) equipped with diffusion reflectance accessories. A small amount of the analysed liquids was placed on the diamond surface of the spectrometer and fixed by the metal penetrator to maintain contact between the sample and the diamond. A total of 64 scans were obtained within a wavenumber of 4000 to 400 cm^{-1}. Baseline correction and band normalisation were carried out using Thermo Scientific OMNIC™ 9.2.106 software.

3.5. Advanced Oxidation Tests

Advanced oxidation tests were performed using a sunlight emission simulation chamber (SUNTEST CPS+ATLAS, Mount Prospect, IL, USA) (Figure 7a) with a Xenon lamp (Mount Prospect, IL, USA) that simulates solar radiation, which can be set from 250 to 765 W/m^2, measured in the range of 400–800 nm. A UVA/UVB light metre 850009 from Sper Scientific (Scottsdale, AZ, USA) was used to determine the light intensity in the different configurations of the experiments.

Two types of reactors were used (Figure 7b): (1) a Raceway Pond Reactor in a 1 L batch RPR (10 cm wide, 27 cm long, and 5 cm in depth) with a stirring speed of 250 rpm (0.5 m/s), and (2) three cylindrical reactors with a 13 cm^2 diameter and liquid depths of 5, 10, and 15 cm. Water temperature in the cylindrical reactor was controlled with a cooling coil connected to a thermostatic bath from PolyScience (Niles, IL, USA).

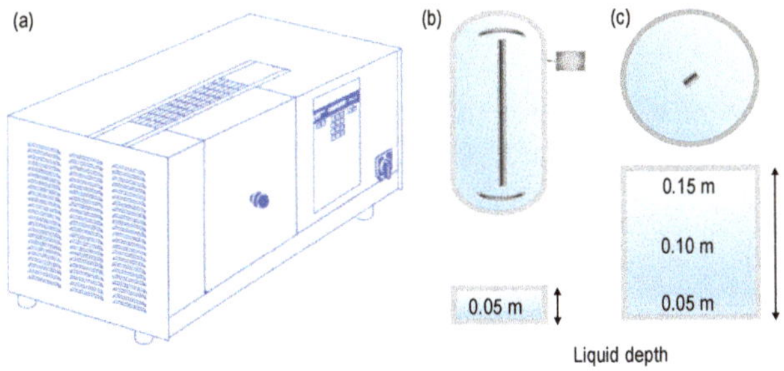

Figure 7. The reaction system for the photocatalytic degradation tests. (**a**) A solar simulator and thermostatic bath, (**b**) a raceway pond reactor with a 5 cm liquid depth, and (**c**) cylindrical reactors with liquid depths of 5, 10, and 15 cm.

3.5.1. Hydroxyl Radical Formation

The Fe_{MS}:Cit and Fe:Cit complexes were used to produce the hydroxyl radical ($HO^{\bullet}$) using the dosimetry technique with salicylic acid (SA), which is based on the principle of aromatic hydroxylation of this compound [47]. The technique indicates that in the presence of hydroxyl radicals, salicylic acid turns into 2,3-dihidroxybenzoic acid and 2,5-dihidroxybenzoic acid; consequently, the formation of the hydroxylated subproducts indicate the formation of hydroxyl radicals (Figure 8).

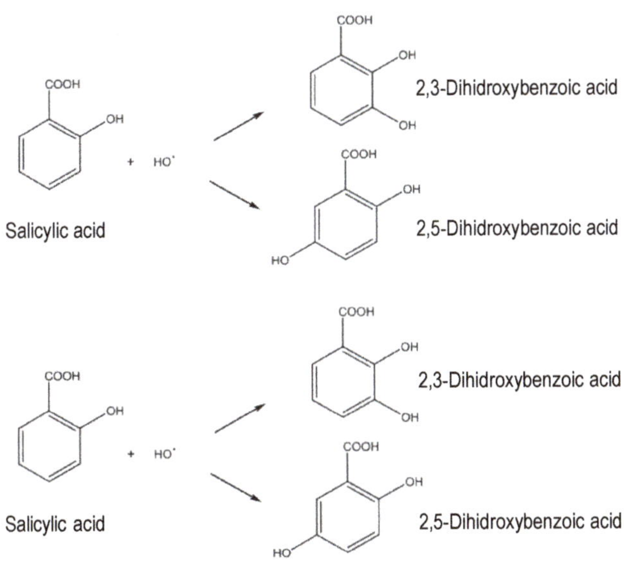

Figure 8. Reaction between salicylic acid and hydroxyl radical [48].

Thus, in order to determine the formation of hydroxyl radicals by the reaction of the supernatant generated in Section 3.4 and hydrogen peroxide under simulated sunlight, a solution of salicylic acid (50 mg/L) in distilled water, whose pH was adjusted to 7, was reacted with hydrogen peroxide (0.88 mM) and an aliquot of the supernatant obtained in 2.1 along with the addition of 0.1 mM of iron to the reactor.

The experiment was performed in a raceway pond reactor (1 L capacity), located inside the solar simulator chamber, whose light intensity was constant at 500 W/m^2. Since the Fe-Cit molar ratio was variable according to the iron lixiviated for each of the days of the lixiviation stage, each one of the supernatants was tested in the $HO^{\bullet}$ production stage by SA hydroxylation following the same procedure.

3.5.2. The Effects of the Iron Source and Liquid Depth on Drug Degradation

Then, the supernatants obtained in Section 3.5.1 were mixed in a single stock, which allowed for the modification of the molar ratio between the iron and the citric acid, resulting in a 1:1 Fe:Cit molar proportion. In order to compare the performance of the complex formed by the slags and the one formed by $Fe_2(SO_4)_3$ in the same molar proportion (1:1), the experiments were performed using 0.1 mM of Fe_{MS}:Cit and 0.88 mM of H_2O_2.

The degradation of DEX, NAP, and KTR was tested with an initial concentration of 10 mg/L each. The drug solution was prepared by adding 1 mL of a stock solution (dissolved in methanol) with a concentration of 10 mg/mL and 15 mg/L of inorganic carbon (from $NaHCO_3$). They were rinsed with distilled water in a 1 L volumetric flask. IC was added to maintain the neutrality of the solution through the reaction, and it was also used in combination with methanol to simulate the scavenger effect.

Oxidation experiments were performed in a set of cylindrical reactors with liquid depths of 5, 10, and 15 cm, irradiated at 500 W/m^2. All of the oxidation tests were performed in duplicate (standard deviation lower than 5%), but only the mean concentration was plotted to prevent oversaturating the figures.

3.5.3. The Effects of the Liquid Depth and Radiation on Drug Degradation

The effects of the changes in the irradiance and liquid depth on drug degradation were tested, and the experiments were carried out using Fe_{MS}:Cit as a catalyst and modifying the radiation in the interval available in the simulation chamber, which means the radiation value was 250, 500, or 750 W/m^2, which are equivalent to 6, 1.6, and 26 mW_{UV}/cm^2, which were the values measured in the wavelength range from 280 to 400 nm on the surface of the liquid. The liquid depth was set at 5, 10, and 15 cm.

Experimental conditions: The oxidant, catalyst, and drug concentrations were the same values as those in Section 3.5.2.

The degradation results obtained in this section were used to determine the reaction order and the rate constant. According to Equation (5), where k is the pseudo-first-order rate constant, C_0 is the initial concentration, and C is the concentration at time t, DXM, NAP, and KTR exhibited a pseudo-first-order reaction.

$$ln\frac{C}{C_0} = k't \quad (5)$$

3.6. *Chemical Analysis*

Throughout each experiment, the hydrogen peroxide concentration was determined by the DIN 38 401 H15 method using titanium (IV) oxysulfate solution. The total leached iron concentration was measured by the Triazine method using a Merck Kit (Spectroquant 1.14761.001, Darmstadt, Alemania), while the o-Phenanthroline method [49] was used to determine the Fe^{2+} concentration. The total organic carbon (TOC) analyses were performed using the standardised 5310B method [50] with a TOC analyser (Shimadzu TOC-L CSH, Kyoto, Japan).

3.7. *Micropollutant Detection and Quantification*

High-performance liquid chromatography (HPLC) was used to determine the micropollutant concentration using a 1100 Agilent chromatography apparatus equipped with a photodiode array detector (Agilent Technologies, Inc., Santa Clara, CA, USA). Two different chromatographic methods were used, one for salicylic acid dosimetry and another for drug degradation. In both of them, the stationary phase was an ACE5 C18-AR, and the injection volume of the samples was 30 μL.

The mobile phase for the salicylic acid dosimetry method was constituted by a mixture of methanol, acetonitrile, and phosphoric acid (0.1%, v/v) at a 10:8:82 ratio (v/v). The eluent flow rate was 1.0 mL/min during the first 11 min and then 1.5 min for the next 12 min. The detection wavelengths (λ) of each compound were the following: λ_1: 236 nm for salicylic acid and 2,5-dihydroxybenzoic acid (2,5-dHBA), and λ_2: 246 nm to detect 2,3-dihydroxybenzoic acid (2,3-dHBA). The retention times were 22, 7, and 9 min, respectively.

For the drug mix, acetonitrile and phosphoric acid (0.1%, v/v) at a 35:65 ratio (v/v) were used, and the eluent flow rate was 1.0 mL/min. The detection wavelengths (λ) of each compound were the following: $\lambda 2$: 245 nm dexamethasone, $\lambda 3$: 310 nm ketorolac, and $\lambda 1$: 230 nm naproxen. The retention times were 6, 9, and 15 min, respectively.

4. Conclusions

The direct formation of an Fe:Cit complex was achieved using copper slag as the iron source and citric acid as acid media and a chelating agent simultaneously, merging the two steps and obtaining 87% iron recovery. The formation of an Fe:Cit complex allows for the operation of the photo-Fenton process in a homogeneous phase, demonstrating $HO^{\bullet}$ formation during salicylic acid hydroxylation.

The Fe:Cit complex formed by the metallurgical slag was compared to the one formed by reactants, with better results regarding drug degradation, which may be related to the trace compounds from the slag that can be lixiviated by the citric acid. These results not only point out that an industrial by-product can be valorised in environmental remediation, but also simplifying the process is possible, which may lead to a diminution in the process cost and a reduction in rough materials.

Thus, the degradation of DEX, NAP, and KTR in a water matrix with a high scavenger effect was demonstrated at some liquid depths and irradiances, showing that by combining both of them, similar degradation rates can be achieved; this information is necessary during the operation of a solar system since the irradiance is not constant, but it can be compensated by modifying the liquid depth.

Author Contributions: All authors contributed to the study conception and design, and all authors read and approved the final manuscript. The individual tasks were divided as follows: S.Y.A.S.: Investigation, Conceptualisation, Methodology, Validation, Formal Analysis, Writing—Original Draft, and Writing—Review and Editing; A.Y.-A.: Investigation and Writing—Review and Editing; R.-M.R.-Z.: Formal Analysis, Writing—Review and Editing, Project Administration, and Funding Acquisition. All authors have read and agreed to the published version of the manuscript.

Funding: Sandra Yazmin Arzate Salgado and Ana Yañez-Aulestia thank DGAPA for their post-doctoral fellowship. The authors acknowledge the financial support granted by PAPIIT, UNAM (IT103723).

Data Availability Statement: The raw data supporting the conclusions of this article will be made available by the authors upon request.

Conflicts of Interest: The authors have no competing interests to declare that are relevant to the content of this article.

References

1. FAO. *Water for Sustainable Food and Agriculture. A Report Produced for the G20 Presidency of Germany*; Food and Agriculture Organization of the United Nations: Rome, Italy, 2017.
2. Ryder, G. *The United Nations World Water Development Report, 2017: Wastewater: The Untapped Resource*; UNESCO World Water Assessment Programme: Paris, France, 2017.
3. Krzeminski, P.; Tomei, M.C.; Karaolia, P.; Langenhoff, A.; Almeida, C.M.R.; Felis, E.; Gritten, F.; Andersen, H.R.; Fernandes, T.; Manaia, C.M.; et al. Performance of Secondary Wastewater Treatment Methods for the Removal of Contaminants of Emerging Concern Implicated in Crop Uptake and Antibiotic Resistance Spread: A Review. *Sci. Total Environ.* **2019**, *648*, 1052–1081. [CrossRef]
4. Giannakis, S.; Gamarra Vives, F.A.; Grandjean, D.; Magnet, A.; De Alencastro, L.F.; Pulgarin, C. Effect of Advanced Oxidation Processes on the Micropollutants and the Effluent Organic Matter Contained in Municipal Wastewater Previously Treated by Three Different Secondary Methods. *Water Res.* **2015**, *84*, 295–306. [CrossRef]

5. Wang, J.; Tian, Z.; Huo, Y.; Yang, M.; Zheng, X.; Zhang, Y. Monitoring of 943 Organic Micropollutants in Wastewater from Municipal Wastewater Treatment Plants with Secondary and Advanced Treatment Processes. *J. Environ. Sci.* **2018**, *67*, 309–317. [CrossRef]
6. Luo, Y.; Guo, W.; Ngo, H.H.; Nghiem, L.D.; Hai, F.I.; Zhang, J.; Liang, S.; Wang, X.C. A Review on the Occurrence of Micropollutants in the Aquatic Environment and Their Fate and Removal during Wastewater Treatment. *Sci. Total Environ.* **2014**, *473–474*, 619–641. [CrossRef]
7. Echevarría, C.; Valderrama, C.; Cortina, J.L.; Martín, I.; Arnaldos, M.; Bernat, X.; De la Cal, A.; Boleda, M.R.; Vega, A.; Teuler, A.; et al. Techno-Economic Evaluation and Comparison of PAC-MBR and Ozonation-UV Revamping for Organic Micro-Pollutants Removal from Urban Reclaimed Wastewater. *Sci. Total Environ.* **2019**, *671*, 288–298. [CrossRef]
8. Li, X.; Shi, H.; Li, K.; Zhang, L. Combined Process of Biofiltration and Ozone Oxidation as an Advanced Treatment Process for Wastewater Reuse. *Front. Environ. Sci. Eng.* **2015**, *9*, 1076–1083. [CrossRef]
9. Bui, X.T.; Vo, T.P.T.; Ngo, H.H.; Guo, W.S.; Nguyen, T.T. Multicriteria Assessment of Advanced Treatment Technologies for Micropollutants Removal at Large-Scale Applications. *Sci. Total Environ.* **2016**, *563–564*, 1050–1067. [CrossRef]
10. Fang, L.L.; Valverde-Pérez, B.; Damgaard, A.; Plósz, B.G.; Rygaard, M. Life Cycle Assessment as Development and Decision Support Tool for Wastewater Resource Recovery Technology. *Water Res.* **2016**, *88*, 538–549. [CrossRef]
11. Ameta, R.; Chohadia, A.; Abhilasha, J.; Punjabi, P.B. Fenton and Photo-Fenton Processes. In *Advanced Oxidation Processes for Waste Water Treatment*; Ameta, S., Ameta, R., Eds.; Academic Press: Cambridge, MA, USA, 2018; pp. 49–87. [CrossRef]
12. Lado Ribeiro, A.R.; Moreira, N.F.F.; Li Puma, G.; Silva, A.M.T. Impact of Water Matrix on the Removal of Micropollutants by Advanced Oxidation Technologies. *Chem. Eng. J.* **2019**, *363*, 155–173. [CrossRef]
13. Miralles-Cuevas, S.; Oller, I.; Sánchez Pérez, J.A.; Malato, S. Removal of Pharmaceuticals from MWTP Effluent by Nanofiltration and Solar Photo-Fenton Using Two Different Iron Complexes at Neutral PH. *Water Res.* **2014**, *64*, 23–31. [CrossRef]
14. Cabrera Reina, A.; López, J.C.; Rubio, M.M.; Jordá, L.S.J.; Sánchez, J.G.; Pérez, J.S. Effects of Environmental Variables on the Photo-Fenton Plant Design. *Chem. Eng. J.* **2014**, *237*, 469–477. [CrossRef]
15. Malato, S.; Blanco, J.; Alarcón, D.C.; Maldonado, M.I.; Fernández-Ibáñez, P.; Gernjak, W. Photocatalytic Decontamination and Disinfection of Water with Solar Collectors. *Catal. Today* **2007**, *122*, 137–149. [CrossRef]
16. Gualda-Alonso, E.; Soriano-Molina, P.; López, J.L.C.; Sánchez, J.L.G.; Plaza-Bolaños, P.; Agüera, A.; Pérez, J.A.S. Large-Scale Raceway Pond Reactor for CEC Removal from Municipal WWTP Effluents by Solar Photo-Fenton. *Appl. Catal. B Environ.* **2022**, *319*, 121908. [CrossRef]
17. Arzate, S.; García Sánchez, J.L.; Soriano-Molina, P.; Casas López, J.L.; Campos-Mañas, M.C.; Agüera, A.; Sánchez Pérez, J.A. Effect of Residence Time on Micropollutant Removal in WWTP Secondary Effluents by Continuous Solar Photo-Fenton Process in Raceway Pond Reactors. *Chem. Eng. J.* **2017**, *316*, 1114–1121. [CrossRef]
18. Soares, P.A.; Batalha, M.; Souza, S.M.A.G.U.; Boaventura, R.A.R.; Vilar, V.J.P. Enhancement of a Solar Photo-Fenton Reaction with Ferric-Organic Ligands for the Treatment of Acrylic-Textile Dyeing Wastewater. *J. Environ. Manag.* **2015**, *152*, 120–131. [CrossRef]
19. Pereira, J.H.O.S.; Queirós, D.B.; Reis, A.C.; Nunes, O.C.; Borges, M.T.; Boaventura, R.A.R.; Vilar, V.J.P. Process Enhancement at near Neutral PH of a Homogeneous Photo-Fenton Reaction Using Ferricarboxylate Complexes: Application to Oxytetracycline Degradation. *Chem. Eng. J.* **2014**, *253*, 217–228. [CrossRef]
20. De Luca, A.; Dantas, R.F.; Esplugas, S. Assessment of Iron Chelates Efficiency for Photo-Fenton at Neutral PH. *Water Res.* **2014**, *61*, 232–242. [CrossRef]
21. Sánchez-Pérez, J.A.; Arzate, S.; Soriano-Molina, P.; García-Sánchez, J.L.; Casas López, J.L.; Plaza-Bolaños, P. Neutral or Acidic pH for the Removal of Contaminants of Emerging Concern in Wastewater by Solar Photo-Fenton? A Techno-Economic Assessment of Continuous Raceway Pond Reactors. *Sci. Total Environ.* **2020**, *736*, 139681. [CrossRef]
22. Kim, B.S.; Jo, S.K.; Shin, D.; Lee, J.C.; Jeong, S.B. A Physico-Chemical Separation Process for Upgrading Iron from Waste Copper Slag. *Int. J. Miner. Process.* **2013**, *124*, 124–127. [CrossRef]
23. Solís-López, M.; Durán-Moreno, A.; Rigas, F.; Morales, A.A.A.; Navarrete, M.; Ramírez-Zamora, R.M. *Assessment of Copper Slag as a Sustainable Fenton-Type Photocatalyst for Water Disinfection*; Elsevier Inc.: Amsterdam, The Netherlands, 2014. [CrossRef]
24. Mercado-Borrayo, B.M.; González-Chávez, J.L.; Ramírez-Zamora, R.M.; Schouwenaars, R. Valorization of Metallurgical Slag for the Treatment of Water Pollution: An Emerging Technology for Resource Conservation and Re-Utilization. *J. Sustain. Metall.* **2018**, *4*, 50–67. [CrossRef]
25. García-Estrada, R.; Esteban García, B.; Ramírez-Zamora, R.M.; Sánchez Pérez, J.A. Micropollutant Degradation by the Heterogeneous Solar Photo-Fenton Process at Circumneutral pH Using Copper Slag. *J. Water Process Eng.* **2020**, *38*, 101562. [CrossRef]
26. Huanosta-Gutiérrez, T.; Dantas, R.F.; Ramírez-Zamora, R.M.; Esplugas, S. Evaluation of Copper Slag to Catalyze Advanced Oxidation Processes for the Removal of Phenol in Water. *J. Hazard. Mater.* **2012**, *213–214*, 325–330. [CrossRef]
27. Ruales-Lonfat, C.; Barona, J.F.; Sienkiewicz, A.; Vélez, J.; Benítez, L.N.; Pulgarín, C. Bacterial Inactivation with Iron Citrate Complex: A New Source of Dissolved Iron in Solar Photo-Fenton Process at near-Neutral and Alkaline PH. *Appl. Catal. B Environ.* **2016**, *180*, 379–390. [CrossRef]
28. Linge, K.L.; Gruchlik, Y.; Busetti, F.; Ryan, U.; Joll, C.A. Use of Micropollutant Indicator Ratios to Characterize Wastewater Treatment Plant Efficiency and to Identify Wastewater Impact on Groundwater. *J. Environ. Manag.* **2024**, *358*, 120822. [CrossRef]

29. Morales-Paredes, C.A.; Rodríguez-Díaz, J.M.; Boluda-Botella, N. Pharmaceutical Compounds Used in the COVID-19 Pandemic: A Review of Their Presence in Water and Treatment Techniques for Their Elimination. *Sci. Total Environ.* **2022**, *814*, 152691. [CrossRef]
30. Musee, N.; Kebaabetswe, L.P.; Tichapondwa, S.; Tubatsi, G.; Mahaye, N.; Leareng, S.K.; Nomngongo, P.N. Occurrence, Fate, Effects, and Risks of Dexamethasone: Ecological Implications Post-COVID-19. *Int. J. Environ. Res. Public Health* **2021**, *18*, 11291. [CrossRef]
31. Lowell, S.; Shields, J.E.; Thomas, M.A.; Thommes, M. *Characterization of Porous Solids and Powders: Surface Area, Pore Size and Density*; Springer Science & Business Media: New York, NY, USA, 2012. [CrossRef]
32. Arzate-Salgado, S.-Y.; Morales-Pérez, A.-A.; Solís-López, M.; Ramírez-Zamora, R.-M. Evaluation of Metallurgical Slag as a Fenton-Type Photocatalyst for the Degradation of an Emerging Pollutant: Diclofenac. *Catal. Today* **2016**, *266*, 126–135. [CrossRef]
33. Silva, A.M.N.; Kong, X.; Parkin, M.C.; Cammack, R.; Hider, R.C. Iron(III) Citrate Speciation in Aqueous Solution. *Dalt. Trans.* **2009**, 8616–8625. [CrossRef]
34. Stuart, B.H. *Infrared Spectroscopy: Fundamentals and Applications*; Wiley: Chichester, England, 2004. Available online: https://onlinelibrary.wiley.com/doi/book/10.1002/0470011149 (accessed on 15 June 2024).
35. Hwang, S.W.; Umar, A.; Dar, G.N.; Kim, S.H.; Badran, R.I. Synthesis and Characterization of Iron Oxide Nanoparticles for Phenyl Hydrazine Sensor Applications. *Sens. Lett.* **2014**, *12*, 97–101. [CrossRef]
36. Clarizia, L.; Russo, D.; Di Somma, I.; Marotta, R.; Andreozzi, R. Homogeneous Photo-Fenton Processes at near Neutral PH: A Review. *Appl. Catal. B Environ.* **2017**, *209*, 358–371. [CrossRef]
37. Pignatello, J.J.; Oliveros, E.; MacKay, A. Advanced Oxidation Processes for Organic Contaminant Destruction Based on the Fenton Reaction and Related Chemistry. *Crit. Rev. Environ. Sci. Technol.* **2006**, *36*, 1–84. [CrossRef]
38. Sánchez Pérez, J.A.; Soriano-Molina, P.; Rivas, G.; García Sánchez, J.L.; Casas López, J.L.; Fernández Sevilla, J.M. Effect of Temperature and Photon Absorption on the Kinetics of Micropollutant Removal by Solar Photo-Fenton in Raceway Pond Reactors. *Chem. Eng. J.* **2017**, *310*, 464–472. [CrossRef]
39. Pulgarin, A.; Giannakis, S.; Pulgarin, C.; Ludwig, C.; Refardt, D. A Novel Proposition for a Citrate-Modified Photo-Fenton Process against Bacterial Contamination of Microalgae Cultures. *Appl. Catal. B Environ.* **2020**, *265*, 118615. [CrossRef]
40. Soriano-Molina, P.; García Sánchez, J.L.; Malato, S.; Pérez-Estrada, L.A.; Sánchez Pérez, J.A. Effect of Volumetric Rate of Photon Absorption on the Kinetics of Micropollutant Removal by Solar Photo-Fenton with Fe^{3+}-EDDS at Neutral pH. *Chem. Eng. J.* **2018**, *331*, 84–92. [CrossRef]
41. Klamerth, N.; Malato, S.; Agüera, A.; Fernández-Alba, A.; Mailhot, G. Treatment of Municipal Wastewater Treatment Plant Effluents with Modified Photo-Fenton as a Tertiary Treatment for the Degradation of Micro Pollutants and Disinfection. *Environ. Sci. Technol.* **2012**, *46*, 2885–2892. [CrossRef]
42. Machado, F.; Teixeira, A.C.S.C.; Ruotolo, L.A.M. Critical Review of Fenton and Photo—Fenton Wastewater Treatment Processes over the Last Two Decades. *Int. J. Environ. Sci. Technol.* **2023**, *20*, 13995–14032. [CrossRef]
43. Rivas, G.; Carra, I.; García Sánchez, J.L.L.; Casas López, J.L.L.; Malato, S.; Sánchez Pérez, J.A.A. Modelling of the Operation of Raceway Pond Reactors for Micropollutant Removal by Solar Photo-Fenton as a Function of Photon Absorption. *Appl. Catal. B Environ.* **2015**, *178*, 210–217. [CrossRef]
44. Pazoki, M.; Parsa, M.; Farhadpour, R. Removal of the Hormones Dexamethasone (DXM) by Ag Doped on TiO_2 Photocatalysis. *J. Environ. Chem. Eng.* **2016**, *4*, 4426–4434. [CrossRef]
45. Bai, R.; Xiao, Y.; Yan, W.; Wang, S.; Ding, R.; Yang, F.; Li, J.; Lu, X.; Zhao, F. Rapid and Efficient Removal of Naproxen from Water by $CuFe_2O_4$ with Peroxymonosulfate. *Environ. Sci. Pollut. Res.* **2020**, *27*, 21542–21551. [CrossRef]
46. Kalariya, P.D.; Raju, B.; Borkar, R.M.; Namdev, D.; Gananadhamu, S.; Nandekar, P.P.; Sangamwar, A.T.; Srinivas, R. Characterization of Forced Degradation Products of Ketorolac Tromethamine Using LC/ESI/Q/TOF/MS/MS and in Silico Toxicity Prediction. *J. Mass Spectrom.* **2014**, *49*, 380–391. [CrossRef]
47. Jen, J.F.; Leu, M.F.; Yang, T.C. Determination of Hydroxyl Radicals in an Advanced Oxidation Process with Salicylic Acid Trapping and Liquid Chromatography. *J. Chromatogr. A* **1998**, *796*, 283–288. [CrossRef]
48. Peralta, E.; Roa, G.; Hernandez-Severin, J.A.; Romero, R.; Balderas, P.; Natividad, R. Hydroxyl Radicals Quantification by UV Spectrophotometry. *Electrochim. Acta* **2014**, *129*, 137–141. [CrossRef]
49. *ISO 6332:1988*; Water Quality—Determination of Iron—Spectrometric Method Using 1,10-Phenanthroline 2nd ed. ISO—International Organization for Standardization: Geneva, Switzerland, 2023.
50. American Public Health Association; American Water Works Association; Water Environment Federation. *Standard Methods for the Examination of Water and Wastewater*, 24th ed.; APHA Press: Washington DC, USA, 2023. Available online: https://www.standardmethods.org/Buy/ (accessed on 15 June 2024).

catalysts

MDPI

Article

Kinetic Study of the Water Quality Parameters during the Oxidation of Diclofenac by UV Photocatalytic Variants

Natalia Villota [1,*], Begoña Echevarria [2], Unai Duoandicoechea [1], Jose Ignacio Lombraña [3] and Ana María De Luis [2]

[1] Department of Environmental and Chemical Engineering, Faculty of Engineering Vitoria-Gasteiz, Nieves Cano 12-01006 Vitoria-Gasteiz, University of the Basque Country UPV/EHU, 12, 01006 Vitoria-Gasteiz, Spain; unai.duoandicoechea@ehu.eus

[2] Department of Environmental and Chemical Engineering, Bilbao School of Engineering, Rafael Moreno "Pitxitxi" 2-48013 Bilbao, University of the Basque Country UPV/EHU, Plaza Ingeniero Torres Quevedo, 1, 48013 Bilbao, Spain; jimenezbego450@gmail.com (B.E.); ana.deluis@ehu.eus (A.M.D.L.)

[3] Department of Chemical Engineering, Faculty of Science and Technology, Barrio Sarriena s/n-48940 Leioa, University of the Basque Country UPV/EHU, P.O. Box 644, 48080 Bilbao, Spain; ji.lombrana@ehu.eus

* Correspondence: natalia.villota@ehu.eus; Tel.: +34-945013248

Abstract: Diclofenac (DCF, $C_{14}H_{11}Cl_2NO_2$) is a widely used non-steroidal anti-inflammatory drug, with a significant occurrence in waste effluents. DCF is especially persistent and difficult to degrade, with numerous toxic effects on aquatic fauna and humans. In 2015, DCF was identified as a priority pollutant (EU Directives on water policy). In this work, UV irradiation and its combination with hydrogen peroxide only or catalyzed by iron salts (photo-Fenton) are analyzed to find the most efficient alternative. DCF aqueous solutions were treated in a stirred 150 W UV photocatalytic reactor. Depending on the case, 1.0 mM H_2O_2 and 0–5.0 mg/L Fe^{2+} catalyst, such as $FeSO_4$, was added. During the reaction, DCF, pH, turbidity, UVA at 254 and 455 nm, dissolved oxygen (DO), and TOC were assessed. The degradation of DCF yields a strong increase in aromaticity because of the rise in aromatic intermediates (mono-hydroxylated (4-hydroxy-diclofenac and 5-hydroxy-diclofenac) and di-hydroxylated products (4,5-dihydroxy-diclofenac), which subsequently generate compounds of a quinoid nature), which are very stable and non-degradable by UV light. Thus, only if H_2O_2 is added can UV completely degrade these aromatic colour intermediates. However, adding ferrous ion (photo-Fenton) the aromaticity remains constant due to iron com-plexes, that generates maximum colour and turbidity at an stoichiometric Fe^{2+} : DCF ratio of 3. As a result of the study, it is concluded that, with UV light only, a strong yellow colour is generated and maintained along the reaction, but by adding H_2O_2, a colourless appearance, low turbidity (<1 NTU), and [DO] = 8.1 mg/L are obtained. Surprisingly, photo-Fenton was found to be unsuitable for degrading DCF.

Keywords: colour; diclofenac; turbidity

1. Introduction

Among pharmaceutical compounds detected at measurable concentrations in aquatic environments are non-steroidal anti-inflammatory drugs. It is estimated that the annual production of these drugs can reach several kilotons. In Germany, for example, 87.5 million prescriptions for these substances were issued in 2001, not considering that many of these medications are purchased over the counter, making the actual consumption even higher [1]. Sodium diclofenac (DCF) is a commonly used NSAID for pain relief. It is estimated that approximately 490 tons of DCF are consumed annually worldwide [2].

DCF has garnered attention due to its toxicity and potential consequences for various organisms in aquatic ecosystems, even at low concentrations [3]. Therefore, it was classified as an emerging contaminant, and it was included in the first watch list established by the European Commission [4]. DCF has been detected in water at concentrations of up to μg/L [5–7], with the highest concentration detected in surface waters being 4.4 μg/L in countries such as the USA, Pakistan, Austria, and Germany [8]. One of the most significant cases of environmental toxicity is the decline in vulture populations in Pakistan due to the ingestion of DCF through food. Gyps vultures in South Asia were considered among the world's most common large raptors in the mid-1980s, but by 2000, three species were classified as critically endangered due to the rapid population decline [9].

Conventional water treatment and natural attenuation are not sufficient to remove these contaminants from wastewater, surface water, and drinking water, leading to their bioaccumulation in aquatic ecosystems and the human body [10]. Since wastewater treatment plants (WWTPs) are not designed to eliminate pharmaceuticals from water, these substances reach irrigation water for vegetables that we consume daily through WWTP effluents. It has been found that in tomatoes, radishes, and lettuce, the bioconcentration factor of diclofenac was the highest among the analyzed compounds, with concentrations in all vegetables ranging from 31 to 118 ng/g [11]. Therefore, it is of the utmost importance to develop efficient treatment techniques to degrade DCF in aquatic systems. Although DCF is a fairly stable compound, it is sensitive to photolysis [12]. Indeed, solar irradiation can cause the photochemical decomposition of DCF in surface waters.

Studies in the literature show the efficacy of treatments based on the use of UV light combined with hydrogen peroxide and/or intensified with ferrous ion salts in the degradation of DCF [13,14]. Reported works analyze the efficacy of these treatments based on the degradation of DCF concentration in water and aim to identify the degradation intermediates generated in treated waters [15–18]. However, although diclofenac is a compound widely studied within this context, these studies have not analyzed or delved into the quality parameters of water once processed. Therefore, this work aims to complement previous studies by analyzing the effect of UV light combined with hydrogen peroxide and iron salts on water quality parameters of an organoleptic nature (colour and turbidity) and other parameters of special relevance, such as dissolved oxygen, pH, and aromaticity, to ensure that treated waters meet environmental regulations. Among the main characteristics of oxidized DCF aqueous solutions, the strong yellow colour of partially oxidized water and the poor solubility of DCF at acidic pH values stand out. This results in an increase in suspended particles, causing high turbidity in the treated waters. Consequently, the primary objective of this work is to analyze these phenomena to determine the operating conditions (the molar ratios of oxidant and catalyst relative to the contaminant) that would allow the application of UV technology to obtain treated waters that are safe and suitable for discharge into natural water systems.

At this stage, the technology based on the use of UV light, enhanced by the oxidative action of hydrogen peroxide and iron salts, is compared in terms of DCF mineralization efficiency and its impact on water colour and turbidity. This comparison involves correlating the molecular structures of the different degradation intermediates generated during DCF oxidation, in the presence and absence of iron, with the formation of colour and turbidity [19–21]. Based on the obtained results and supported by DCF degradation mechanisms published in the literature [22–25], the aim is to theoretically deduce the DCF degradation pathways that lead to the formation of intermediates responsible for high turbidity and colour, as well as to propose models of molecular structures responsible for the intense colour in the water.

2. Results and Discussion

2.1. Kinetic Analysis of DCF Oxidation

2.1.1. Parameters Indicating the Quality of Treated Water

Figure 1 shows the degradation of DCF through the photo-Fenton treatment. As depicted in the graph, during the oxidation of DCF, high turbidity is generated in the water, but the photo-Fenton treatment is capable of degrading some of the turbidity-causing species. It is also observed that the total dissolved solids (TDS) increase exponentially over time until they reach a stable value that persists in the treated water. To explain the notable increase in TDS, it is necessary to consider that the solubility of DCF depends on the pH of the water, such that at values lower than pH > 4.0, DCF becomes insoluble as the pH becomes more acidic. Consequently, it is experimentally observed that the water acquires a whitish colour, and a whitish substance floats on the water surface when operating at pH = 2.0. This whitish substance is the insoluble DCF, which remains suspended in the water, causing high TDS and turbidity in the water. Since this effect has a direct influence on the turbidity and TDS of the water, it is necessary to adjust the initial pH of the water when studying these parameters.

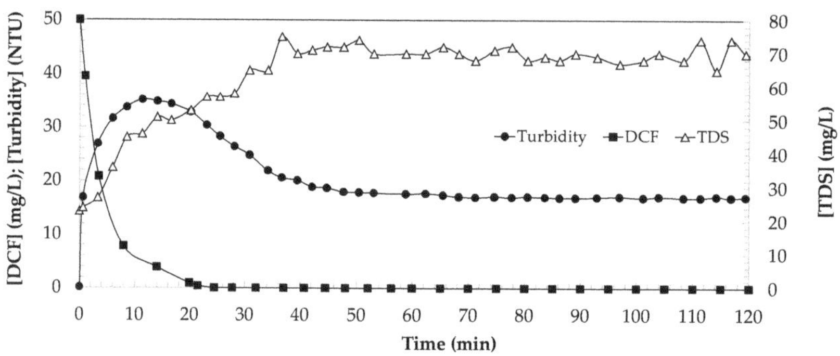

Figure 1. Kinetics of DCF concentration, turbidity, and total dissolved solids (TDS) during the oxidation of DCF using a photo-Fenton treatment. Experimental conditions: $[DCF]_0$ = 50.0 mg/L; $[pH]_0$ = 4.4; [UV] = 150 W; $[H_2O_2]_0$ = 1.0 mM; $[Fe^{2+}]_0$ = 1.0 mg/L; [T] = 30 °C.

Figure 2 shows the kinetic evolution of aromaticity and colour during DCF oxidation. As verified, DCF degrades to intermediates of a strongly aromatic nature until it reaches a maximum value, which would correspond to the maximum concentration of these species. Once these species are formed, the photo-Fenton treatment is capable of degrading part of the aromatic charge generated in the water. The evolution of colour maintains a trend parallel to the formation of aromaticity, so it is possible to deduce that aromatic species are responsible for the yellow-brown colour generated in the treated waters.

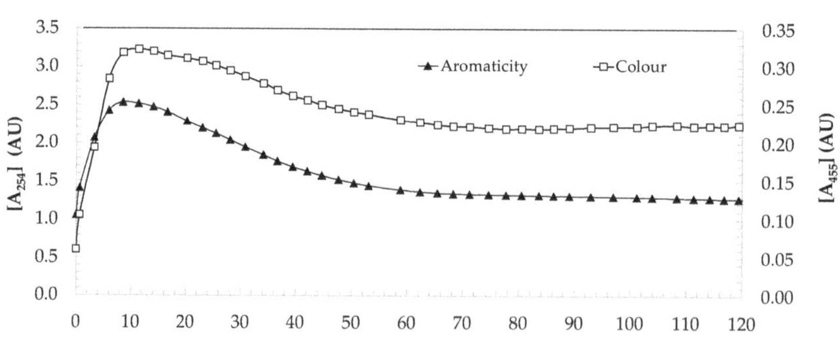

Figure 2. Kinetics of aromaticity (A_{254}) and colour (A_{455}) during the oxidation of DCF using a photo-Fenton treatment. Experimental conditions: $[DCF]_0$ = 50.0 mg/L; $[pH]_0$ = 4.4; [UV] = 150 W; $[H_2O_2]_0$ = 1.0 mM; $[Fe^{2+}]_0$ = 1.0 mg/L; [T] = 30 °C.

2.1.2. Kinetic Modelling of Pseudo-First Order for DCF Oxidation

Figure 3 shows the degradation kinetics of DCF when carrying out the treatment using UV light, and combining it with hydrogen peroxide, and with ferrous sulphate salts (photo-Fenton). As can be seen, DCF is completely oxidized during the first 20 min of the reaction in all cases. It is important to highlight that no significant dependence of the DCF oxidation rate was observed depending on the treatment, or the operating conditions used in the different tests carried out.

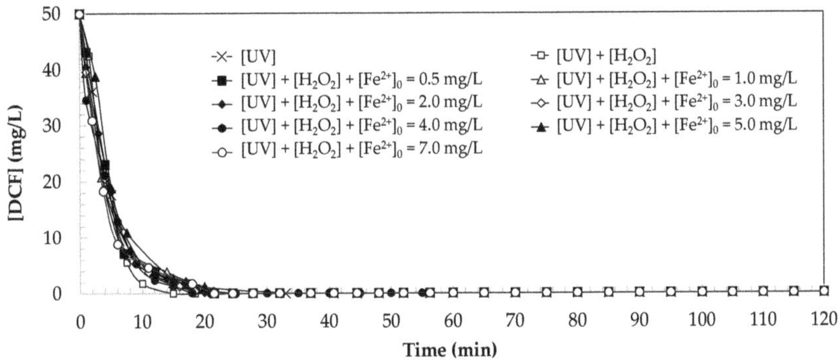

Figure 3. DCF kinetics as a function of UV treatment combined with hydrogen peroxide and iron. Experimental conditions: $[DCF]_0$ = 50.0 mg/L; $[pH]_0$ = 5.5; $[H_2O_2]_0$ = 1.0 mM; [UV] = 150 W; [T]=30 °C.

Based on the results obtained, the values of the DCF oxidation kinetic constants were graphically estimated (Figure 4) considering a pseudo-first order kinetic model (Equation (1)), where DCF degrades to intermediates of degradation according to a kinetic constant k_{DCF} (1/min) according to the kinetic equation shown in Equation (2).

Mass balance:

$$d[DCF]/dt = -k_{DCF}[DCF] \tag{1}$$

First order kinetic equation for DCF oxidation:

$$[DCF] = [DCF]_0 exp(-k_{DCF}t) \tag{2}$$

where:

$[DCF]_0$: initial concentration of DCF (= 50.0 mg/L);
[DCF]: concentration of DCF (mg/L);
k_{DCF}: pseudo-first-order kinetic constant for DCF oxidation (1/min).

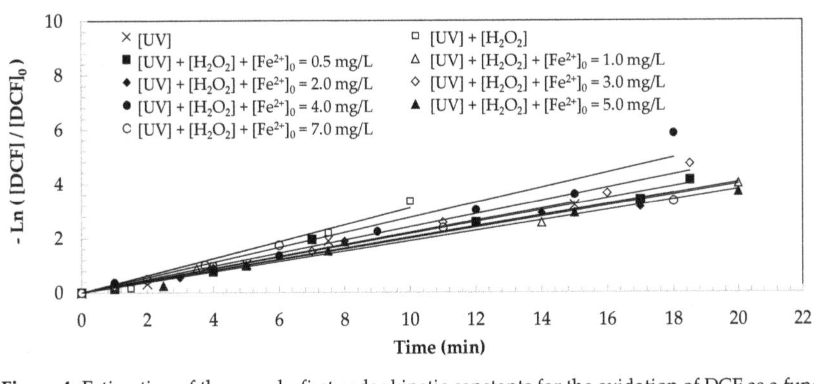

Figure 4. Estimation of the pseudo-first-order kinetic constants for the oxidation of DCF as a function of UV treatment combined with hydrogen peroxide and iron catalyst. Experimental conditions: $[DCF]_0$ = 50.0 mg/L; $[pH]_0$ = 5.5; $[H_2O_2]_0$ = 1.0 mM; [UV] = 150 W; [T] = 30 °C.

Table 1 shows the estimated kinetic values based on the operating conditions used, verifying that in the case of operating with a UV/H_2O_2 treatment, a higher degradation rate is obtained (k_{DCF} = 0.3104 1/min) than those obtained using only the action of UV light or with the photo-Fenton treatment (average k_{DCF} value = 0.2184 1/min). It should be noted that the values are very close, which indicates that k_{DCF} is not strongly influenced by the dose of iron, and therefore an average value was estimated between the experiments carried out.

Table 1. Estimated pseudo-first-order kinetic constants for DCF oxidation. Experimental conditions: $[DCF]_0$ = 50.0 mg/L; $[pH]_0$ = 5.5; [T] = 30 °C.

[UV] (W)	$[H_2O_2]_0$ (mM)	$[Fe^{2+}]_0$ (mg/L)	k_{DCF} (1/min)	r^2
150	0	0	0.2201	0.9879
150	1.0	0	0.3104	0.9709
150	1.0	0.5	0.2164	0.9808
150	1.0	1.0	0.1995	0.9867
150	1.0	2.0	0.2022	0.9817
150	1.0	3.0	0.2403	0.9918
150	1.0	4.0	0.2755	0.9489
150	1.0	5.0	0.1896	0.9932
150	1.0	7.0	0.2036	0.9431

2.2. Kinetic Study of the Parameters Indicating the Quality of the Treated Water

2.2.1. Formation of Aromatic Intermediates

Figure 5 shows the formation of aromaticity in water during the degradation of DCF using the different UV techniques tested. It should be noted that when adding iron to the aqueous DCF solution, the iron is adjusted to pH = 2.0 so that it remains in the form of a ferrous ion. By adding different volumes of the ferrous ion solution to water, the pH value of the water decreases, affecting the concentration of DCF that remains dissolved in water and in insoluble form (suspended), thus affecting the initial absorbance of DCF analyzed.

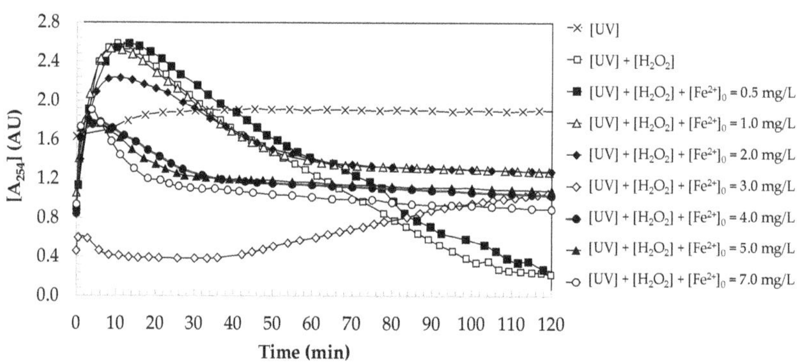

Figure 5. Formation of aromaticity in water during the oxidation of DCF as a function of UV treatment combined with hydrogen peroxide and iron catalyst. Experimental conditions: $[DCF]_0$ = 50.0 mg/L; $[pH]_0$ = 5.5; $[H_2O_2]_0$ = 1.0 mM; [UV] = 150 W; [T] = 30 °C.

The results obtained allow us to verify that UV light degrades DCF to aromatic intermediates that persist in water. When DCF is oxidized through a process activated by ultraviolet (UV) light, the aromaticity of the water containing DCF does not vary throughout the oxidation reaction and the water is found not to mineralize. This is because the action of UV light is not oxidative enough to affect the aromatic bonds of the DCF molecules.

When intensifying the treatment by combining UV light with hydrogen peroxide, DCF degrades to strongly aromatic intermediates, generating a maximum aromaticity value in the water ($[A_{254}]_{max}$ = 1.913 AU), which corresponds to the maximum concentration of aromatic intermediates. which is capable of generating 50.0 mg/L of DCF ($[A_{254}]_0$ = 0.91 AU) (see Figure 6a). Studies have identified several of these intermediates, which can vary slightly depending on the specific process conditions. In general, DCF is degraded through hydroxylation reactions by the addition of hydroxyl radicals to the aromatic rings of DCF, leading to the generation of mono-hydroxylated (4-hydroxy-diclofenac and 5-hydroxy-diclofenac) and di-hydroxylated products (4,5-dihydroxy-diclofenac), which subsequently generate compounds of a quinoid nature (diclofenac-2,5-iminoquinone, 2,6-dichlorobenzoquinone, and 2-[(2-amino-3-chlorophenyl)-hydroxymethyl]-6-chlorobenzoquinone) [18]. These intermediates continue to react under the action of UV light and the oxidant, finally leading to the opening of the aromatic rings, leading to the formation of diclofenac acetic acid, and the subsequent decarboxylation of acetic acid [17], mineralizing the DCF contained in the water to lower molecular weight products, such as CO_2, H_2O, and inorganic salts characterized by being compounds of low aromaticity, and obtaining treated waters with aromaticity values of $[A_{254}]_{final}$ = 0.224 AU (see Figure 6a) and mineralization yields of the final [TOC] = 86% (see Figure 6b).

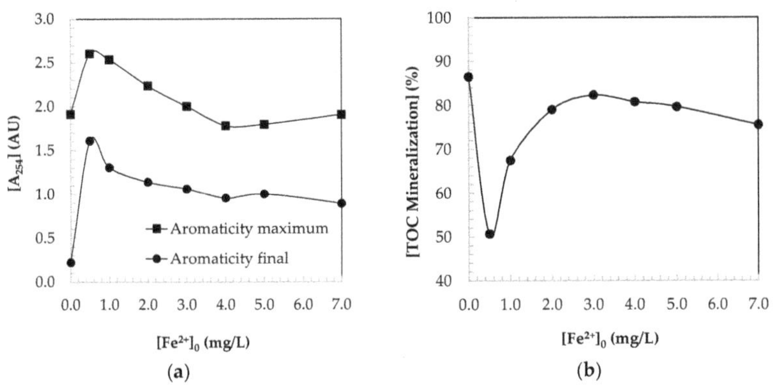

Figure 6. Effect of the dosage of iron catalyst on the quality parameters analyzed in the treated water. (**a**) Aromaticity. (**b**) Total Organic Carbon Mineralization (%). Experimental conditions: $[DCF]_0$ = 50.0 mg/L; $[pH]_0$ = 5.5; $[H_2O_2]_0$ = 1.0 mM; [UV] = 150 W; [T] = 30 °C; $[TOC]_0$ = 25.4 mg/L.

On the other hand, the photo-Fenton technology was tested operating with mass concentrations of iron in the range $[Fe^{2+}]_0$ = 0.5–7.0 mg/L. The results shown in Figure 5 indicate that when operating with iron concentrations of $[Fe^{2+}]_0$ = 0.5 mg/L, the highest aromaticity values are generated in the water ($[A_{254}]_{max}$ = 2.604 AU). As shown in Figure 6a, the maximum turbidity generated in the water decreases throughout the reaction, maintaining a considerable recalcitrant aromatic load in the treated waters. It is proven that by increasing the concentration of iron catalyst, the efficiency of removal of aromatic DCF degradation species improves, until reaching an optimal mineralization value (82% TOC removal yield) when operating with ferrous ion concentrations of $[Fe^{2+}]_0$ = 3.0 mg/L, which corresponds to an approximate molar ratio of 3 mol DCF:1 mol Fe^{2+}. Operating under these conditions, the aromatic load contained in the water is oxidized to values of the order of $[A_{254}]_{final}$ = 1.06 AU (Figure 6a). However, it is proven that the UV/H_2O_2 treatment allows us to obtain higher degrees of mineralization and elimination of aromatic load from the water than the photo-Fenton treatment and UV light. These results suggest that iron, given its high reactivity, could react with the organic load contained in the water, leading to the formation of stable metal complexes of 1 mol of iron with 3 mol DCF, which cause a decrease in the effectiveness of the photo-Fenton treatment.

2.2.2. Colour Formation in Water

Figure 7 shows the formation of a yellow-brown colour in the aqueous solutions of oxidized DCF. It should be noted that when DCF is oxidized with UV light, colour is generated during the first 20 min of reaction until reaching a maximum colour value ($[A_{455}]_{max}$ = 0.282 AU) that lasts in the treated water ($[A_{455}]_{final}$ = 0.282 AU) (see Figure 8). When intensifying the treatment by combining UV light with hydrogen peroxide, the maximum colour is generated after 10 min of reaction ($[A_{455}]_{max}$ = 0.216 AU), but the treatment is capable of degrading the reaction intermediates causing the colour, until treated waters with a colourless appearance are obtained ($[A_{455}]_{final}$ = 0.053 AU), as shown in Figure 8a. Using the photo-Fenton treatment, the maximum colour generated is of greater intensity than that observed in the previous treatments, with the treated water acquiring brown tones of up to $[A_{455}]_{max}$ = 0.412 AU when operating with ferrous ion concentrations of $[Fe^{2+}]_0$ = 4.0 mg/L, which corresponds to molar ratios of 2 mol DCF:1 mol Fe^{2+}. This residual colour generated in treated water reduces its quality and restricts its reuse and use. Even at low environmental levels, diclofenac and its degradation products can pose significant ecological risks, particularly to aquatic life and birds. Its persistence and the potential effects of its degradation products further complicate the issue. Addressing these challenges requires strict regulatory measures, improved wastewater treatment technologies, and an increased public awareness of the proper disposal of pharmaceuticals.

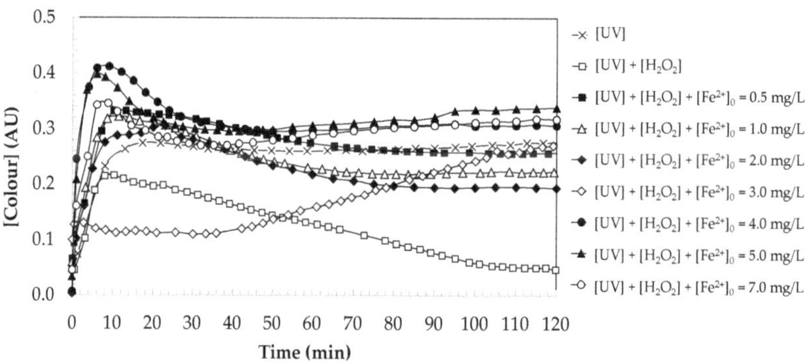

Figure 7. Colour formation in water during the oxidation of DCF as a function of UV treatment combined with hydrogen peroxide and iron catalyst. Experimental conditions: $[DCF]_0$ = 50.0 mg/L; $[pH]_0$ = 5.5; $[H_2O_2]_0$ = 1.0 mM; [UV] = 150 W; [T] = 30 °C.

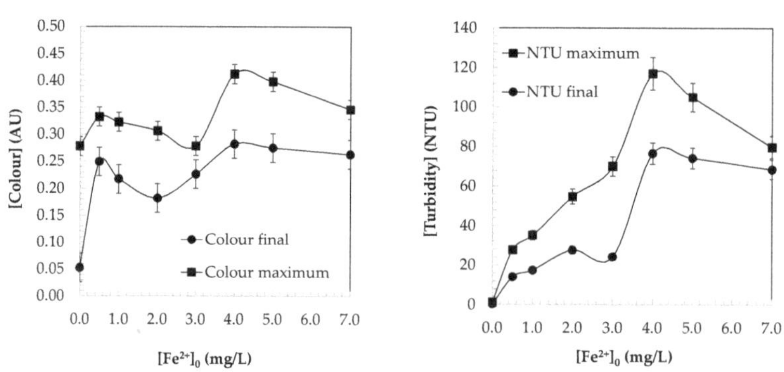

Figure 8. Quality parameters analyzed in treated water. (**a**) Colour. (**b**) Turbidity. Experimental conditions: $[DCF]_0$ = 50.0 mg/L; $[pH]_0$ = 5.5; $[H_2O_2]_0$ = 1.0 mM; [UV] = 150 W; [T] = 30 °C.

The colour generated in water during the oxidation of DCF is mainly due to the formation of oxidation intermediates, many of which are aromatic compounds containing structures with conjugated double-bond systems that can absorb light in the visible range [18]. These intermediates can be formed through different DCF degradation processes, including photodegradation, biodegradation, and chemical oxidation. The presence of these intermediates in water can cause a yellow or brown colouration, which is often an indication of the transformation of DCF into other, less known, and potentially less studied compounds in terms of toxicity and environmental behaviour.

Within these compounds, a variety of intermediates can be included with colours that vary between yellow and brown, such as quinones, aminoquinones (they can be formed from secondary reactions of quinones), and substituted anilines (they are formed as a consequence of the breaking of bonds in the original DCF molecule). Nitroaromatic compounds (yellowish colour) and chlorinated phenols (which tend to have yellowish or brown colours) can also be generated. Specifically, some of these intermediates, such as 2,6-dichloro-1,4-benzoquinone [22–24], can impart a deep yellow or brown colour to the water (see Figure 9).

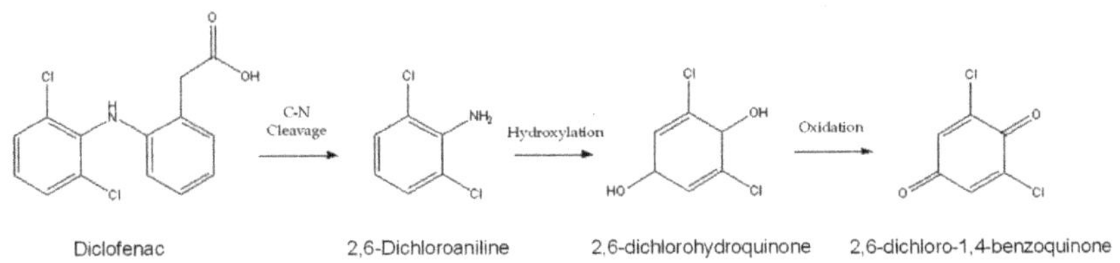

Figure 9. Proposed mechanism of degradation of DCF to 2,6-dichloro-1,4-benzoquinone (species causing an intense yellow-brown colour in water).

Some of the coloured products still remain even after extensive UV/H_2O_2 treatment because coloured products are very recalcitrant and persistent. This is because many of the coloured compounds have aromatic structures that allow electronic conjugation. This conjugation is responsible for the absorption of light in the visible spectrum, which gives rise to colouration. On the other hand, the presence of functional groups such as amino, nitro, hydroxyl, and carboxyl can influence the stability and optical behaviour of these compounds. The formation and stability of coloured compounds during the degradation of diclofenac have important environmental implications. Stable compounds can persist in the environment, affecting water quality and aquatic life. On the other hand, less stable compounds can degrade more quickly, reducing their long-term environmental impact.

2.2.3. Turbidity Formation

Figure 10 shows the turbidity changes in waters containing DCF when they are oxidized in the presence of UV light. As can be seen, when oxidizing DCF with UV light or combining it with hydrogen peroxide, turbidity is not generated in the water, but it remains constant throughout the reaction at values less than 1 NTU. However, when carrying out the oxidation by adding iron (photo-Fenton), high turbidity is generated in the water during the first 10 min of the reaction until a maximum value is reached. As shown in Figure 8b, the maximum turbidity generated in the water depends on the concentration of iron catalyst dosed, so that the turbidity increases with the concentration of iron, until the maximum intensity of turbidity corresponding to 117 NTU is formed when operating with $[Fe^{2+}]_0$ = 4.0 mg/L of ferrous ion, which corresponds to a molar ratio of 2 mol DCF : 1 mol Fe^{2+}. This is because iron can form iron hydroxides $Fe(OH)_3$ during the reaction, which are suspended particles that generate turbidity in the water. When operating with iron concentrations higher than $[Fe^{2+}]_0$ > 4.0 mg/L, turbidity decreases linearly with the ferrous ion concentration dosed, because the precipitation of ferric hydroxide particles occurs.

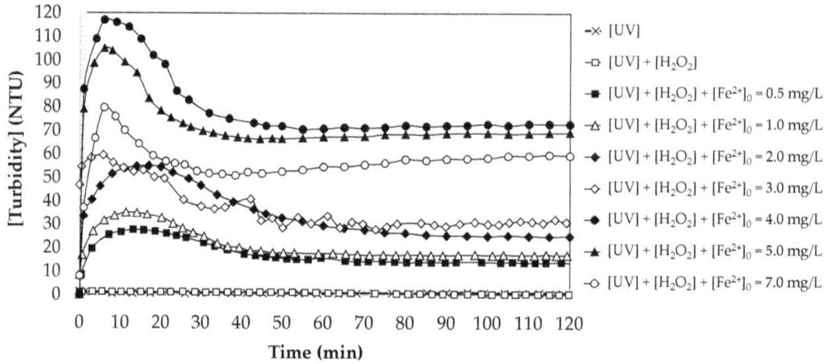

Figure 10. Formation of turbidity in water during the oxidation of DCF as a function of UV treatment combined with hydrogen peroxide and iron catalyst. Experimental conditions: $[DCF]_0$ = 50.0 mg/L; $[pH]_0$ = 5.5; $[H_2O_2]_0$ = 1.0 mM; [UV] = 150 W; [T] = 30 °C.

Once the maximum formation of turbidity in the water is reached, turbidity decreases exponentially until it reaches a stable value that persists in the treated waters (see Figure 9b). The photo-Fenton process generates hydroxyl radicals which are highly reactive species that can fragment diclofenac molecules into smaller compounds. These fragments can aggregate or interact with each other, forming larger particles that contribute to turbidity.

On the other hand, it should be considered that during the oxidation of DCF, several intermediate and final products are formed, such as carboxylic acids, aldehydes, ketones, and other degradation products that may have low solubility in water, due to the solubility of these functional groups decreasing with molecular weight (the number of carbons in their chains). As a consequence of their low solubility, they can precipitate or form colloids, increasing the turbidity of the treated water. In addition, it must be taken into account that the oxidation products of diclofenac can form coordination complexes with iron, which would contribute to the turbidity of the treated water. Two examples of these complexes are the 4′-hydroxydiclofenac complex [25], where the hydroxyl group can coordinate with the Fe^{3+} ion (see Figure 11a); and the complex of diclofenac and acridine with Fe^{2+} (see Figure 11b). These complexes may have properties that make them difficult to remove from water.

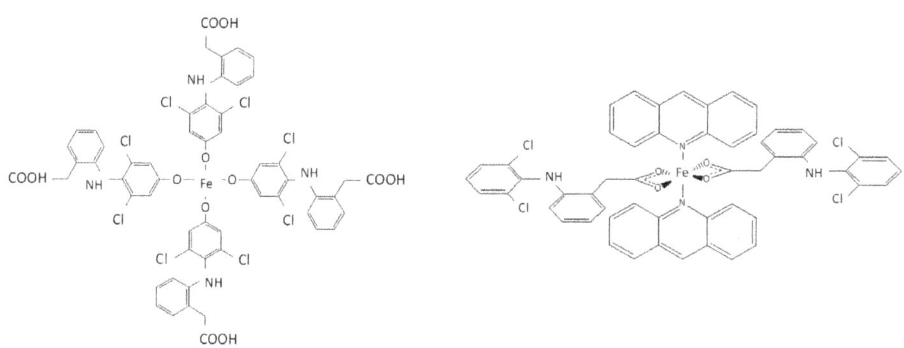

(**a**) (**b**)

Figure 11. Possible iron coordination complexes with DCF degradation intermediates. (**a**) Complex between 4-hydroxydiclofenac and Fe^{3+}. (**b**) Complex between diclofenac, acridine, and Fe^{2+}.

2.2.4. Dissolved Oxygen Concentration in Water

Figure 12 shows the changes in the concentration of dissolved oxygen in aqueous samples containing DCF during their oxidation with different treatments based on the use of UV light. As can be seen, the experiments begin with a different initial DO. This effect is due to the fact that an initial decrease in dissolved oxygen occurs in the aqueous solutions of diclofenac when ferrous ion (Fe^{2+}) is added, due to the oxidation of the ferrous ion to ferric ion (Fe^{3+}), a process which consumes oxygen dissolved in water.

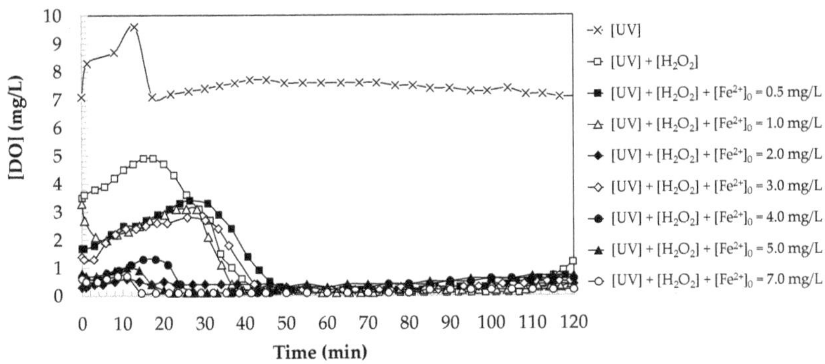

Figure 12. Dissolved oxygen concentration in water during DCF oxidation as a function of UV treatment combined with hydrogen peroxide and iron catalyst. Experimental conditions: $[DCF]_0$ = 50.0 mg/L; $[pH]_0$ = 5.5; $[H_2O_2]_0$ = 1.0 mM; [UV] = 150 W; [T] = 30 °C.

As verified, the direct photolysis process of DCF with UV light causes an increase in the concentration of DO during the first 13 min of the reaction, until it reaches a maximum value $[DO]_{max}$ = 9.6 mg/L. This effect is due to the photolysis of DCF generating hydroxyl radicals that react with the oxygen in water, decreasing its concentration and generating hydrogen peroxide (Equation (3)). This hydrogen peroxide can, in turn, decompose again into water and oxygen (Equation (4)), so that it compensates for the decrease in oxygen, and allows a net increase in the DO concentration analyzed in the water to be observed. The subsequent formation and decomposition reactions of hydrogen peroxide finally stabilize the DO concentration of the water at values of the order of $[DO]_{final}$ = 7.1 mg/L.

$$OH^{\bullet} + O_2 \rightarrow H_2O_2 + O_2^{\bullet -} \tag{3}$$

$$2H_2O_2 \rightarrow 2H_2O + O_2 \tag{4}$$

During the oxidation of DCF with UV/H_2O_2, an initial increase in the DO concentration takes place. This effect is due to the fact that despite the initial consumption of oxygen in the formation of hydroxyl radicals (Equation (5)), which are highly reactive agents that oxidize DCF consuming DO (Equation (6)), the production of molecular oxygen takes place from the photolysis of hydrogen peroxide (Equation (7)). This reaction compensates for the consumption of DO, allowing a net increase in DO concentration to be observed, until reaching a maximum value of $[DO]_{max}$ = 4.9 mg/L 15 min after oxidation. Once this maximum value is reached, the DO concentration begins to decrease as the initial load of hydrogen peroxide used in the treatment is consumed, finally stabilizing at values of the order of $[DO]_{final}$ = 0.7 mg/L.

$$H_2O_2 + h \rightarrow 2OH^{\bullet} \tag{5}$$

$$OH^{\bullet} + DCF \rightarrow \text{Oxidized products} \tag{6}$$

$$H_2O_2 \rightarrow H_2O + \frac{1}{2}O_2 \tag{7}$$

During the oxidation of DCF using a photo-Fenton treatment, the DO concentration decreases slightly during the first 5 min of the oxidation, due to the consumption of oxygen in the reactions of the generation of hydroxyl radicals and the formation of hydrogen peroxide (Equations (3) and (5)). Next, the DO in the water begins to increase because the photodegradation of hydrogen peroxide releases molecular oxygen (Equation (7)), which causes a net increase in the DO concentration. The continuous photodegradation of hydrogen peroxide under the action of UV light continues to release molecular oxygen. As proven, in the photo-Fenton process, the maximum DO value generated in the water is a function of the concentration of iron catalyst, due to the effect of exposure to UV light, which facilitates the regeneration of ferrous ions from ferric ions releasing oxygen (Equation (8)). Finally, as the concentration of hydrogen peroxide initially supplied in the treatment is depleted, the concentration of DO decreases until it stabilizes at values of the order of $[DO]_{final}$ = 0.6 mg/L.

$$Fe^{3+} + O_{2\bullet}^{-} \rightarrow Fe^{2+} + O_2 \quad (8)$$

3. Materials and Methods

To carry out the oxidation reactions of the DCF-doped aqueous solutions, a 1.0 L photocatalytic reactor containing a 150 W medium-pressure mercury lamp (Heraeus, 85.8 V, 148.8 W, 1.79 A, 95% transmission between 300 and 570 nm) was used, which was agitated by means of a magnetic stirrer. The reactor, operating in batch mode, was fed with the aqueous samples of diclofenac sodium (Fragon, >99.5%) of concentration $[DCF]_0$ = 50.0 mg/L. Then, the concentration of ferrous sulphate catalyst estimated for each test $[Fe^{2+}]_0$ = 0–7.0 mg/L and oxidant $[H_2O_2]_0$ = 1.0 mM was added. All reactions were carried out at the natural pH of water (around $[pH]_0$ = 6.5). During the reaction, the temperature of the reaction mixture was kept within a range of 30 °C using a heating bath (Frigiterm-P Selecta). During the 120 min of the test, which, considering the total energy absorbed by the water, corresponds to a UV dose per unit volume of water of 1,026,000,000 J/m^3, the following variables were analyzed: pH was measured with a pH meter (Kent EIL9142).

Water turbidity was measured in nephelometric turbidity units (NTU) with a turbidity meter (Model HI88703, Hanna Instruments S.L., Eibar, Spain). Aromaticity and colour were analyzed using a UV/Vis spectrophotometer (Model V-630, Jasco, Madrid, Spain) measuring at λ = 254 and 455 nm. Dissolved oxygen and temperature were measured with a portable oxygen meter (Model HI 9142, Hanna Instruments S.L., Eibar, Spain). A TOC analyser (Shimadzu TOC-V, Shimadzu Corporation, Kyoto, Japan) was used to measure total organic carbon (TOC, mg/L).

DCF concentration was measured using High-Performance Liquid Chromatography (Model 2695, Waters Cromatografía S.A., Cerdanyola del Vallès, Spain) with a Dual λ Absorbance Detector (Model 2487, Waters Cromatografía S.A., Cerdanyola del Vallès, Spain). A ZORBAX Eclipse PAH column (150 mm, 4.6 mm, particle size 5 μm) and a guard column ZORBAX Eclipse PAH (4.6 mm, 12.5 mm) supplied by Agilent (Santa Clara, California, USA) were used. The mobile phase consisted of water and acetonitrile (ACN), at a flow rate of 0.8 mL/min. Initial gradient conditions were set at 20% ACN and increased from 45% *v/v* ACN during 3 min, then held during 6 min, and finally decreased to 20% *v/v* ACN at 1 min. Total run time was 10 min. Injection volume was 50 μL and all separations were performed at room temperature. The identification of DCF was performed by comparison of standard. Detection was carried out at the following wavelength: 275 nm.

4. Conclusions

The results obtained in the study allow us to assert the superiority of the combined UV/H_2O_2 treatment for the mineralization and elimination of aromatic load and colour in the oxidation of DCF contained in water, in comparison with UV light treatments and photo-Fenton. This study shows that UV light degrades DCF to aromatic intermediates, but does not achieve complete mineralization due to its insufficient oxidizing power. The combination of UV light with hydrogen peroxide significantly improves the degradation of DCF, achieving high mineralization (a mineralization efficiency of 86%) by transforming aromatic intermediates into lower molecular weight products such as CO_2, H_2O, and inorganic salts. The photo-Fenton treatment allows us to obtain a mineralization efficiency of 82% using molar ratios of 3 mol DCF : 1 mol Fe^{2+}. During the oxidation of DCF with UV light, a yellow colour is generated during the first 20 min that remains in the treated waters. By combining UV with hydrogen peroxide, the maximum colour is reached in 10 min but treated water is obtained with a colourless appearance. When using the photo-Fenton treatment, a more intense residual colour (yellow-brown) is generated that affects the quality of the treated water. Turbidity remains at values less than 1 NTU by oxidizing DCF with UV or combined with hydrogen peroxide. The photo-Fenton treatment generates high initial turbidity (a maximum of 117 NTU with molar ratios of 2 mol DCF : 1 mol Fe^{2+}), which decreases with the precipitation of ferric hydroxide and stabilizes over time.

Finally, it should be noted that this study lays out the foundations for future work, in which it would be advisable to work at the scale of a pilot plant and with real water from the inflows and effluents treated in wastewater treatment plants to study the change in scaling in the results obtained. When working with real water, it will be necessary to consider that the concentrations of diclofenac are much lower, which will require equipment for concentrating emerging contaminants, to treat smaller flows of water than the flow of water entering the treatment plant.

Author Contributions: Conceptualization, N.V. and A.M.D.L.; methodology, B.E. and U.D.; software, B.E., U.D. and N.V.; validation, N.V.; formal analysis, B.E., U.D. and N.V.; investigation, N.V., U.D. and B.E.; resources, B.E., U.D., N.V. and A.M.D.L.; data curation, B.E., U.D. and N.V.; writing—original draft preparation, N.V.; writing—review and editing, N.V.; visualization, N.V., A.M.D.L., B.E. and U.D.; supervision, A.M.D.L. and J.I.L.; project administration, A.M.D.L. and J.I.L.; funding acquisition, A.M.D.L. and J.I.L. All authors have read and agreed to the published version of the manuscript.

Funding: This research was funded by PIBA_2023_01_0032- Basic and/or Applied Research Project. Financing entity: Basque Government 2023.

Data Availability Statement: The data supporting the findings of this study can be found within the article.

Acknowledgments: The authors thank the Department of Environmental and Chemical Engineering and the Department of Chemical Engineering of the University of the Basque Country UPV/EHU, for their support.

Conflicts of Interest: The authors declare no conflicts of interest.

References

1. Cleuvers, M. Mixture toxicity of the anti-inflammatory drugs diclofenac, ibuprofen, naproxen, and acetylsalicylic acid. *Ecotoxicol. Environ. Saf.* **2003**, *59*, 309–315. [CrossRef] [PubMed]
2. Buser, H.R.; Poiger, T.; Müller, M.D. Occurrence and fate of the pharmaceutical drug diclofenac in surface waters: Rapid photodegradation in a lake. *Environ. Sci. Technol.* **1998**, *32*, 3449–3456. [CrossRef]
3. Schwaiger, J.; Ferling, H.; Mallow, U.; Wintermayr, H.; Negele, R. Toxic effects of the non-steroidal anti-inflammatory drug diclofenac. *Aquat. Toxicol.* **2004**, *68*, 141–150. [CrossRef]
4. Commission Implementing Decision (EU) 2015/495 of 20 March 2015 Establishing a Watch List of Substances for Union-Wide Monitoring in the Field of Water Policy Pursuant to Directive 2008/105/EC of the European Parliament and of the Council (Notified under Document C(2015) 1756) Text with EEA Relevance. Available online: http://data.europa.eu/eli/dec_impl/2015/495/oj (accessed on 24 August 2024).
5. Sousa, J.C.G.; Ribeiro, A.R.; Barbosa, M.O.; Ribeiro, C.; Tiritan, M.E.; Pereira, M.F.R.; Silva, A.M. Monitoring of the 17 EU Watch List contaminants of emerging concern in the Ave and the Sousa Rivers. *Sci. Total Environ.* **2019**, *649*, 1083–1095. [CrossRef]
6. Lonappan, L.; Brar, S.K.; Das, R.K.; Verma, M.; Surampalli, R.Y. Diclofenac and its transformation products: Environmental occurrence and toxicity — A review. *Environ. Int.* **2016**, *96*, 127–138. [CrossRef] [PubMed]
7. Vieno, N.; Sillanpää, M. Fate of diclofenac in municipal wastewater treatment plant — A review. *Environ. Int.* **2014**, *69*, 28–39. [CrossRef] [PubMed]
8. Yu, H.; Nie, E.; Xu, J.; Yan, S.; Cooper, W.J.; Song, W. Degradation of Diclofenac by Advanced Oxidation and Reduction Processes: Kinetic Studies, Degradation Pathways and Toxicity Assessments. *Water Res.* **2013**, *47*, 1909–1918. [CrossRef]

9. Gilbert, M.; Watson, R.T.; Virani, M.Z.; Oaks, J.L.; Ahmed, S.; Chaudhry, M.J.I.; Arshad, M.; Mahmood, S.; Ali, A.; Khan, A.A. Rapid population declines and mortality clusters in three Oriental white-backed vulture *Gyps bengalensis* colonies in Pakistan due to diclofenac poisoning. *Oryx* **2006**, *40*, 388–399. [CrossRef]
10. Ruhí, A.; Acuña, V.; Barceló, D.; Huerta, B.; Mor, J.R.; Rodríguez-Mozaz, S.; Sabater, S. Bioaccumulation and trophic magnification of pharmaceuticals and endocrine disruptors in a Mediterranean river food web. *Sci. Total. Environ.* **2016**, *540*, 250–259. [CrossRef]
11. Kovacs, E.D.; Silaghi-Dumitrescu, L.; Kovacs, M.H.; Roman, C. Determination of the Uptake of Ibuprofen, Ketoprofen, and Diclofenac by Tomatoes, Radishes, and Lettuce by Gas Chromatography-Mass Spectrometry (GC-MS). *Anal. Lett.* **2021**, *54*, 314–330. [CrossRef]
12. Pizzichetti, R.; Martín-Gamboa, M.; Pablos, C.; Reynolds, K.; Stanley, S.; Dufour, J.; Marugán, J. Environmental life cycle assessment of UV-C LEDs vs. mercury lamps and oxidant selection for diclofenac degradation. *Sustain. Mater. Technol.* **2024**, *41*, e01002. [CrossRef]
13. Pizzichetti, R.; Reynolds, K.; Pablos, C.; Casado, C.; Moore, E.; Stanley, S.; Marugán, J. Removal of diclofenac by UV-B and UV-C light-emitting diodes (LEDs) driven advanced oxidation processes (AOPs): Wavelength dependence, kinetic modelling and energy consumption. *Chem. Eng. J.* **2023**, *471*, 144520. [CrossRef]
14. Rehman, F.; Ahmad, W.; Sayed, M. Mechanistic investigations on the removal of diclofenac sodium by $UV/S_2O_8{}^{2-}/Fe^{2+}$, $UV/HSO_5{}^{-}/Fe^{2+}$ and $UV/H_2O_2/Fe^{2+}$-based advanced oxidation processes. *Environ. Technol.* **2020**, *42*, 3995–4005. [CrossRef]
15. Dhawle, R.; Mantzavinos, D.; Lianos, P. UV/H_2O_2 degradation of diclofenac in a photocatalytic fuel cell. *Appl. Catal. B: Environ.* **2021**, *299*, 120706. [CrossRef]
16. Hong, M.; Wang, Y.; Lu, G. UV-Fenton degradation of diclofenac, sulpiride, sulfamethoxazole and sulfisomidine: Degradation mechanisms, transformation products, toxicity evolution and effect of real water matrix. *Chemosphere* **2020**, *258*, 127351. [CrossRef]
17. Zhang, L.; Liu, Y.; Fu, Y. Degradation kinetics and mechanism of diclofenac by UV/peracetic acid. *RSC Adv.* **2020**, *10*, 9907–9916. [CrossRef]
18. Li, X.; Zhou, M.; Pan, Y. Degradation of diclofenac by H_2O_2 activated with pre-magnetization Fe^0: Influencing factors and degradation pathways. *Chemosphere* **2018**, *212*, 853–862. [CrossRef]
19. Villota, N.; Cruz-Alcalde, A.; Ferreiro, C.; Lombraña, J.I.; Esplugas, S. Changes in solution turbidity and color during paracetamol removal in laboratory and pilot-scale semicontinuous ozonation reactors. *Sci. Total. Environ.* **2023**, *854*, 158682. [CrossRef]
20. Villota, N.; Jankelevitch, S.; Lomas, J.M. Kinetic modelling of colour and turbidity formation in aqueous solutions of sulphamethoxazole degraded by UV/H_2O_2. *Environ. Technol.* **2022**, *45*, 349–359. [CrossRef]
21. Villota, N.; Ferreiro, C.; Qulatein, H.A.; Lomas, J.M.; Camarero, L.M.; Lombraña, J.I. Colour Changes during the Carbamazepine Oxidation by Photo-Fenton. *Catalysts* **2021**, *11*, 386. [CrossRef]
22. Muelas-Ramos, V.; Sampaio, M.; Silva, C.; Bedia, J.; Rodriguez, J.; Faria, J.; Belver, C. Degradation of diclofenac in water under LED irradiation using combined $g\text{-}C_3N_4/NH_2$-MIL-125 photocatalysts. *J. Hazard. Mater.* **2021**, *416*, 126199. [CrossRef] [PubMed]
23. Kosaka, K.; Nakai, T.; Hishida, Y.; Asami, M.; Ohkubo, K.; Akiba, M. Formation of 2,6-dichloro-1,4-benzoquinone from aromatic compounds after chlorination. *Water Res.* **2017**, *110*, 48–55. [CrossRef]
24. Pan, Z.; Zhu, X.; Li, G.; Wang, Y.; Li, M.; Sun, S.; Jia, R.; Hou, L. Degradation of 2,6-dichloro-1,4-benzoquinone by advanced oxidation with UV, H_2O_2, and O_3: Parameter optimization and model building. *J. Water Supply Res. Technol.* **2021**, *70*, 1159–1169. [CrossRef]
25. Bouju, H.; Nastold, P.; Beck, B.; Hollender, J.; Corvini, P.F.X.; Wintgens, T. Elucidation of biotransformation of diclofenac and 4′hydroxydiclofenac during biological wastewater treatment. *J. Hazard. Mater.* **2016**, *301*, 443–452. [CrossRef] [PubMed]

catalysts

MDPI

Article

Peroxymonosulfate Activation by Fe/C Composites for Paracetamol Degradation: Performance Evaluation and Mechanism Insight

Yujun Zhuo [1], Hong Meng [1], Yongqing Zhang [1,2,*], Yu Chen [1] and Jiaqi Cui [1]

1 School of Environment and Energy, South China University of Technology, Guangzhou 510006, China; zyj19806785427@163.com (Y.Z.); 202110188969@mail.scut.edu.cn (H.M.); cy720316@163.com (Y.C.); jqcui0617stu@163.com (J.C.)
2 The Key Lab of Pollution Control and Ecosystem Restoration in Industry Clusters, Ministry of Education, Guangzhou 510006, China
* Correspondence: zhangyq@scut.edu.cn

Abstract: Peroxymonosulfate-based advanced oxidation processes (PMS-AOPs) relying on non-radical pathways offer advantages such as resistance to interference, efficient oxidant utilization, and selective degradation of pollutants. In this study, an Fe, N co-doped activator (Fe-N-$C_{1.5}$) was synthesized using a simple mixed solvent pyrolysis method. The Fe-N-$C_{1.5}$ exhibited excellent PMS activation activity. A total of 100% of paracetamol (PCT, 10 ppm) was degraded in the Fe-N-$C_{1.5}$/PMS system in 7 min. Furthermore, this oxidation system maintained effective PCT removal even in the presence of background ions and in real water matrices. In addition, the leached Fe concentration after 60 min was only 0.084 mg/L, and 94% of PCT could still be removed during the fourth cyclic use of the catalyst. Quenching experiments, electron paramagnetic resonance (EPR), and electrochemical analysis revealed that the Fe-N-$C_{1.5}$/PMS/PCT system predominantly relies on non-radical pathways, including singlet oxygen ($^1O_2$) and catalyst-interface-mediated electron transfer process (ETP). X-ray photoelectron spectroscopy (XPS) analysis and KSCN toxicity experiment confirmed that the graphitic N, carbonyl (C=O), and Fe-N_x were the main PMS activation sites. This study provides an understanding of degradation mechanisms of the Fe-N-$C_{1.5}$/PMS/PCT system and offers insights into the design of iron–carbon composite catalysts that carry out non-radical PMS activation.

Keywords: peroxymonosulfate; paracetamol; singlet oxygen; electron transfer process; Fe/C composites

1. Introduction

Paracetamol (PCT) is a widely used analgesic and antipyretic owing to its safety, tolerability, and minimal side effects [1,2]. However, approximately 58–68% of ingested PCT is not absorbed and utilized by the human body [3,4] and is excreted directly into the municipal wastewater treatment system [5]. Due to its low biodegradability, traditional biological treatment processes do not efficiently treat PCT [3]. According to reports, in municipal wastewater treatment plants of various regions, the PCT concentrations remained between 0.27 and 300 μg/L [6–8]. If PCT is not adequately treated, its release into the environment will present toxicological risks to aquatic organisms, including hepatotoxicity, genotoxicity, and endocrine disruption [8]. Furthermore, PCT bioaccumulation, through the food chain, poses a threat to human health. Therefore, developing a rapid and effective PCT degradation treatment strategy is urgent.

Peroxymonosulfate-based advanced oxidation processes (PMS-AOPs) are a promising remediation technology for PCT degradation, due to their rapid generation of various reactive oxygen species (ROS) with high oxidation-reduction potentials [9–11]. Radical-based PMS-AOPs show significant cost-effectiveness and high efficiency, as they do not require additional energy input while enabling the rapid generation of strongly oxidizing radicals [12]. However, its application is limited by natural organic matter, background ions, and pH [13]. In addition, the reaction of halide ions with free radicals results in the formation of halogen radicals, which further react with organics to form more toxic organohalides [14]. In contrast, the non-radical-based PMS-AOPs exhibit a broader pH operational range, higher selectivity for pollutants, more improved oxidant utilization efficiency, and more sustained degradation capacity [15]. They can be carried out on the catalyst surface to directly oxidize organics with electron-rich groups without additional oxidant consumption and radical generation [16,17]. Moreover, non-radical-based PMS-AOPs hinder the formation of halogen free radicals, minimizing the production of halogenated disinfection by-products. Therefore, developing suitable catalysts that efficiently activate PMS for selectively degrading organic pollutants through non-radical pathway is crucial for facilitating the efficient PCT removal by the PMS-AOPs system.

Most carbon materials, such as biochar [18], carbon nanotubes [19], and graphene oxide [20], exhibit poor PMS activation activities and stabilities due to their low degree of graphitization and high impedance [21]. In contrast, metal materials exhibit better activation capacity and stability [22]; however, metal ion leaching during the reaction process poses an additional environmental threat [23]. To improve the stability and activation ability of materials and reduce metal leaching, researchers have proposed metal composite materials, which incorporate both metal and non-metal elements into carbon materials [24,25]. On one hand, the doping of metal and non-metal elements can break the structural inertness of the carbon network, enhancing the electron transfer capacity of materials while also inducing the formation of structural defects that increase the number of active sites [26]. For instance, doping carbon materials with electron-rich nitrogen can create Lewis basic sites, which tend to bind with transition metal atoms, forming strong M-Nx coordination sites with high catalytic activity (M, representing Lewis acids) [27,28]. On the other hand, carbon materials disperse and anchor metal atoms, preventing particle aggregation, avoiding the loss of metal active sites, and reducing metal leaching [24].

Iron is an excellent candidate for the preparation of metal composites due to its abundant natural reserves and environmental friendliness [29,30]. A large number of Fe/C composite catalysts have been designed and applied to AOPs. Wu et al. developed a carbon-dot-modified iron-based MOF material as persulfate (PS) activator for PCT degradation. In total, 100% of PCT was removed in the iron-based MOF/PS system in 20 min [31]. Zhu et al. synthesized an Fe-N-C catalyst by anchoring Fe onto a covalent triazine framework. The Fe-N-C exhibited efficient PMS activation, and the Fe-N-C/PMS system achieved 100% PCT degradation across a broad pH range (2.6–10.67) [32]. However, iron–carbon composite materials reported in previous studies still face several challenges on their practical applications. Zhuo et al. developed a magnetic biochar catalyst with dual active sites of Fe_3C and Fe_4N, but the synthesis process required high energy consumption [33]. Huang et al. designed an Fe(BDC)(DMF,F)-OA/PS system that exhibited significant iron leaching (up to 1.51 mg/L) after 120 min of reaction [34]. Furthermore, due to limitations in quencher selection and experimental design, the mechanistic understanding of high-valent metal oxides, singlet oxygen, and electron transfer pathways remains controversial in previous studies [24,35]. Therefore, it is necessary to conduct further research for enhancing the understanding of the oxidation mechanisms in Fe/C composite/persulfate systems.

In this study, an Fe-N-C catalyst was synthesized through a simple and low-energy mixed-solvent pyrolysis method to construct a non-radical-dominated PMS oxidation system for PCT degradation. By regulating the Fe/C mass ratio in the precursor, the catalytic performance of Fe-N-C was optimized while achieving extremely low Fe leaching. This work systematically explored the operational conditions affecting the performance of the Fe-N-C/PMS system and identified the main reactive species responsible for PCT degradation and the active sites for PMS activation in the Fe-N-C/PMS system through quenching experiments, EPR, electrochemical tests, XPS analysis, and KSCN toxicity experiment. Additionally, this work proposed possible degradation pathways for PCT and evaluated the toxicity of the intermediates. In conclusion, this study provides a comprehensive understanding of the PCT degradation mechanism in the Fe-N-C/PMS system and presents a new approach for developing efficient and stable Fe-N-C catalysts.

2. Result and Discussion

2.1. Characterization of Materials

The Fe-N-C_x microstructure was observed using scanning electron microscopy (SEM). Fe-N-C_6, with the lowest Fe:C mass ratio, exhibited a unique carbon block structure with a smooth and dense surface (Figure 1a,b). As the Fe:C mass ratio increased to 2:3, the carbon block surface gradually became rough, fragmented, and more porous (Figure 1c,d), indicating that the Fe:C mass ratio in the precursor had an impact on the catalyst structure. Fe may promote the graphitization of carbon material during pyrolysis [36], thereby affecting the microstructure of the carbon material. Additionally, the rough surface morphology facilitates exposure of additional active sites and enhances reactant accessibility, thereby significantly improving catalytic activity [37]. Interestingly, granular protrusions were observed on the Fe-N-$C_{0.25}$ surface (Figure S1), which could be attributed to the formation of encapsulated iron oxides from the excess Fe [38]. The energy-dispersive X-ray spectroscopy (EDS) elemental mapping images (Figure 1e) of Fe-N-$C_{1.5}$ showed uniform dispersion of C, N, O, and Fe on the catalyst's surface.

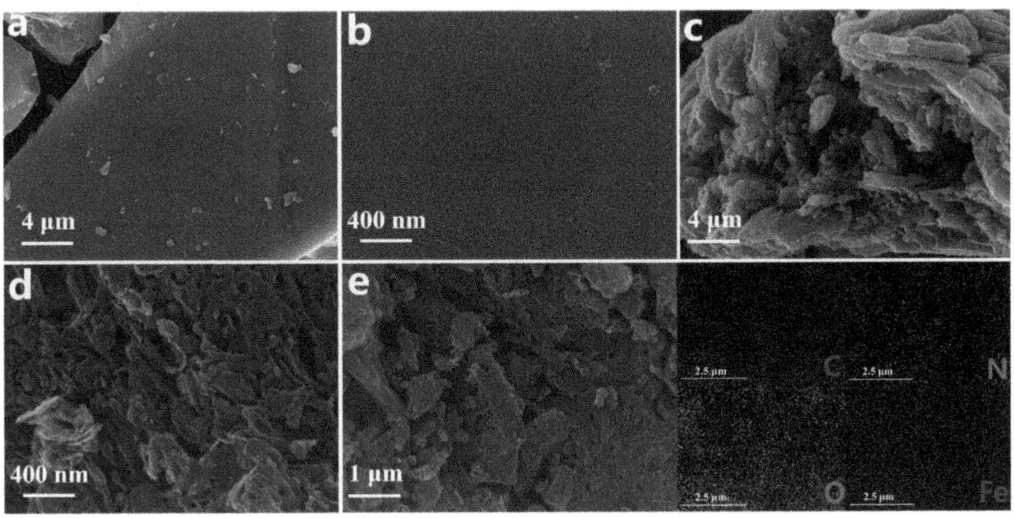

Figure 1. SEM images of (**a**,**b**) Fe-N-C_6 and (**c**,**d**) Fe-N-$C_{1.5}$ at different scales; (**e**) EDS images of Fe-N-$C_{1.5}$.

The crystalline phase composition of the catalyst was obtained using X-ray diffraction (XRD). No characteristic peaks for Fe or its oxides were observed for Fe-N-$C_{0.75}$, Fe-N-$C_{1.5}$, Fe-N-C_3, and Fe-N-C_6. This may be due to the low Fe-doping level, which coordinated with C and N during pyrolysis, resulting in its high dispersion within the material [31]. The broad peak at 24° was from the 002 plane, and the weak diffraction peak at 43.8° was from the 101 plane of graphite (Figure 2a), indicating that the synthesized carbon structure was amorphous [39]. In contrast, the higher Fe content in Fe-N-$C_{0.25}$ led to the formation of iron oxides (Fe_3O_4, PDF 97-015-8743) during pyrolysis. Additionally, the XRD pattern of Fe-N-$C_{0.25}$ (Figure 2b) showed characteristic peaks for elemental sulfur (S_8, PDF 97-020-0453). The presence of sulfur could be explained by the decomposition of excess sulfate ions into sulfur dioxide (SO_2) at high temperatures, which were then reduced to elemental sulfur by the carbon [40].

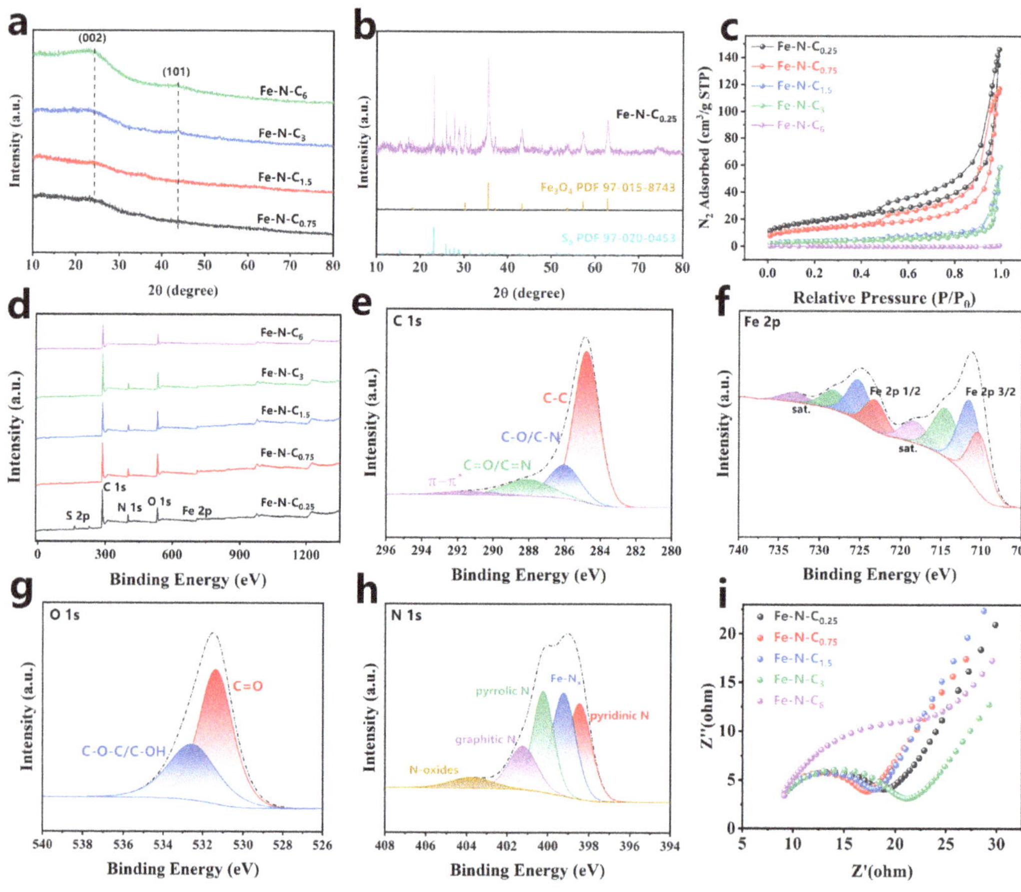

Figure 2. (**a**,**b**) XRD patterns, (**c**) N_2 adsorption/desorption isotherms, (**d**) XPS survey spectra of Fe-N-C_x; XPS C 1s (**e**), Fe 2p (**f**), O 1s (**g**), and N 1s (**h**) of Fe-N-$C_{1.5}$; (**i**) The Nyquist plots of Fe-N-C_x.

The N_2 adsorption–desorption isotherm of the prepared catalysts (Figure 2c) exhibited typical type-IV isotherms with H_3-type hysteresis loops, indicating the presence of mesoporous structures (2–50 nm) in the catalyst [41]. The Barrett–Joyner–Halenda (BJH) curve (Figure S2) revealed that the prepared catalysts pore structures were primarily composed of mesopores and macropores, with most of the pores having a diameter of around 50 nm.

Furthermore, as the Fe:C mass ratio in the precursor increased, the pyrolyzed catalyst had a larger Brunauer–Emmett–Teller (BET) surface area and smaller average pore diameter (Table S2), suggesting that Fe promoted porous structure formation in the material during pyrolysis. In addition, Fe-N-C_6 did not exhibit a type-IV isotherm, which was attributed to its low Fe-doping level, resulting in a smooth and flat structure with a smaller specific surface area. The specific surface area has a significant impact on the catalytic activity. Figure S3a shows the relationship between the BET surface area of catalyst and degradation rate constant (K_{obs}). Compared to Fe-N-C_3 and Fe-N-C_6, Fe-N-$C_{0.25}$ and Fe-N-$C_{0.75}$, which had larger surface areas, exhibited higher K_{obs} values. This might be attributed to the larger surface area providing more active sites that facilitate the adsorption of PMS and pollutants, thereby enhancing catalytic performance [39].

The Fe content in the catalyst was determined by ICP-OES (Table S3). The Fe loading increased with the Fe:C mass ratio in the precursor, but when the Fe:C mass ratio exceeded 2:3, the Fe loading did not significantly increase. High Fe loading in the iron–carbon composite catalyst provided additional active sites. Figure S3b illustrates an approximately positive correlation between Fe content and K_{obs}, which might be due to the higher Fe content leading to the formation of more Fe active sites after pyrolysis. These active sites played a crucial role in PMS activation.

High-resolution X-ray photoelectron spectroscopy (XPS) was used to analyze the surface elements and chemical composition of the catalyst. XPS analysis (Figure 2d) revealed that the catalyst surface primarily contains C, N, O, and Fe, which was consistent with the EDS analysis results. Table S4 presents the surface composition results and their relative contents for the different catalysts. The C 1s spectrum (Figure 2e) was deconvoluted into four peaks, C-C bonds (284.8 eV), C-O bonds (286.02 eV), C=O bonds (288.07 eV), and π-π* shake-up satellite (291.44 eV) [17]. Compared to Fe-N-C_6, catalysts with higher Fe content (Fe-N-$C_{0.75}$ and Fe-N-$C_{1.5}$) showed a higher proportion of C-O and C=O bonds. This result showed that Fe promoted carbon matrix reconstruction during pyrolysis, leading to the formation of more oxygen-containing functional groups [42]. Previous studies have demonstrated that C=O is the active center for PMS activation on catalysts to produce $^1O_2$, with higher C=O content significantly enhancing catalytic activation performance [43].

The Fe 2p spectrum (Figure 2f) could be deconvoluted into Fe^{2+} (710.21 eV, 723.13 eV) and Fe^{3+} (711.39 eV, 725.19 eV) [32]. Additionally, four satellite peaks were observed: two corresponding to Fe^{2+} at 714.36 eV and 728.23 eV, and two corresponding to Fe^{3+} at 718.27 eV and 733.04 eV. The O 1s spectrum (Figure 2g) could be deconvoluted into C=O (531.37 eV) and C-O-C/C-OH (532.54 eV). Notably, no discernible peaks corresponding to Fe-O bonds (~529.5–530.5 eV) were observed in the O 1s spectrum, suggesting negligible contributions from iron oxides (Fe-O) or iron–carbonate (Fe-O-C) species. This absence aligns with the catalyst preparation protocol, which includes an acid-washing step to remove unstable metallic aggregates, thereby significantly reducing oxygenated iron phases. Previous studies on Fe/C composites have similarly reported that acid-treated catalysts predominantly exhibit oxygen functionalities associated with carbon matrices (C=O and C-O) rather than Fe-bonded oxygen [44].

The N 1s spectrum was deconvoluted into five peaks (Figure 2h), corresponding to pyridinic N (398.47 eV), Fe-N_x (399.25 eV), pyrrolic N (400.25 eV), graphite N (401.25 eV), and N-oxide (403.75 eV) [31,32]. As shown in Table S4, the proportions of Fe-N_x on the surface of the catalyst gradually increased as the Fe:C mass ratio increased, while the proportions of Fe-N_x decreased when the Fe:C mass ratio exceeded 0.75. It was reported that the Fe-N_x site was favorable to alter the electronic distribution on the catalyst surface, enhancing electron transfer capability [31]. Electrochemical impedance spectroscopy

(EIS) was used to analyze the interfacial charge transfer resistance of the catalysts. The Nyquist plot (Figure 2i) showed that Fe-N-$C_{0.75}$ had the smallest semicircular arc diameter, indicating it had the lowest charge transfer resistance and the highest electron transfer efficiency compared to other prepared catalysts [45]. This allowed it to serve more effectively as an electron transfer medium, facilitating electron transfer from the pollutants to PMS.

2.2. The Performance of Catalysts

The PMS activation performance of the catalyst was investigated using PCT as the target pollutant. The synthesized catalysts showed negligible PCT adsorption (less than 5%) (Figure S4). In the PMS-only system, PCT removal was almost nonexistent within 30 min (Figure S5), indicating that PMS alone could not degrade PCT through self-decomposition, which is consistent with previous studies [39]. However, when the catalyst was introduced, PCT degradation was significantly enhanced. As shown in Figure 3a, both the Fe-N-$C_{0.75}$/PMS and Fe-N-$C_{1.5}$/PMS systems removed 99% of PCT in 10 min. The K_{obs} for the Fe-N-C_3/PMS system was 0.118 min^{-1} (Figure 3b), while the K_{obs} of the Fe-N-$C_{0.25}$/PMS system was approximately three times higher. In contrast, the Fe-N-C_6/PMS system removed only 10% of PCT in 30 min. These results demonstrate that the catalyst performance is proportional to the Fe-doping ratio within a certain range, with higher Fe content generating more active sites. Moreover, the decreased performance of Fe-N-$C_{0.25}$ might be attributed to the formation of excessive unstable iron clusters during pyrolysis, which were removed during acid washing, leading to a reduction in active sites. The Fe-N-C_x (x = 0.75–6) systems exhibited low iron leaching after a 60 min reaction (0.007–0.268 mg/L) (Figure 3c), all within the emission standards for drinking water, surface water, and ground water (GB 5749-2006 [46], GB 3838-2002 [47], and GB/T 14848-2017 III [48]). Among the synthesized catalysts, the reaction rates for Fe-N-$C_{0.75}$ and Fe-N-$C_{1.5}$ were similar, and the iron leaching in the Fe-N-$C_{1.5}$ system was only one-third of the Fe-N-$C_{0.75}$ system. Additionally, the Fe-N-$C_{1.5}$ also could efficiently active PDS. As shown in Figure S5, the Fe-N-$C_{1.5}$/PDS system removed 80% of PCT in 30 min. Compared to some previously reported catalysts (Table S5), the Fe-N-$C_{1.5}$/PMS system demonstrated the highest PMS activation, with the lowest iron leaching. Therefore, Fe-N-$C_{1.5}$ was selected as the PMS activator for subsequent experiments.

2.3. Effect of Different Factors on PCT Degradation

Previous studies reported that catalyst dosage had a significant impact on PCT degradation [49]. Therefore, the PCT degradation rates at various catalyst dosages were examined. As shown in Figures 3d and S6a, the PCT degradation rate was accelerated with increasing catalyst dosage, exhibiting a linear correlation (Figure S7). This indicated that increasing the catalyst dosage provides more available active sites, thereby facilitating PMS activation.

The effect of PMS dosage on PCT degradation was also investigated. The K_{obs} increased significantly from 0.5957 min^{-1} to 0.9002 min^{-1} as the PMS dosage increased from 0.5 mM to 2 mM (Figure S6b). However, when the PMS dosage exceeded 2 mM, the enhancement of PCT degradation became negligible (Figure 3e). This result suggested that 2 mM PMS was sufficient for the available active sites on the 0.02 g/L Fe-N-$C_{1.5}$.

Furthermore, the effect of PCT concentration was examined (Figure 3f). As shown in Figure S6c, the K_{obs} gradually decreased as the PCT concentration increased, suggesting that the Fe-N-$C_{1.5}$/PMS system effectively treated PCT concentrations less than 20 mg/L.

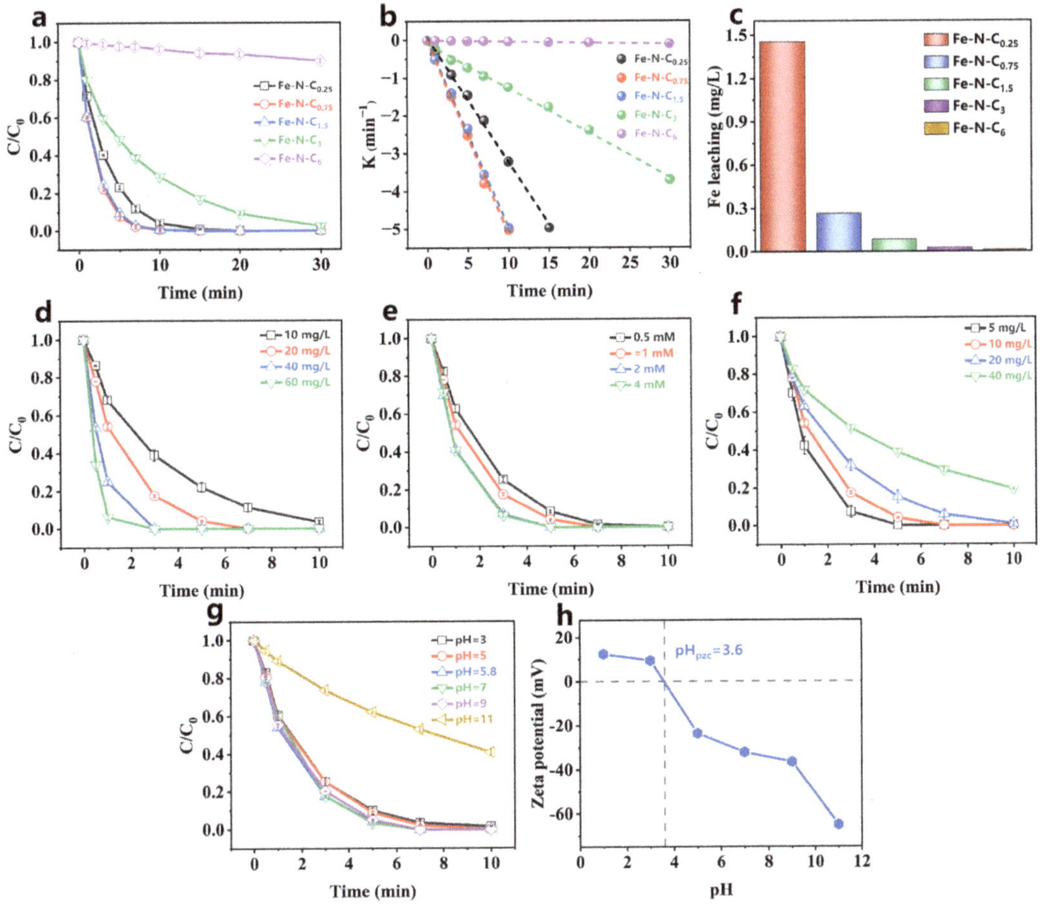

Figure 3. (**a**) Activation performance, (**b**) the corresponding rate constants, and (**c**) the Fe leaching at 60 min for Fe-N-C_x. ([catalyst] = 30 mg/L, [PMS] = 1.5 mM, [PCT] = 50 mg/L, pH_0 = 5.8 (without pH adjustment), temperature = 25–30 °C); PCT degradation in Fe-N-$C_{1.5}$/PMS system under different (**d**) catalyst dosage, (**e**) PMS dosage, (**f**) PCT concentration, and (**g**) initial pH; (**h**) Zeta potential of Fe-N-$C_{1.5}$. [catalyst] = 20 mg/L, [PCT] = 10 mg/L, [PMS] = 1 mM, pH_0 = 5.8 (without pH adjustment), temperature = 25–30 °C.

pH is a critical factor limiting the practical application of advanced oxidation technology. Therefore, the effect of different initial pHs on PCT degradation by the Fe-N-$C_{1.5}$/PMS system was explored. The oxidation system maintained efficient PCT removal over an initial pH from 3 to 9 (Figure 3g). However, PCT degradation was significantly inhibited when the initial pH was 11. To further explore the mechanism of pH impact on PMS activation, the Fe-N-$C_{1.5}$ surface charge characteristics at different pH levels were investigated. Figure 3h shows that the zero point of charge (pH_{zpc}) of Fe-N-$C_{1.5}$ was 3.6, and the zeta potential decreased as the initial pH increased. This suggested that Fe-N-$C_{1.5}$ carried negative charge at pH $\geq$ 3.6. When the initial pH of the solution was 3–9, the Fe-N-$C_{1.5}$/PMS system pH rapidly stabilized in the range of 2.94–3.39 after PMS addition (Table S6). In this pH range, the Fe-N-$C_{1.5}$ carried a positive charge, facilitating the adsorption of the negatively charged HSO_5^-. In contrast, when the initial pH was 11, the pH of the Fe-N-$C_{1.5}$/PMS system decreased to 10.11 after PMS addition. At this pH, Fe-N-$C_{1.5}$ carried a negative charge, and PCT underwent deprotonation (pKa = 9.86) [50]. Thus, at initial pH

≥ 11, electrostatic repulsion existed between the Fe-N-$C_{1.5}$ and PMS that hindered PMS activation, leading to a decrease in the PCT degradation rate.

2.4. Feasibility of Practical Application of Catalysts

Many background ions (Cl^-, NO_3^-, SO_4^{2-}, CO_3^{2-}, HCO_3^-) and humic acid (HA) exist in environmental water. The impacts of these background ions on PCT degradation in the Fe-N-$C_{1.5}$/PMS system were investigated. As shown in Figure 4a–c, Cl^-, NO_3^-, and SO_4^{2-} showed insignificant inhibition for PCT degradation, while CO_3^{2-} and HCO_3^- significantly inhibited PCT degradation (Figure 4d,e), as both ions raise the pH, which was unfavorable for PMS activation. Moreover, Figure 4f shows that HA inhibited PCT degradation, likely due to the fact that HA competes with PCT for the active sites and active species of the catalyst. Despite the varying degrees of inhibition caused by adding these interfering ions, PCT was still largely removed in 30 min. This result demonstrated that the Fe-N-$C_{1.5}$/PMS system had good resistance to ion interference.

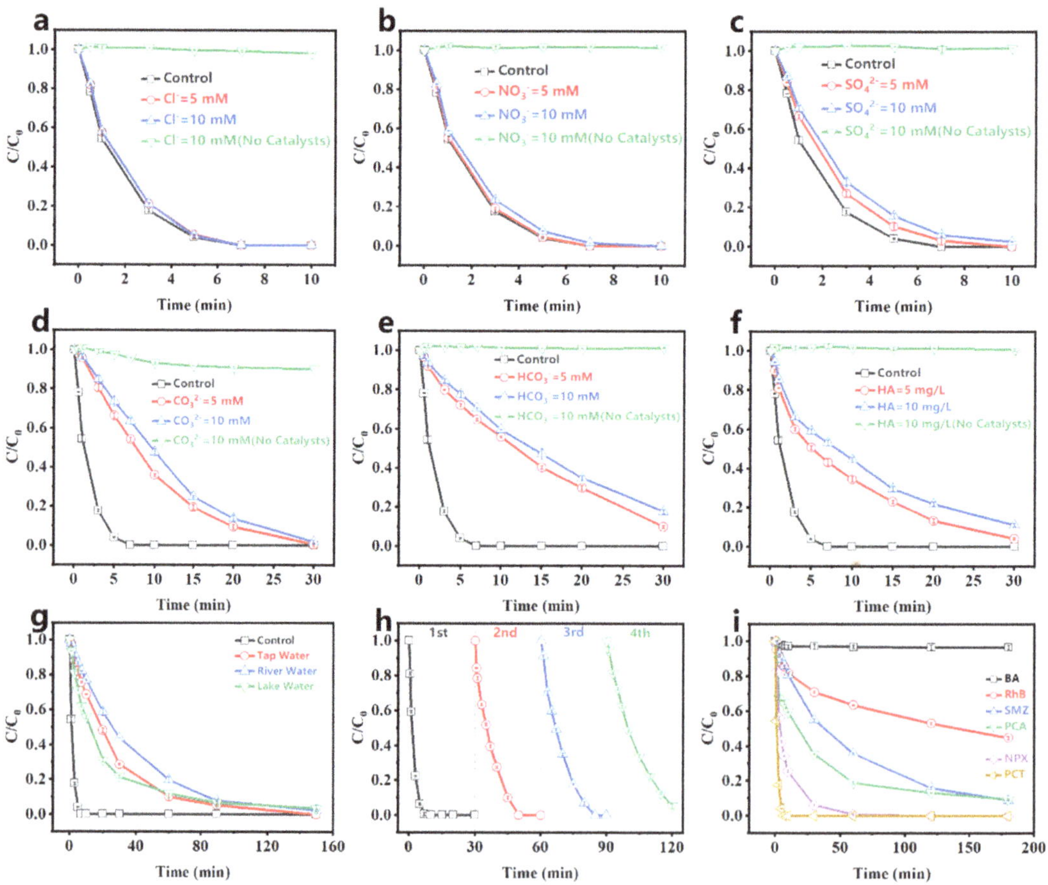

Figure 4. Effect of (**a**) Cl^-, (**b**) NO_3^-, (**c**) SO_4^{2-}, (**d**) CO_3^{2-}, (**e**) HCO_3^-, (**f**) HA on the degradation of PCT; (**g**) PCT degradation by the Fe-N-$C_{1.5}$/PMS system in different real water bodies; (**h**) cycle experiments; (**i**) degradation of different pollutants. [catalyst] = 20 mg/L, [PCT] = 10 mg/L, [PMS] = 1 mM, pH_0 = 5.8 (without pH adjustment), temperature = 25–30 °C.

To more comprehensively assess the practical applicability of the Fe-N-$C_{1.5}$/PMS system, its degradation performance was evaluated in tap water (South China University of Technology), artificial lake water (South China University of Technology), and river water (the Zhujiang River). As shown in Figure 4g, 97% of PCT was removed in 150 min in the natural water matrix, which was lower than in DI water. This was attributed to the presence of abundant interfering ions and organic matter in natural waters, which compete with PCT for active sites and consume reactive oxygen species. Additionally, the mineralization capability of the Fe-N-$C_{1.5}$/PMS system was assessed, showing a TOC removal rate of 16.14% at 120 min (Figure S8).

The stability of the Fe-N-$C_{1.5}$/PMS system was investigated through cycle experiments. Figure 4h shows the degradation performance of the system over four cycles. In the first cycle, PCT was completely degraded in 10 min, and all subsequent cycles showed slower reaction rates than the previous one. In the fourth cycle, only 94% of PCT was degraded in 30 min, which might be due to the large amount of intermediate product adsorption occupying the catalyst active sites and hindering PMS reacting at those sites [32]. The absence of significant changes in the Fe-N-$C_{1.5}$ XRD pattern after four cycles (Figure S9) further demonstrated the excellent stability of Fe-N-$C_{1.5}$.

The degradation of different pollutants via the Fe-N-$C_{1.5}$/PMS system was evaluated. Due to their varying chemical structures and properties, the degradation mechanisms and rates differed significantly. As shown in Figure 4i, the Fe-N-$C_{1.5}$/PMS system demonstrated varying degradation rates for benzoic acid (BA), rhodamine B (RhB), p-chloroaniline (PCA), sulfamethazine (SMZ), naproxen (NPX), and paracetamol (PCT). Among them, the degradation rates of electron-rich pollutants (PCT, NPX) with electron-donating groups such as hydroxyl and alkoxy groups were faster, while there was basically no degradation for BA with electron-withdrawing groups. This was likely due to the lower ionization potential of the electron-donating groups, such as hydroxyl and alkoxy groups, which were more susceptible to electron capture by the oxidation of activated PMS [51]. Furthermore, the cycle experiments for PCT, NPX, PCA, and SMZ (Figure S10) revealed that the system maintains excellent removal efficiency (>98%) for both PCT and NPX after four consecutive cycles, while the removal efficiency for PCA and SMZ decreased to approximately 65%. These results demonstrated that the Fe-N-$C_{1.5}$/PMS system exhibits selective degradation capability towards various contaminants while maintaining superior stability.

2.5. Mechanism Discussion for PCT Degradation

2.5.1. Identification of Reactive Species

Quenching experiments were conducted to identify the reactive oxygen species (ROS) generated in the Fe-N-$C_{1.5}$/PMS system. Ethanol (EtOH) and tert-butyl alcohol (TBA) are commonly used to identify hydroxyl radicals ($^{\bullet}OH$) and sulfate radicals ($SO_4^{\bullet-}$) in the system [52]. EtOH is an effective quencher for both $^{\bullet}OH$ and $SO_4^{\bullet-}$ ($K_{EtOH/\bullet OH} = 1.9 \times 10^9$ $M^{-1}s^{-1}$, $K_{EtOH/SO4\bullet-} = 7.7 \times 10^7$ $M^{-1}s^{-1}$) [53,54], while TBA reacts much faster with $^{\bullet}OH$ than $SO_4^{\bullet-}$ ($K_{TBA/\bullet OH} = 6.0 \times 10^8$ $M^{-1}s^{-1}$, $K_{TBA/SO4\bullet-} = 7.6 \times 10^5$ $M^{-1}s^{-1}$) [15]. As shown in Figure 5a, both EtOH and TBA insignificantly inhibited PCT degradation, indicating PCT degradation was not primarily dependent on $^{\bullet}OH$ or $SO_4^{\bullet-}$. 5,5-dimethyl-1-pyrroline N-oxide (DMPO), a radical scavenger, was used to capture $^{\bullet}OH$ and $SO_4^{\bullet-}$ in the system [55]. As shown in Figure 5b, the appearance of only a weak 1:2:2:1 signal peak further confirmed that $^{\bullet}OH$ and $SO_4^{\bullet-}$ were not the primary active species in the system.

$O_2^{\bullet-}$ is a common radical in PMS systems. Since p-benzoquinone (p-BQ) reacts with $O_2^{\bullet-}$ at a high rate ($K_{p-BQ/O2\bullet-} = 0.9 - 1.0 \times 10^9$ $M^{-1}s^{-1}$) [56], it was used as a quencher to identify $O_2^{\bullet-}$ in the system. After adding 1 mM p-BQ, PCT degradation was significantly inhibited (Figure 5c). However, it has been reported that p-BQ can directly consume PMS [57], resulting in a reduced degradation rate. To further identify the $O_2^{\bullet-}$ contribution, the $O_2^{\bullet-}$ was detected by monitoring the absorption intensity of nitro blue tetrazolium (NBT) at 560 nm in the UV–visible spectrum [41,58]. Figure 5d shows that no NBT absorption peak appeared after 30 min of reaction, confirming the absence of $O_2^{\bullet-}$ in the Fe-N-$C_{1.5}$/PMS system.

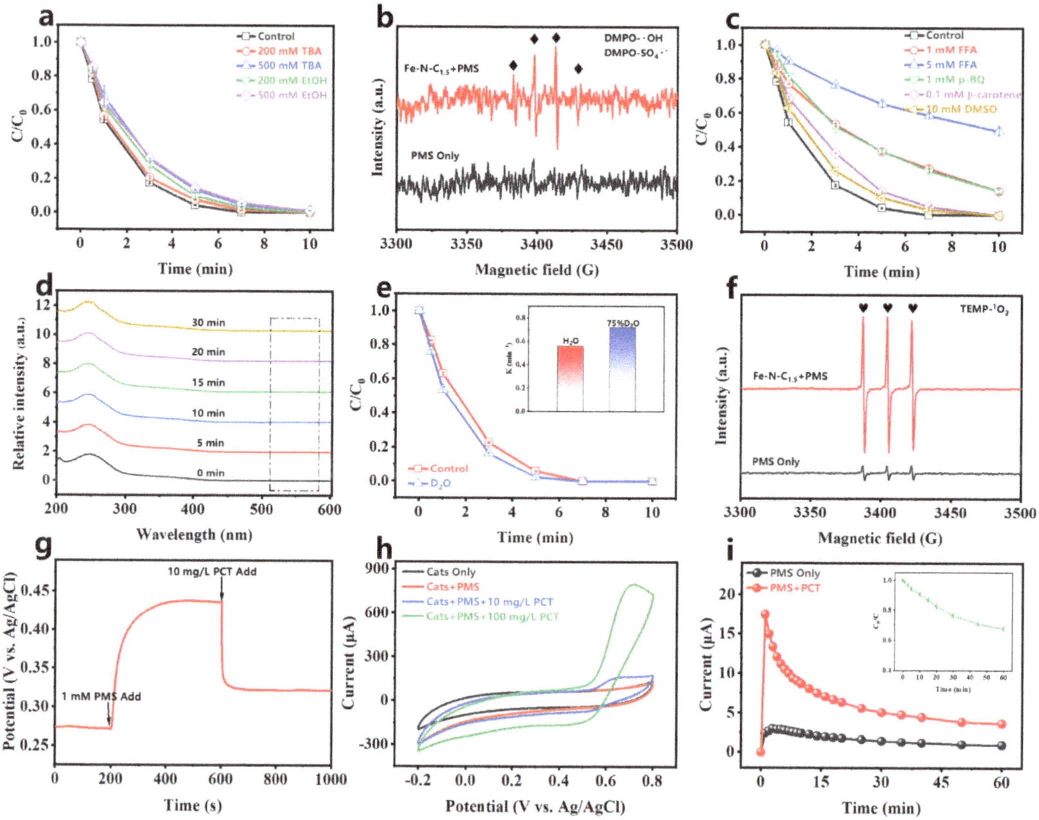

Figure 5. (**a**) Quenching experiments of EtOH and TBA; (**b**) EPR in the presence of DMPO; (**c**) quenching experiments of FFA, p-BQ, β-carotene, and DMSO; (**d**) NBT experiments; (**e**) effect of reaction solvents (H_2O and D_2O); (**f**) EPR in the presence of TEMP; (**g**) OCPT, (**h**) CV; and (**i**) GOP experiment for the Fe-N-$C_{1.5}$/PMS/PCT system. [catalyst] = 20 mg/L, [PCT] = 10 mg/L, [PMS] = 1 mM, pH_0 = 5.8 (without pH adjustment), temperature = 25–30 °C.

Furfuryl alcohol (FFA) was used for quenching singlet oxygen ($^1O_2$) ($K_{FFA/1O2} = 1.2 \times 10^8\ M^{-1}s^{-1}$) [15]. As shown in Figure 5c, after adding FFA, PCT degradation in the system was significantly inhibited, and the inhibitory effect increased as the FFA concentration increased. In the presence of 5 mM FFA, only 50% of PCT was removed in 10 min. However, there is debate that high FFA concentrations may directly react with PMS [59], reducing the reaction rate. β-carotene, a $^1O_2$-quenching agent that does not consume PMS, was used to further evaluate the $^1O_2$ contribution. Figure 5c shows that even at low concentrations (β-carotene: PMS = 0.2:1), the addition of β-carotene significantly inhibited PCT degradation. This result indicated that $^1O_2$ was a crucial ROS in PCT degradation. Since $^1O_2$ has a longer lifetime in D_2O (60 μs) than in H_2O (3.5 μs), a solvent exchange experiment was conducted to further confirm its presence [60]. Figure 5e shows an accelerated PCT degradation rate in D_2O (from 0.5601 min^{-1} to 0.7211 min^{-1}), further confirming the generation of $^1O_2$ in the Fe-N-$C_{1.5}$/PMS system and its involvement in PCT degradation. 2,2,6,6-tetramethylpiperidine (TEMP) was used as a spin trap agent for $^1O_2$. As shown in Figure 5f, a weak TEMPO triplet signal (1:1:1) appeared in the PMS-only system. In contrast, the intensity of the triplet signal increased significantly in the Fe-N-$C_{1.5}$/PMS system, providing further evidence for the generation of $^1O_2$ in the system.

High-valent metal oxide species are commonly present in metal composite catalyst systems and participate in the direct oxidation of pollutants. To verify the generation and role of high-valent metal oxide species (Fe(IV)=O, Fe(V)=O), dimethyl sulfoxide (DMSO) was introduced as a quencher. The addition of 10 mM DMSO had negligible effects on PCT degradation (Figure 5c). This suggested that Fe(IV)=O and Fe(V)=O were not the primary ROS for PCT degradation.

Based on the above results, the electron transfer process (ETP) likely plays a crucial role in PCT degradation in the Fe-N-$C_{1.5}$/PMS system. $K_2Cr_2O_7$, a solution-phase ETP scavenger [61,62], was employed to identify solution-phase ETP's contribution in the system. Figure S11 shows that 100% of PCT was degraded in 30 min, even with the addition of 10 mM $K_2Cr_2O_7$, suggesting that the solution-phase ETP was an insignificant role for PCT removal in this system. The PMS decomposition experiments were performed (Figure S12). The PMS concentration decreased slightly (approximately 8%) after 10 min when either the catalyst or PCT were present, while it significantly decreased 22% when both the catalyst and PCT were present. This result implied a catalyst interface-mediated ETP [39]. Electrochemical tests can directly detect electron transfer in the system to further identify the role of catalyst-interface-mediated ETPs. As shown in Figure 5g, the addition of PMS to the catalyst-coated carbon electrode system led to a rapid increase in potential, reaching a stable plateau, indicating the catalyst interacts with PMS to elevate its potential. After adding PCT, the potential quickly decreased, suggesting the high-potential catalyst directly captures electrons from PCT, thereby proving electron transfer [63,64]. Similar results were observed in the chronoamperometry (i-t) experiment (Figure S13), where the current density showed significant changes upon the sequential addition of PMS and PCT. Cyclic voltammetry (CV) clearly showed the redox processes of the pollutants on the catalyst surface [65]. After adding PMS, the curve shows no significant changes, but after adding PCT, a pair of redox peaks appeared, with the peak current increasing as the PCT concentration increased (Figure 5h), further illustrating that the system captured the electrons from PCT through electron transfer. The galvanic oxidation process (GOP) system developed by Huang et al. provides a more intuitive demonstration of the electron transfer [66]. As shown in Figure 5i, after adding PMS and PCT to the two half-cells, respectively, the current in the external circuit increased, and the PCT began to degrade gradually. This confirmed that the system degraded the PCT via catalyst-mediated electron transfer.

In catalyst-interface-mediated ETP systems, the PMS usually interacts with the surface sites of the catalyst, forming surface-bonded active complexes that have higher redox potentials [67]. Fourier-transform infrared (FTIR) spectroscopy (Figure S14) confirmed the formation of the metastable complex in the Fe-N-$C_{1.5}$/PMS system. Compared to PMS alone, the FTIR spectrum of the system showed a red shift in the S-O absorption peak from 1120 cm^{-1} to 1132 cm^{-1}. This shift was attributed to the interaction between Fe-N-$C_{1.5}$ and PMS, which weakened the electron density of the S-O bond in the metastable complex [68].

Furthermore, ionic strength experiments were conducted to explore whether the interaction between the catalyst, PMS, and PCT during electron transfer occurs through an inner-sphere or outer-sphere process. The ionic strength of the system was adjusted by introducing varying $NaClO_4$ concentrations to modulate the interference from background ions [15]. When the $NaClO_4$ concentration ranged from 1 to 50 mM, the Fe-N-$C_{1.5}$/PMS system was able to completely remove PCT in 7 min (Figure S15), indicating that ionic strength had minimal impact on PCT degradation. This result demonstrated that the electron transfer in the system predominantly follows an inner-sphere mechanism.

The above experimental results indicated that the Fe-N-$C_{1.5}$/PMS system primarily degrades PCT through two non-radical pathways: (i) $^1O_2$- and (ii) catalyst-interface-mediated ETPs.

2.5.2. Identification of Reactive Sites and Analysis of PCT Degradation Mechanisms

The XPS results and the K_{obs} of Fe-N-C_x showed the relative content of Fe-N_x, graphitic N, and C=O on the catalyst surface positively correlated with K_{obs} (Table S4 and Figure 6a). Specifically, the K_{obs} gradually increased as the relative content of Fe-N_x, graphitic N, and C=O increased, indicating that these sites play important roles in PMS activation. To further explore the primary active sites for PMS activation, the catalyst surface before and after the reaction was analyzed by XPS. As shown in Figure 6b–d, the relative content of Fe-N_x decreased from 28.72% to 26.15%, graphitic N decreased from 14.54% to 12.67%, and C=O content declined from 70.14% to 61.28% after the reaction. These results further demonstrated that Fe-N_x, graphitic N, and C=O sites play significant roles for PMS activation. It was reported that N atoms could alter the electronic arrangement of the carbon network, introducing hybrid nitrogen sites such as pyridinic N, pyrrolic N, and graphitic N. Pyrrolic N and pyridinic N, with lone pairs of electrons, could coordinate with the Fe atom, forming Fe-N_x sites that have a strong affinity for PMS [69]. Due to differences in electronegativity between elements, local polarization electric fields exist at Fe-N_x sites, which promote electron transfer on the Fe-N-$C_{1.5}$ electronic bridge [17], thereby facilitating the rapid activation of PMS. Additionally, graphitic N also serves as a reactive site in the catalytic reaction. The more electronegative N in graphitic N has a stronger electron-withdrawing ability, which induces the adjacent carbon atoms to become positively charged, generating electron-deficient C active sites that promote nucleophilic attack by PMS [70]. When one electron from PMS transfers to the electron-deficient C^+ site, a peroxyl radical intermediate ($SO_5^{\bullet -}$) is generated, which easily reacts to form $^1O_2$ [39]. Similarly, the carbonyl (C=O), with its electrophilic nature, also induces nucleophilic attack by PMS, leading to the generation of $^1O_2$ [43]. To further identify the contribution of Fe-N_x for PMS activation, a KSCN toxicity experiment was conducted. As shown in Figure 6e, the addition of KSCN significantly inhibited PCT degradation, confirming that Fe-N_x served as a critical site for PMS activation.

2.6. PCT Degradation Pathway and the Intermediates Toxicity Assessment

The intermediates generated during PCT degradation in the Fe-N-$C_{1.5}$/PMS system were detected using HPLC-MS (detailed information is provided in Table S8 and Figure S16), based on which two possible degradation pathways were proposed (Figure 7). It was reported that the radical pathway contributes to the formation of biphenyl and diphenyl ether [71]. In Pathway I, PCT was attacked by $^\bullet OH$ and $SO_4^{\bullet -}$, leading to the formation of phenoxy radicals, which subsequently produced compound A (2,2′-dihydroxy-5,5′-diacetyldiaminebiphenyl, m/z = 300) or compound B (2-hydroxy-4′,5-diacetamido-diphenyl ether, m/z = 300) [31,41]. In Pathway II, PCT underwent hydrogen abstraction, forming compound C (N-acetyl-p-benzoquinone imine, m/z = 149) [72]. Compound C, after hydroxylation and hydrogen abstraction, was oxidized to form compound D (2-hydroxy-5-acetamido-1,4-benzoquinone, m/z = 181) [73]. Due to the instability of their quinoid structure, compounds C and D were prone to ring-opening reactions [71,74,75], generating small aliphatic hydrocarbon derivatives, such as compound E (4-aminobutane-1,2,3-triol, m/z = 121), compound F (3-aminopropanol, m/z = 75), and compound G (acetamide, m/z = 59). Finally, these substances were further mineralized to CO_2 and H_2O.

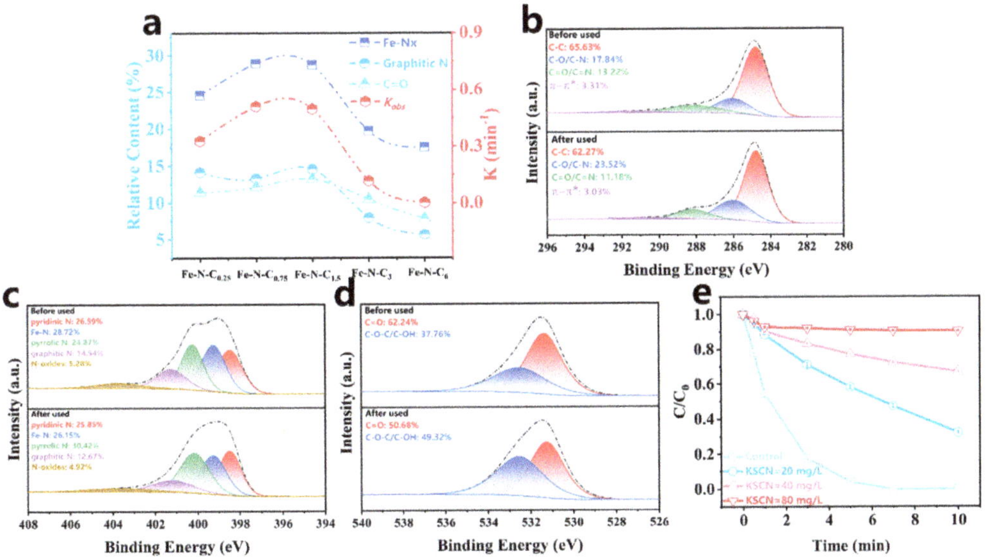

Figure 6. (**a**) Structure–property relationships between different species and the Kobs; XPS C 1s (**b**), N 1s (**c**), O 1s (**d**) of the Fe-N-$C_{1.5}$ before and after use; (**e**) effects of different concentrations of KSCN on Fe-N-$C_{1.5}$/PMS/PCT system.

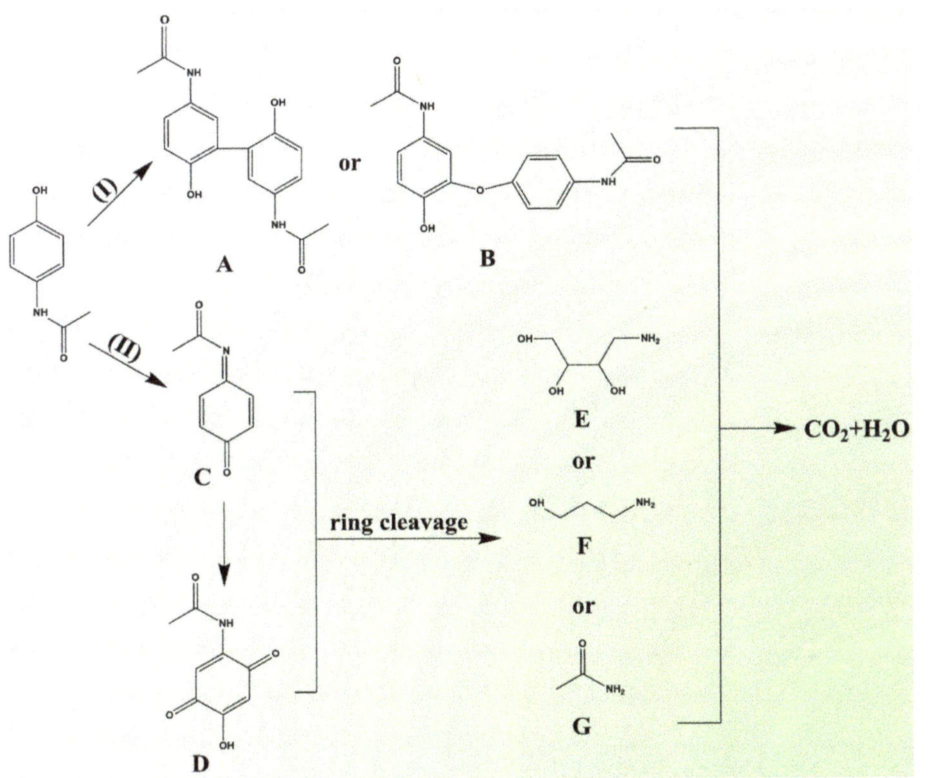

Figure 7. The proposed PCT degradation pathways in the Fe-N-$C_{1.5}$/PMS system.

The toxicity of the PCT degradation intermediates was assessed using a Toxicity Estimation Software Tool (T.E.S.T version 5.1.2) based on the QASR method to evaluate their environmental impact. Four indicators were selected for evaluation: acute toxicity LC_{50} to Daphnia magna, developmental toxicity, bioconcentration factor, and mutagenicity (Figure 8) [3,76,77]. The acute toxicity LC_{50} of PCT to Daphnia magna was 27.14 mg/L, classifying it as a low-toxicity substance (LC_{50} > 10 mg/L) according to the aquatic toxicity classification. The other intermediates were also found to be low-toxicity substances, with most intermediates showing an increase in LC_{50} relative to PCT, indicating that the degradation effectively reduced the pollutants' acute toxicity in the system. However, for developmental toxicity, all intermediates except compound E exhibited higher toxicity. Additionally, intermediates B and C were found to be mutagenic, indicating that PCT degradation in the Fe-N-$C_{1.5}$/PMS system produced toxic intermediates. Fortunately, all intermediates showed lower bioconcentration factors, suggesting that although some intermediates possess higher toxicity, their accumulation is much lower than that of PCT. As a result, the system's overall toxicity may decrease over the course of degradation. However, to ensure the safety of treated water, further oxidation and mineralization are required to eliminate these toxic intermediates.

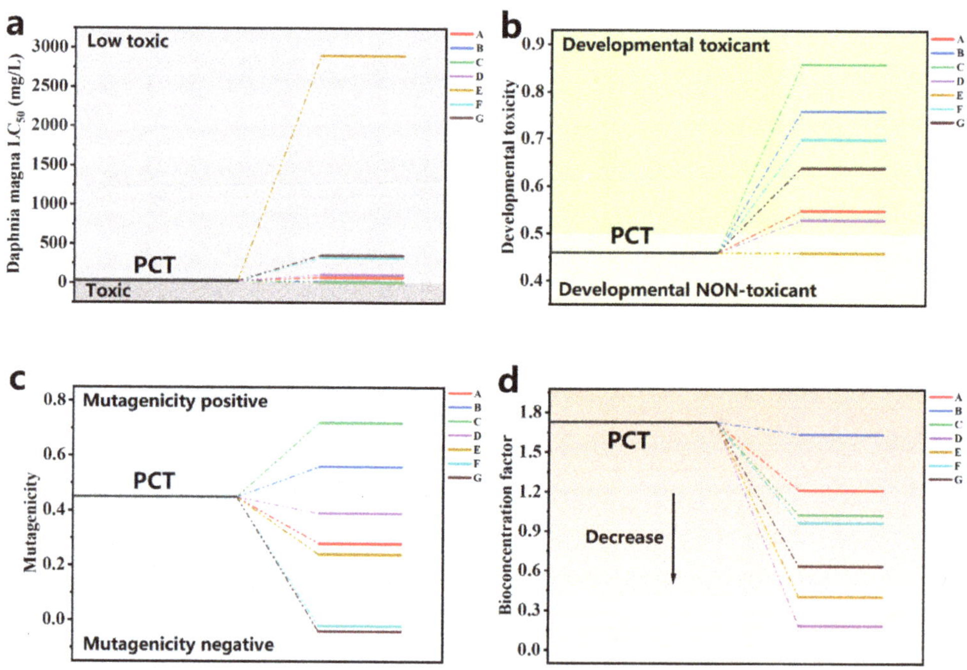

Figure 8. (**a**) Daphnia magna LC_{50}, (**b**) developmental toxicity, (**c**) mutagenicity, and (**d**) bioaccumulation factor of PCT and degradation intermediates in the Fe-N-$C_{1.5}$/PMS system.

3. Materials and Methods

3.1. Chemicals and Catalyst Synthesis

Details of the chemicals used in this work are provided in the Text S1. Fe-N-C_x was synthesized using a simple co-pyrolysis method. Ferrous sulfate heptahydrate (1 mmol) was added to 25 mL of deionized water to obtain solution A. 1,10-phenanthroline (x mmol, x = 0.25, 0.75, 1.5, 3, 6) was added to 25 mL of ethanol to obtain solution B. Solution A was added dropwise to stirred solution B. The mixed solution was dried under vacuum at

60 °C for 48 h to produce the Fe-N precursor. The Fe-N precursor was calcined at 600 °C for 2 h under a N_2 atmosphere to obtain Fe, N co-doped metal composites. Finally, the metal composites were washed with 1 M HCl solution for 2 h to remove unstable metal particles which aggregated on the surface of the metal composites. The acid-treated material was dried at 60 °C for 24 h to obtain stable composites, referred to as Fe-N-C_x.

3.2. Characterizing Methods

The characterization details are provided in the Text S2–S4.

3.3. Batch Experiments

All degradation experiments were conducted at room temperature in a 250 mL beaker with magnetic stirring. The degradation process was initiated by adding catalyst (30 mg/L) and PMS (1.5 mM) to a 50 mg/L PCT solution. At set intervals, a 0.5 mL sample was withdrawn from the reaction mixture and filtered through a 0.45 µm filter, then transferred into a vial containing 0.5 mL of sodium thiosulfate (0.05 M) to quench the oxidation reaction. Adsorption experiments followed the same procedure as above but without the PMS addition. All experiments were conducted in duplicate, and the mean values with their corresponding standard deviations are presented.

The pseudo-first-order kinetic degradation rate constant for the degradation experiments was calculated using the following equation:

$$-ln\left(\frac{C_t}{C_0}\right) = K_{obs} \times t \tag{1}$$

where C_t represents the residual pollutant concentration and C_0 the initial pollutant concentration, while K_{obs} is the rate constant for the target pollutant degradation.

3.4. Analytical Methods

The concentration of the target pollutant was determined via HPLC (Agilent1200, Agilent Technologies, Santa Clara, CA, USA), with details listed in Table S1. The leached Fe ion concentration was determined using a UV–Vis spectrophotometer (UV-2550, Shimadzu, Kyoto, Japan) via the 1,10-phenanthroline colorimetric method. The zeta potential was obtained with a nanoparticle size and zeta potential analyzer (Malvern Zetasizer Nano ZS90, Malvern Panalytical, Worcestershire, UK). The concentration of PMS in the PMS decomposition experiments was determined using the 2,2′-azino-bis (3-ethylbenzothiazoline-6-sulfonic acid) diammonium salt (ABTS) colorimetric method, using the detailed procedures provided in Text S5. The radicals generated in the system were measured via a paramagnetic resonance spectrometer (EPR, Bruker EMXPlus-10/12, Bruker, Rheinstetten, Germany), using DMPO and TEMP as radical scavengers. The total organic carbon (TOC) was measured using a TOC analyzer (Jena Multi N/C 3100, Analytik Jena GmbH + Co. KG., Jena, Germany). Electrochemical analyses were performed using an electrochemical workstation (CHI760E, Shanghai Chenhua Instrument Co., Ltd., Shanghai, China), with detailed methods given (Text S6). The galvanic oxidation process (GOP) experiment (details provided in Text S7) was conducted to further confirm electron transfer, with the current measured with an ammeter. Degradation intermediates were analyzed using an Agilent 1290 ultra-high-performance liquid chromatography–mass spectrometry system (HPLC-MS, Bruker maXis impact, Bruker, Bremen, Germany), with detailed information presented in Text S8.

4. Conclusions

In this study, an efficient and stable Fe-N-$C_{1.5}$ catalyst was synthesized via a simple mixed solvent–pyrolysis method to activate PMS for PCT degradation. The Fe-N-$C_{1.5}$/PMS system exhibited exceptional performance, achieving 100% PCT (10 mg/L) removal in 7 min with low catalyst doses (20 mg/L) and PMS (1 mM). Moreover, the system demonstrated outstanding anti-interference capabilities and practical applicability, effectively degrading PCT in the presence of various inorganic anions and in different real environmental water matrices. The cycle experiments further confirmed the stability of the Fe-N-$C_{1.5}$/PMS system. Mechanistic analysis revealed that PCT degradation primarily occurred via two non-radical pathways: $^1O_2$ and catalyst-interface-mediated ETP. PMS was rapidly activated at Fe-N_x, graphitic nitrogen, and C=O, generating singlet oxygen or forming an active PMS complex species, which subsequently oxidized PCT. In addition, two degradation pathways for PCT were proposed, and the toxicity of the potential intermediates was assessed. The results indicated that some intermediates exhibited more ecological toxicity than PCT, highlighting the need for further mineralization to eliminate these harmful byproducts. In summary, this study deepens an understanding of the PCT degradation mechanism by the Fe-N-$C_{1.5}$/PMS system and offers a promising strategy for organic contaminant removal from water.

Supplementary Materials: The following supporting information can be downloaded at: https://www.mdpi.com/article/10.3390/catal15030217/s1, Text S1. Chemicals. Text S2. Characterizing methods. Text S3. The determination of Fe content in activators by ICP-OES. Text S4. The experimental method of electrochemical impedance spectroscopy (EIS). Text S5. Measurement of PMS concentration by the ABTS colorimetric method. Text S6. Electrochemical measurements. Text S7. The experimental method of galvanic oxidation processes (GOP). Text S8. Analytical method of LC-MS. Table S1. HPLC analysis parameters for different pollutants. Table S2. BET surface area, pore volume and pore size of Fe-N-C_x. Table S3. ICP-OES test of different catalysts. Table S4. The surface element chemical states of different catalysts. (XPS analysis) Table S5. Comparison of reported iron-based catalysts. Table S6. Comparison of the solution pH before and after adding PMS. Table S7. Comparison of the surface element chemical states before and after reaction of Fe-N-$C_{1.5}$. (XPS analysis). Table S8. The intermediate products of PCT degradation. Figure S1. (a, b) SEM images of Fe-N-$C_{0.25}$. Figure S2. Pore size distribution of Fe-N-$C_{0.25}$ (a), Fe-N-$C_{0.75}$ (b), Fe-N-$C_{1.5}$ (c), Fe-N-C_3 (d) and Fe-N-C_6 (e). Figure S3. Structure-property relationships between the Kobs and (a) S_{BET}, (b) Fe content. Figure S4. Adsorption performance of Fe-N-C_x. [catalyst] = 30 mg/L, [PCT] = 50 mg/L, pH_0 = 5.8 (without pH adjustment), room temperature. Figure S5. PCT degradation in different systems. [catalyst] = 20 mg/L, [PCT] = 10 mg/L, [PMS] = 1 mM, [PDS] = 1 mM, pH_0 = 5.8 (without pH adjustment), room temperature. Figure S6. The first-order reaction rate constants of PCT degradation under different catalyst dosage (a), PMS dosage (b), PCT concentration (c). Figure S7. Catalyst dosages and linear fitting of first-order reaction rate constants. Figure S8. TOC removal of PCT degradation in the Fe-N-$C_{1.5}$/PMS system. Figure S9. XRD patterns of Fe-N-$C_{1.5}$ before and after use. Figure S10. Cycle experiments for PCT, NPX, PCA, and SMZ. (reaction time per cycle: 180 min). Figure S12. PMS decomposition in different systems. [catalyst] = 20 mg/L, [PCT] = 10 mg/L, [PMS] = 1 mM, pH_0 = 5.8 (without pH adjustment), room temperature. Figure S13. I-t experiment of Fe-N-$C_{1.5}$/PMS/PCT system. Figure S14. FTIR of Fe-N-$C_{1.5}$ only, PMS only and Fe-N-$C_{1.5}$/PMS. Figure S15. Effects of different concentrations of $NaClO_4$ on Fe-N-$C_{1.5}$/PMS/PCT system. [catalyst] = 20 mg/L, [PCT] = 10 mg/L, [PMS] = 1 mM, pH_0 = 5.8 (without pH adjustment), room temperature. Figure S16. Chromatogram and mass spectrum of substances at 5 min. Refs. [32,34,66,78–85] are cited in the Supplementary Materials.

Author Contributions: Y.Z. (Yujun Zhuo): Writing—original draft, investigation, formal analysis, methodology; H.M.: formal analysis, methodology, writing—review and editing; Y.Z. (Yongqing Zhang): writing—review and editing, supervision, funding acquisition; Y.C.: investigation, methodology, validation; J.C.: investigation, validation. All authors have read and agreed to the published version of the manuscript.

Funding: This work was financially supported by the National Natural Science Foundation of China (51572089), the National Key Research and Development Program of China (2016YFC0400708) and Guangdong Basic and Applied Basic Research Foundation (2023A1515012378).

Data Availability Statement: The data that support the findings of this study are available from the corresponding author upon reasonable request.

Conflicts of Interest: The authors declare no conflicts of interest.

References

1. Chidiac, A.S.; Buckley, N.A.; Noghrehchi, F.; Cairns, R. Paracetamol (acetaminophen) overdose and hepatotoxicity: Mechanism, treatment, prevention measures, and estimates of burden of disease. *Expert Opin. Drug Metab. Toxicol.* **2023**, *19*, 297–317. [CrossRef]
2. Lee, W.M. Acetaminophen (APAP) hepatotoxicity—Isn't it time for APAP to go away? *J. Hepatol.* **2017**, *67*, 1324–1331. [CrossRef]
3. Meng, H.; Gong, Z.; Xiang, X.; Zhu, Y.; Wu, X.; Chen, Y.; Zhang, Y. Influence of entropy on catalytic performance of high-entropy oxides (NiMgZnCuCoOx) in peroxymonosulfate-mediated acetaminophen degradation. *Chemosphere* **2024**, *362*, 142610. [CrossRef]
4. Chacón, F.J.; Cayuela, M.L.; Sánchez-Monedero, M.A. Paracetamol degradation pathways in soil after biochar addition. *Environ. Pollut.* **2022**, *307*, 119546. [CrossRef] [PubMed]
5. Kalantzi, L.; Reppas, C.; Dressman, J.B.; Amidon, G.L.; Junginger, H.E.; Midha, K.K.; Shah, V.P.; Stavchansky, S.A.; Barends, D.M. Biowaiver monographs for immediate release solid oral dosage forms: Acetaminophen (paracetamol). *J. Pharm. Sci.* **2006**, *95*, 4–14. [CrossRef] [PubMed]
6. Wu, J.-l.; Liu, Z.-h.; Ma, Q.-g.; Dai, L.; Dang, Z. Occurrence, removal and risk evaluation of ibuprofen and acetaminophen in municipal wastewater treatment plants: A critical review. *Sci. Total Environ.* **2023**, *891*, 164600. [CrossRef] [PubMed]
7. Ghanbari, F.; Giannakis, S.; Lin, K.-Y.A.; Wu, J.; Madihi-Bidgoli, S. Acetaminophen degradation by a synergistic peracetic acid/UVC-LED/Fe(II) advanced oxidation process: Kinetic assessment, process feasibility and mechanistic considerations. *Chemosphere* **2021**, *263*, 128119. [CrossRef] [PubMed]
8. Parolini, M. Toxicity of the Non-Steroidal Anti-Inflammatory Drugs (NSAIDs) acetylsalicylic acid, paracetamol, diclofenac, ibuprofen and naproxen towards freshwater invertebrates: A review. *Sci. Total Environ.* **2020**, *740*, 140043. [CrossRef] [PubMed]
9. Liu, Z.; Ren, X.; Duan, X.; Sarmah, A.K.; Zhao, X. Remediation of environmentally persistent organic pollutants (POPs) by persulfates oxidation system (PS): A review. *Sci. Total Environ.* **2023**, *863*, 160818. [CrossRef] [PubMed]
10. Ding, Y.; Wang, X.; Fu, L.; Peng, X.; Pan, C.; Mao, Q.; Wang, C.; Yan, J. Nonradicals induced degradation of organic pollutants by peroxydisulfate (PDS) and peroxymonosulfate (PMS): Recent advances and perspective. *Sci. Total Environ.* **2021**, *765*, 142794. [CrossRef] [PubMed]
11. Matzek, L.W.; Carter, K.E. Activated persulfate for organic chemical degradation: A review. *Chemosphere* **2016**, *151*, 178–188. [CrossRef]
12. Tian, K.; Hu, L.; Li, L.; Zheng, Q.; Xin, Y.; Zhang, G. Recent advances in persulfate-based advanced oxidation processes for organic wastewater treatment. *Chin. Chem. Lett.* **2022**, *33*, 4461–4477. [CrossRef]
13. Lee, J.; von Gunten, U.; Kim, J.-H. Persulfate-Based Advanced Oxidation: Critical Assessment of Opportunities and Roadblocks. *Environ. Sci. Technol.* **2020**, *54*, 3064–3081. [CrossRef]
14. Zhang, T.; Zhu, H.; Croué, J.-P. Production of Sulfate Radical from Peroxymonosulfate Induced by a Magnetically Separable $CuFe_2O_4$ Spinel in Water: Efficiency, Stability, and Mechanism. *Environ. Sci. Technol.* **2013**, *47*, 2784–2791. [CrossRef] [PubMed]
15. Ren, W.; Cheng, C.; Shao, P.; Luo, X.; Zhang, H.; Wang, S.; Duan, X. Origins of Electron-Transfer Regime in Persulfate-Based Nonradical Oxidation Processes. *Environ. Sci. Technol.* **2022**, *56*, 78–97. [CrossRef] [PubMed]
16. Tang, L.; Liu, Y.; Wang, J.; Zeng, G.; Deng, Y.; Dong, H.; Feng, H.; Wang, J.; Peng, B. Enhanced activation process of persulfate by mesoporous carbon for degradation of aqueous organic pollutants: Electron transfer mechanism. *Appl. Catal. B Environ.* **2018**, *231*, 1–10. [CrossRef]
17. Liu, Y.; Zhang, Y.; Zhang, L.; Hayat, W.; Wu, X.; Meng, H.; Wu, Q. Selective degradation of organic pollutants by highly dispersed Fe, Ni anchored on nitrogen carbon activating persulfate system: Electron transfer promotion, the impact of pollutant redox and adsorption properties. *Chem. Eng. J.* **2024**, *490*, 151585. [CrossRef]

18. Wang, H.; Guo, W.; Liu, B.; Wu, Q.; Luo, H.; Zhao, Q.; Si, Q.; Sseguya, F.; Ren, N. Edge-nitrogenated biochar for efficient peroxydisulfate activation: An electron transfer mechanism. *Water Res.* **2019**, *160*, 405–414. [CrossRef] [PubMed]
19. Cheng, X.; Guo, H.; Zhang, Y.; Wu, X.; Liu, Y. Non-photochemical production of singlet oxygen via activation of persulfate by carbon nanotubes. *Water Res.* **2017**, *113*, 80–88. [CrossRef]
20. Wang, R.; Zhai, H.; Luo, W.; Liu, W.; Zhuang, Z.; Ji, M. Multifunctional sites on reduced graphene oxide synergistically improving the degradation of diclofenac in peroxydisulfate systems. *J. Environ. Chem. Eng.* **2022**, *10*, 108251. [CrossRef]
21. Huang, W.; Xiao, S.; Zhong, H.; Yan, M.; Yang, X. Activation of persulfates by carbonaceous materials: A review. *Chem. Eng. J.* **2021**, *418*, 129297. [CrossRef]
22. Liang, J.; Duan, X.; Xu, X.; Chen, K.; Zhang, Y.; Zhao, L.; Qiu, H.; Wang, S.; Cao, X. Persulfate Oxidation of Sulfamethoxazole by Magnetic Iron-Char Composites via Nonradical Pathways: Fe(IV) Versus Surface-Mediated Electron Transfer. *Environ. Sci. Technol.* **2021**, *55*, 10077–10086. [CrossRef] [PubMed]
23. Zheng, X.; Niu, X.; Zhang, D.; Lv, M.; Ye, X.; Ma, J.; Lin, Z.; Fu, M. Metal-based catalysts for persulfate and peroxymonosulfate activation in heterogeneous ways: A review. *Chem. Eng. J.* **2022**, *429*, 132323. [CrossRef]
24. Yan, H.; Lai, C.; Liu, S.; Wang, D.; Zhou, X.; Zhang, M.; Li, L.; Li, X.; Xu, F.; Nie, J. Metal-carbon hybrid materials induced persulfate activation: Application, mechanism, and tunable reaction pathways. *Water Res.* **2023**, *234*, 119808. [CrossRef] [PubMed]
25. Oyekunle, D.T.; Gendy, E.A.; Ifthikar, J.; Chen, Z. Heterogeneous activation of persulfate by metal and non-metal catalyst for the degradation of sulfamethoxazole: A review. *Chem. Eng. J.* **2022**, *437*, 135277. [CrossRef]
26. Liu, S.; Lai, C.; Li, B.; Liu, X.; Zhou, X.; Zhang, C.; Qin, L.; Li, L.; Zhang, M.; Yi, H.; et al. Heteroatom doping in metal-free carbonaceous materials for the enhancement of persulfate activation. *Chem. Eng. J.* **2022**, *427*, 131655. [CrossRef]
27. Wang, G.; Nie, X.; Ji, X.; Quan, X.; Chen, S.; Wang, H.; Yu, H.; Guo, X. Enhanced heterogeneous activation of peroxymonosulfate by Co and N codoped porous carbon for degradation of organic pollutants: The synergism between Co and N. *Environ. Sci. Nano* **2019**, *6*, 399–410. [CrossRef]
28. Ma, M.; Chen, Y.; Su, R.; Liu, Z.; He, J.; Zhou, W.; Gu, M.; Yan, M.; Li, Q. In situ synthesis of Fe-N co-doped carbonaceous nanocomposites using biogas residue as an effective persulfate activator for remediation of aged petroleum contaminated soils. *J. Hazard. Mater.* **2022**, *435*, 128963. [CrossRef] [PubMed]
29. Zhao, G.; Zou, J.; Chen, X.; Liu, L.; Wang, Y.; Zhou, S.; Long, X.; Yu, J.; Jiao, F. Iron-based catalysts for persulfate-based advanced oxidation process: Microstructure, property and tailoring. *Chem. Eng. J.* **2021**, *421*, 127845. [CrossRef]
30. Xiao, S.; Cheng, M.; Zhong, H.; Liu, Z.; Liu, Y.; Yang, X.; Liang, Q. Iron-mediated activation of persulfate and peroxymonosulfate in both homogeneous and heterogeneous ways: A review. *Chem. Eng. J.* **2020**, *384*, 123265. [CrossRef]
31. Wu, X.; Zhu, J.; Hayat, W.; Zhang, Y.; Huang, S.; Jiang, R. Fe-N coordination moieties regulate the defect formation in carbon nanomaterial for efficient peroxydisulfate activation: Significant role of surface complex. *J. Environ. Chem. Eng.* **2023**, *11*, 111585. [CrossRef]
32. Zhu, J.; Wang, L.; Hayat, W.; Zhang, Y.; Huang, S.; Zhang, X.; Zhou, S. The efficient degradation of paracetamol using covalent triazine framework-derived Fe-N-C activated peroxymonosulfate via a non-radical pathway: Analysis of high-valent iron oxide, singlet oxygen and electron transfer. *Sep. Purif. Technol.* **2023**, *310*, 123034. [CrossRef]
33. Zhuo, S.-N.; Sun, H.; Wang, Z.-Y.; Ren, H.-Y.; Xing, D.-F.; Ren, N.-Q.; Liu, B.-F. A magnetic biochar catalyst with dual active sites of Fe_3C and Fe_4N derived from floc: The activation mechanism for persulfate on degrading organic pollutant. *Chem. Eng. J.* **2023**, *455*, 140702. [CrossRef]
34. Huang, M.; Wang, Y.; Wan, J.; Ma, Y.; Chi, H.; Xu, Y.; Qiu, S. Facile construction of highly reactive and stable defective iron-based metal organic frameworks for efficient degradation of Tetrabromobisphenol A via persulfate activation. *Environ. Pollut.* **2020**, *256*, 113399. [CrossRef] [PubMed]
35. Zhang, X.; Wei, J.; Wang, C.; Wang, L.; Guo, Z.; Song, Y. Recent advance of Fe-based bimetallic persulfate activation catalysts for antibiotics removal: Performance, mechanism, contribution of the key ROSs and degradation pathways. *Chem. Eng. J.* **2024**, *487*, 150514. [CrossRef]
36. Sun, Z.; Yao, D.; Cao, C.; Zhang, Z.; Zhang, L.; Zhu, H.; Yuan, Q.; Yi, B. Preparation and formation mechanism of biomass-based graphite carbon catalyzed by iron nitrate under a low-temperature condition. *J. Environ. Manag.* **2022**, *318*, 115555. [CrossRef]
37. Peng, H.; Xiong, W.; Yang, Z.; Tong, J.; Jia, M.; Xiang, Y.; Sun, S.; Xu, Z. Fe_3O_4-supported N-doped carbon channels in wood carbon form etching and carbonization: Boosting performance for persulfate activating. *Chem. Eng. J.* **2023**, *457*, 141317. [CrossRef]
38. Li, X.; Jia, Y.; Zhou, M.; Su, X.; Sun, J. High-efficiency degradation of organic pollutants with Fe, N co-doped biochar catalysts via persulfate activation. *J. Hazard. Mater.* **2020**, *397*, 122764. [CrossRef] [PubMed]
39. Wu, Q.; Zhang, Y.; Meng, H.; Wu, X.; Liu, Y.; Li, L. Cu/N co-doped biochar activating PMS for selective degrading paracetamol via a non-radical pathway dominated by singlet oxygen and electron transfer. *Chemosphere* **2024**, *357*, 141858. [CrossRef] [PubMed]

40. Jin, H.; Zhou, H.; Li, W.; Wang, Z.; Yang, J.; Xiong, Y.; He, D.; Chen, L.; Mu, S. In situ derived Fe/N/S-codoped carbon nanotubes from ZIF-8 crystals as efficient electrocatalysts for the oxygen reduction reaction and zinc–air batteries. *J. Mater. Chem. A* **2018**, *6*, 20093–20099. [CrossRef]
41. Zhang, L.; Zhang, Y.; Meng, H.; Liu, Y.; Hayat, W.; Wu, X. Efficient degradation of acetaminophen by activated peroxymonosulfate using Mn/C composites: Performance and mechanism. *Sep. Purif. Technol.* **2024**, *341*, 126768. [CrossRef]
42. Liang, G.; Yang, Z.; Wang, Z.; Cai, X.; Zhang, X.; Xie, X. Relying on the non-radical pathways for selective degradation organic pollutants in Fe and Cu co-doped biochar/peroxymonosulfate system: The roles of Cu, Fe, defect sites and ketonic group. *Sep. Purif. Technol.* **2021**, *279*, 119697. [CrossRef]
43. Shao, P.; Tian, J.; Yang, F.; Duan, X.; Gao, S.; Shi, W.; Luo, X.; Cui, F.; Luo, S.; Wang, S. Identification and Regulation of Active Sites on Nanodiamonds: Establishing a Highly Efficient Catalytic System for Oxidation of Organic Contaminants. *Adv. Funct. Mater.* **2018**, *28*, 1705295. [CrossRef]
44. Jiang, N.; Xu, H.; Wang, L.; Jiang, J.; Zhang, T. Nonradical Oxidation of Pollutants with Single-Atom-Fe(III)-Activated Persulfate: Fe(V) Being the Possible Intermediate Oxidant. *Environ. Sci. Technol.* **2020**, *54*, 14057–14065. [CrossRef]
45. Chen, H.; Guo, P.; Huang, Z.; Sun, J.; Lei, Y.; Xu, J. Enhanced stability and conductivity of montmorillonite and sucrose loaded Fe-MOFs for degradation of chlortetracycline hydrochloride via electrochemically activated persulfate. *Appl. Clay Sci.* **2024**, *249*, 107231. [CrossRef]
46. *GB 5749-2006*; Standards for Drinking Water Quality. Ministry of Health of China: Beijing, China, 2006.
47. *GB 3838-2002*; Environmental Quality Standards for Surface Water. Ministry of Ecology and Environment of the PRC: Beijing, China, 2002.
48. *GB/T 14848-2017*; Standard for Groundwater Quality. State Bureau of Technical Supervision: Beijing, China, 2017.
49. Wang, L.; Zhu, J.; Zhang, Y.; Zhou, L.; Huang, S.; Zhou, S. MOFs-derived Mn/C composites activating peroxymonosulfate for efficient degradation of sulfamethazine: Kinetics, pathways, and mechanism. *Surf. Interfaces* **2023**, *36*, 102551. [CrossRef]
50. Moussavi, G.; Momeninejad, H.; Shekoohiyan, S.; Baratpour, P. Oxidation of acetaminophen in the contaminated water using UVC/S2O82− process in a cylindrical photoreactor: Efficiency and kinetics of degradation and mineralization. *Sep. Purif. Technol.* **2017**, *181*, 132–138. [CrossRef]
51. Huang, R.; Gao, P.; Zhu, J.; Zhang, Y.; Chen, Y.; Huang, S.; Wang, G.; Yu, Z.; Zhao, S.; Zhou, S. Insights into the pollutant electron property inducing the transformation of peroxymonosulfate activation mechanisms on manganese dioxide. *Appl. Catal. B Environ.* **2022**, *317*, 121753. [CrossRef]
52. Yi, C.; He, Z.; Hu, Y.; Liang, D.; Zhang, Y.; Chen, Y. FeOOH@MoS2 as a highly effective and stable activator of peroxymonosulfate-based advanced oxidation processes for pollutant degradation. *Surf. Interfaces* **2021**, *27*, 101465. [CrossRef]
53. Zhou, X.; Luo, H.; Sheng, B.; Chen, X.; Wang, Y.; Chen, Q.; Zhou, J. Cu2+/Cu+ cycle promoted PMS decomposition with the assistance of Mo for the degradation of organic pollutant. *J. Hazard. Mater.* **2021**, *411*, 125050. [CrossRef]
54. Yuan, R.; Jiang, M.; Gao, S.; Wang, Z.; Wang, H.; Boczkaj, G.; Liu, Z.; Ma, J.; Li, Z. 3D mesoporous α-$Co(OH)_2$ nanosheets electrodeposited on nickel foam: A new generation of macroscopic cobalt-based hybrid for peroxymonosulfate activation. *Chem. Eng. J.* **2020**, *380*, 122447. [CrossRef]
55. Wang, C.; Jia, S.; Zhang, Y.; Nian, Y.; Wang, Y.; Han, Y.; Liu, Y.; Ren, H.; Wu, S.; Yao, K.; et al. Catalytic reactivity of Co_3O_4 with different facets in the hydrogen abstraction of phenol by persulfate. *Appl. Catal. B Environ.* **2020**, *270*, 118819. [CrossRef]
56. Yang, S.; Wu, P.; Liu, J.; Chen, M.; Ahmed, Z.; Zhu, N. Efficient removal of bisphenol A by superoxide radical and singlet oxygen generated from peroxymonosulfate activated with Fe0-montmorillonite. *Chem. Eng. J.* **2018**, *350*, 484–495. [CrossRef]
57. Yun, E.-T.; Lee, J.H.; Kim, J.; Park, H.-D.; Lee, J. Identifying the Nonradical Mechanism in the Peroxymonosulfate Activation Process: Singlet Oxygenation Versus Mediated Electron Transfer. *Environ. Sci. Technol.* **2018**, *52*, 7032–7042. [CrossRef] [PubMed]
58. Yin, Y.; Ren, Y.; Lu, J.; Zhang, W.; Shan, C.; Hua, M.; Lv, L.; Pan, B. The nature and catalytic reactivity of UiO-66 supported Fe_3O_4 nanoparticles provide new insights into Fe-Zr dual active centers in Fenton-like reactions. *Appl. Catal. B Environ.* **2021**, *286*, 119943. [CrossRef]
59. Li, H.; Tian, J.; Xiao, F.; Huang, R.; Gao, S.; Cui, F.; Wang, S.; Duan, X. Structure-dependent catalysis of cuprous oxides in peroxymonosulfate activation via nonradical pathway with a high oxidation capacity. *J. Hazard. Mater.* **2020**, *385*, 121518. [CrossRef] [PubMed]
60. Yang, J.-C.E.; Zhu, M.-P.; Duan, X.; Wang, S.; Yuan, B.; Fu, M.-L. The mechanistic difference of 1T-2H MoS2 homojunctions in persulfates activation: Structure-dependent oxidation pathways. *Appl. Catal. B Environ.* **2021**, *297*, 120460. [CrossRef]
61. Yu, J.; Tang, L.; Pang, Y.; Zeng, G.; Feng, H.; Zou, J.; Wang, J.; Feng, C.; Zhu, X.; Ouyang, X.; et al. Hierarchical porous biochar from shrimp shell for persulfate activation: A two-electron transfer path and key impact factors. *Appl. Catal. B Environ.* **2020**, *260*, 118160. [CrossRef]
62. Meng, H.; Zhou, J.; Zhang, Y.; Cui, J.; Chen, Y.; Zhong, W.; Chen, Y.; Jia, C.Q. Single-atom Co-N_3 sites induce peroxymonosulfate activation for acetaminophen degradation via nearly 100% internal electron transfer process. *Appl. Catal. B Environ. Energy* **2025**, *366*, 125038. [CrossRef]

63. Wang, Y.; Zhang, Z.; Yin, Z.; Liu, Z.; Liu, Y.; Yang, Z.; Yang, W. Adsorption and catalysis of peroxymonosulfate on carbocatalysts for phenol degradation: The role of pyrrolic-nitrogen. *Appl. Catal. B Environ.* **2022**, *319*, 121891. [CrossRef]
64. Ren, W.; Zhang, Q.; Cheng, C.; Miao, F.; Zhang, H.; Luo, X.; Wang, S.; Duan, X. Electro-Induced Carbon Nanotube Discrete Electrodes for Sustainable Persulfate Activation. *Environ. Sci. Technol.* **2022**, *56*, 14019–14029. [CrossRef] [PubMed]
65. Zuo, S.; Guan, Z.; Xia, D.; Yang, F.; Xu, H.; Huang, M.; Li, D. Polarized heterogeneous CuO-CN for peroxymonosulfate nonradical activation: An enhancement mechanism of mediated electron transfer. *Chem. Eng. J.* **2021**, *420*, 127619. [CrossRef]
66. Huang, K.Z.; Zhang, H. Direct Electron-Transfer-Based Peroxymonosulfate Activation by Iron-Doped Manganese Oxide (δ-MnO_2) and the Development of Galvanic Oxidation Processes (GOPs). *Environ. Sci. Technol.* **2019**, *53*, 12610–12620. [CrossRef]
67. Li, W.; Nie, C.; Wang, X.; Ye, H.; Li, D.; Ao, Z. Alkaline lignin-derived N-doped biochars as peroxymonosulfate activators for acetaminophen degradation: Performance and catalytic bridging mediated Electron-Transfer mechanism. *Sep. Purif. Technol.* **2023**, *323*, 124418. [CrossRef]
68. Tan, W.; Ren, W.; Wang, C.; Fan, Y.; Deng, B.; Lin, H.; Zhang, H. Peroxymonosulfate activated with waste battery-based Mn-Fe oxides for pollutant removal: Electron transfer mechanism, selective oxidation and LFER analysis. *Chem. Eng. J.* **2020**, *394*, 124864. [CrossRef]
69. Gao, Y.; Wu, T.; Yang, C.; Ma, C.; Zhao, Z.; Wu, Z.; Cao, S.; Geng, W.; Wang, Y.; Yao, Y.; et al. Activity Trends and Mechanisms in Peroxymonosulfate-Assisted Catalytic Production of Singlet Oxygen over Atomic Metal-N-C Catalysts. *Angew. Chem. Int. Ed.* **2021**, *60*, 22513–22521. [CrossRef] [PubMed]
70. Gao, Y.; Zhu, Y.; Lyu, L.; Zeng, Q.; Xing, X.; Hu, C. Electronic Structure Modulation of Graphitic Carbon Nitride by Oxygen Doping for Enhanced Catalytic Degradation of Organic Pollutants through Peroxymonosulfate Activation. *Environ. Sci. Technol.* **2018**, *52*, 14371–14380. [CrossRef] [PubMed]
71. Zhou, X.; Lai, C.; Liu, S.; Li, B.; Qin, L.; Liu, X.; Yi, H.; Fu, Y.; Li, L.; Zhang, M.; et al. Activation of persulfate by swine bone derived biochar: Insight into the specific role of different active sites and the toxicity of acetaminophen degradation pathways. *Sci. Total Environ.* **2022**, *807*, 151059. [CrossRef]
72. Li, J.; Ye, Q.; Gan, J. Degradation and transformation products of acetaminophen in soil. *Water Res.* **2014**, *49*, 44–52. [CrossRef]
73. Nematollahi, D.; Shayani-Jam, H.; Alimoradi, M.; Niroomand, S. Electrochemical oxidation of acetaminophen in aqueous solutions: Kinetic evaluation of hydrolysis, hydroxylation and dimerization processes. *Electrochim. Acta* **2009**, *54*, 7407–7415. [CrossRef]
74. Alvandi, M.; Nourmoradi, H.; Nikoonahad, A.; Aghayani, E.; Abbas Mirzaee, S. LED visible light assisted photo-oxidation of acetaminophen using one-step synthesis of Cu,Fe@g-C_3N_4 nanosheet—Activated persulfate system in aqueous solutions. *Arab. J. Chem.* **2023**, *16*, 105251. [CrossRef]
75. Zhuo, S.-N.; Ren, H.-Y.; Cao, G.-L.; Xie, G.-J.; Xing, D.-F.; Ren, N.-Q.; Liu, B.-F. Highly efficient activation of persulfate by encapsulated nano-Fe0 biochar for acetaminophen degradation: Rich electron environment and dominant effect of superoxide radical. *Chem. Eng. J.* **2022**, *440*, 135947. [CrossRef]
76. Cai, Z.; Hao, X.; Sun, X.; Du, P.; Liu, W.; Fu, J. Highly active WO_3@anatase-SiO_2 aerogel for solar-light-driven phenanthrene degradation: Mechanism insight and toxicity assessment. *Water Res.* **2019**, *162*, 369–382. [CrossRef] [PubMed]
77. Zhao, W.; Duan, Z.; Zheng, Z.; Li, B. Efficient diclofenac removal by superoxide radical and singlet oxygen generated in surface Mn(II)/(III)/(IV) cycle dominated peroxymonosulfate activation system: Mechanism and product toxicity. *Chem. Eng. J.* **2022**, *433*, 133742. [CrossRef]
78. Shi, J.; Dai, B.; Shen, X.; Xu, L.; Zhang, Y.; Gan, L. Wood induced preparation of Fe3C decorated biochar for peroxymonosulfate activation towards bisphenol a degradation with low ion leaching. *J. Environ. Manag.* **2023**, *340*, 117978. [CrossRef] [PubMed]
79. Mohammadi, S.A.; Najafi, H.; Asasian-Kolur, N.; Ebrahimian Pirbazari, A.; Sharifian, S. The efficiency of a granular Fe-based metal-organic framework for the persulfate oxidative degradation of levofloxacin, an emerging antibiotic in wastewater. *J. Mol. Liq.* **2024**, *405*, 125119. [CrossRef]
80. Zhu, K.; Bin, Q.; Shen, Y.; Huang, J.; He, D.; Chen, W. In-situ formed N-doped bamboo-like carbon nanotubes encapsulated with Fe nanoparticles supported by biochar as highly efficient catalyst for activation of persulfate (PS) toward degradation of organic pollutants. *Chem. Eng. J.* **2020**, *402*, 126090. [CrossRef]
81. Liu, J.; Zhu, J.; Zeng, X.; Yang, Y.; Zhang, G.; Sun, Y.; Fan, G. Enhanced activation of persulfate by bimetal and nitrogen co-doped biochar for efficient degradation of refractory organic contaminants: The role of the second metal. *J. Phys. Chem. Solids* **2024**, *193*, 112191. [CrossRef]
82. Peng, L.; Duan, X.; Shang, Y.; Gao, B.; Xu, X. Engineered carbon supported single iron atom sites and iron clusters from Fe-rich Enteromorpha for Fenton-like reactions via nonradical pathways. *Appl. Catal. B Environ.* **2021**, *287*, 119963. [CrossRef]
83. Li, X.; Chen, J.; Liu, Z.; He, C.; Pang, J.; Zhang, L.; Tang, F.; Yang, X. Recognizing the relevance of non-radical peroxymonosulfate activation by co-doped Fe metal-organic framework for the high-efficient degradation of acetaminophen: Role of singlet oxygen and the enhancement of redox cycle. *Chem. Eng. J.* **2024**, *499*, 156081. [CrossRef]

84. Wang, T.; Liang, Q.; Xing, Y.; Sun, M.; Luo, H. Magnetic Fe/N-codoped carbon derived from modified Fe-base MOFs: Synergism of multiple active sites for peroxymonosulfate activation. *J. Environ. Chem. Eng.* **2023**, *11*, 109905. [CrossRef]
85. Chen, C.; Ma, T.; Shang, Y.; Gao, B.; Jin, B.; Dan, H.; Li, Q.; Yue, Q.; Li, Y.; Wang, Y.; et al. In-situ pyrolysis of Enteromorpha as carbocatalyst for catalytic removal of organic contaminants: Considering the intrinsic N/Fe in Enteromorpha and non-radical reaction. *Appl. Catal. B Environ.* **2019**, *250*, 382–395. [CrossRef]

catalysts

MDPI

Article

Amorphous MnO_2 Supported on CN@SiO_2 for Levofloxacin Degradation via a Non-Radical Pathway by PMS Activation

Longfei Xia, Xilin Wang, Jiahui Li and Dongyan Xu *

School of Chemical Engineering, Qingdao University of Science & Technology, Qingdao 266042, China; 4023011089@mails.qust.edu.cn (L.X.); 2021030062@mails.qust.edu.cn (X.W.)
* Correspondence: xdy0156@qust.edu.cn

Abstract: Mn-based catalysts have been extensively studied in advanced oxidation processes based on peroxymonosulfate (PMS) oxidants, demonstrating their significant potential for treating antibiotic-contaminated wastewater. In this study, an amorphous MnO_2-based composite catalyst (MnO_2/CN@SiO_2) was prepared and used to activate PMS for degrading levofloxacin (LEV). The effects of reaction conditions, such as reaction temperature, catalyst dosage, PMS concentration, and solution pH, on LEV degradation were comprehensively investigated. The interference of water components, e.g., NO_3^-, SO_4^{2-}, Cl^-, CO_3^{2-}, and humic acid, on the degradation efficiency of LEV was analyzed, and the stability of the catalysts was explored by cycling experiments. Finally, radical quenching experiments and electron paramagnetic resonance spectroscopy were employed to elucidate the contribution of active species to the degradation reaction process. A non-radical-based pathway for LEV degradation was proposed based on these results.

Keywords: amorphous MnO_2; levofloxacin; PMS; Fenton-like

1. Introduction

Antibiotics are widely used in the pharmaceutical and aquaculture industries due to their excellent bacteriostatic properties. With the increasing demand for antibiotics in recent years, substantial amounts of antibiotic effluent from hospitals, animal husbandry, and aquaculture have been discharged into aquatic environments [1,2]. Levofloxacin (LEV), a third-generation quinolone with broad-spectrum antimicrobial properties, is commonly used in the treatment of urinary, gastrointestinal, and respiratory diseases [3,4]. Typically, LEV cannot be fully absorbed and utilized in humans and animals, leading to unmetabolized drugs entering ecosystems. This phenomenon not only exacerbates water pollution, but also accelerates the development of bacterial resistance, thereby posing a great threat to human health [5–7]. Therefore, the treatment of antibiotic-containing wastewater has become an urgent priority. However, the chemical stability of antibiotics makes traditional wastewater treatment processes ineffective in removing them. Advanced oxidation processes (AOPs), which possess superior oxidizing capabilities, hold greater promise for treating antibiotic-containing wastewater [8,9].

In advanced oxidation techniques, free radicals with strong oxidizing properties (e.g., $\bullet OH$, $\bullet SO_4^-$, and $\bullet O_2^-$) can effectively degrade antibiotics and achieve a high degree of mineralization of organic pollutants in wastewater [10,11]. Peroxymonosulfate (PMS)-based advanced oxidation processes (SR-AOPs) are commonly used for the treatment of antibiotic pollutants in water due to the high redox potential and long half-life of $SO_4^{\bullet-}$ [12–14]. Compared to H_2O_2, solid PMS is easier to store and transport [15]. In summary, SR-AOPs show great promise in antibiotic degradation. Transition metals, such as copper, iron, cobalt, and manganese, can provide electrons to promote the breaking of S–O bonds and are commonly used to activate PMS in heterogeneous systems [16–19]. Among these metals, Mn is usually used for PMS activation in the degradation of antibiotic wastewater because of its abundant reserves, low environmental toxicity, rich valence composition, and controllable crystal morphology [20]. The different arrangements of the basic structural unit [MnO_6] allow for the existence of a variety of crystals of MnO_2, including α, β, γ, and δ–MnO_2 [21]. In addition, differences in crystalline structure and manganese ion valence states in various MnO_2 types lead to the generation of distinct reactive oxygen species (ROS), which in turn exhibit different catalytic activities [22,23]. Recently, it has been reported that amorphous MnO_2 can rapidly activate PMS to degrade pollutants via a non-radical pathway [24]. Generally, highly crystalline MnO_2 needs to be synthesized under the conditions of hydrothermal, microwave, or roasting methods, which are energy-intensive and complex [25,26]. In contrast, amorphous MnO_2 shows great advantages in preparation. It can be obtained by simple redox reactions involving MnO_4^- or Mn^{2+} at ambient temperature and pressure [27]. However, MnO_2 particles tend to agglomerate during the preparation process. This agglomeration seriously hinders the exposure of active sites, thereby diminishing the activation efficiency of PMS. Consequently, enhancing the performance of PMS activation necessitates the introduction of additional components to effectively disperse MnO_2.

Graphitic-phase carbon nitride (g–C_3N_4) is a non-metallic organic polymer with high thermal and chemical stability, commonly used as a carrier in material preparation [28]. The bulk structure of g–C_3N_4 consists of a large number of stacked, flattened layers, which inherently lead to a relatively low specific surface area and limited surface functional groups. Thermal exfoliation has been demonstrated to effectively reduce the stacking degree of these layered structures, thereby significantly increasing the specific surface area [29,30]. The –OH content on the surface of Mn-based catalysts has a significant effect on the activation of PMS [31]. SiO_2, as a hydrophilic material rich in –OH groups, provides binding sites for interfacial interactions between materials and enhances the dispersion stability of composites in aqueous solutions [32,33]. As far as we know, there are relatively few studies on improving the activation efficiency of PMS by compositing SiO_2 with MnO_2. SiO_2 rich in –OH groups can accelerate the binding rate of PMS to the catalyst. Furthermore, the stabilized g–C_3N_4 provides a support matrix for the effective dispersion of MnO_2 particles. Therefore, the composite of MnO_2 with SiO_2 and g–C_3N_4 is expected to solve the problems of easy agglomeration of MnO_2 and slow activation of PMS.

In this study, MnO_2/CN@SiO_2 (MCS) ternary composites were successfully synthesized and used for activating PMS to degrade LEV. The physicochemical properties of the catalysts were comprehensively analyzed by XRD, XPS, and SEM techniques. The catalytic performance of catalyst compositions for LEV degradation by activating PMS was systematically compared. The effects of catalyst dosage, reaction temperature, solution pH, and inorganic anion interference on PMS activation efficiency were thoroughly investigated. Furthermore, the stability of the catalyst was evaluated via cycling experiments. Radical quenching experiments and electron paramagnetic resonance (EPR) spectroscopy were employed to unveil the primary degradation pathways during the reaction process. Based on these results, a non-radical pathway-based mechanism for LEV degradation was proposed. The experimental results showed that the MnO_2/CN@SiO_2 catalyst had the ability to activate PMS efficiently.

2. Results

2.1. Characterizations of $MnO_2/CN@SiO_2$

Figure 1 shows the XRD patterns of MnO_2/SiO_2, MnO_2/CN, MCS, and $MnO_2/g-C_3N_4@SiO_2$. The broad diffraction peaks at 20–30° can be attributed to amorphous SiO_2 [34]. The peak observed at 27.6° corresponds to the (002) crystallographic plane of $g-C_3N_4$ (PDF#87-1526) [35]. In addition, the weak diffraction peaks at 37° and 66° are related to amorphous MnO_2 [36]. The presence of SiO_2, CN, and MnO_2 diffraction peaks confirms the successful preparation of the composite. By comparing the XRD patterns of MCS and $MnO_2/g–C_3N_4@SiO_2$, a significant difference was identified in the intensity of the main diffraction peak at around 23°. This discrepancy may arise from the controlled growth of SiO_2 and MnO_2 after thermal stripping of $g–C_3N_4$.

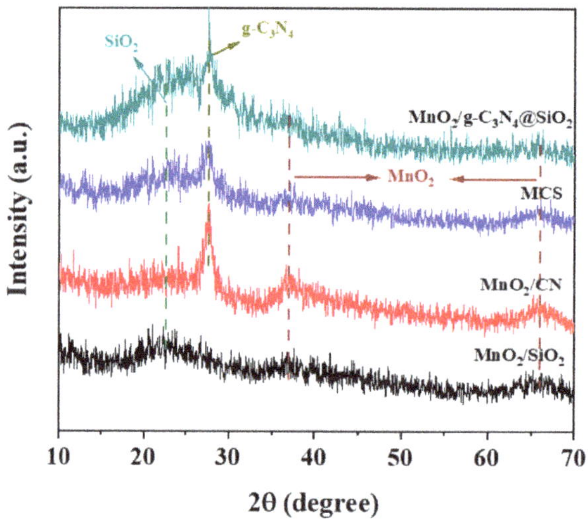

Figure 1. XRD patterns of $MnO_2/CN@SiO_2$ samples.

The surface elements and valence states of the MCS catalyst before and after the reaction were analyzed using XPS. The full XPS spectra of both the fresh catalyst and the recycled catalysts confirm the presence of Mn, O, N, C, and Si elements (Figure 2a). In Figure 2b, the two peaks located at approximately 642 eV and 654 eV correspond to the Mn $2p_{3/2}$ and Mn $2p_{1/2}$ orbitals, respectively [37]. The high-resolution Mn 2p spectra can be deconvoluted into four peaks: those at 641.3 and 641.6 eV are attributed to Mn^{3+}, while those at 642.5 and 642.7 eV are assigned to Mn^{4+} [38]. As presented in Figure 2c, the high-resolution C 1s spectrum of the fresh MCS catalyst shows three peaks located at 284.8 eV, 286.1 eV, and 288.6 eV, corresponding to C–C, C–N, and the N–C=N bonds, respectively [39]. The high-resolution N 1s spectra (Figure 2d) can be deconvoluted into three peaks located at 399.1 eV, 399.8 eV, and 401.7 eV, representing sp^2 hybridized N atoms (N–C=N), quaternary N atoms ($(C)_3$–N), and N atoms in the amino group (C–NH_x), respectively [38,40]. The high-resolution O 1s spectra of the fresh MCS catalyst (Figure 2e) can be divided into three distinct peaks at 529.8 eV, 531.3 eV, and 532.3 eV, which are attributed to the lattice oxygen (Mn–O) in the metal oxides, adsorbed hydroxyl oxygen (–OH), and oxygen in chemisorbed water on the surface of the material [41]. As shown in Figure 2f, the characteristic peak at 101.9 eV for the fresh catalyst corresponds to the 2p electron orbital of the Si element [42].

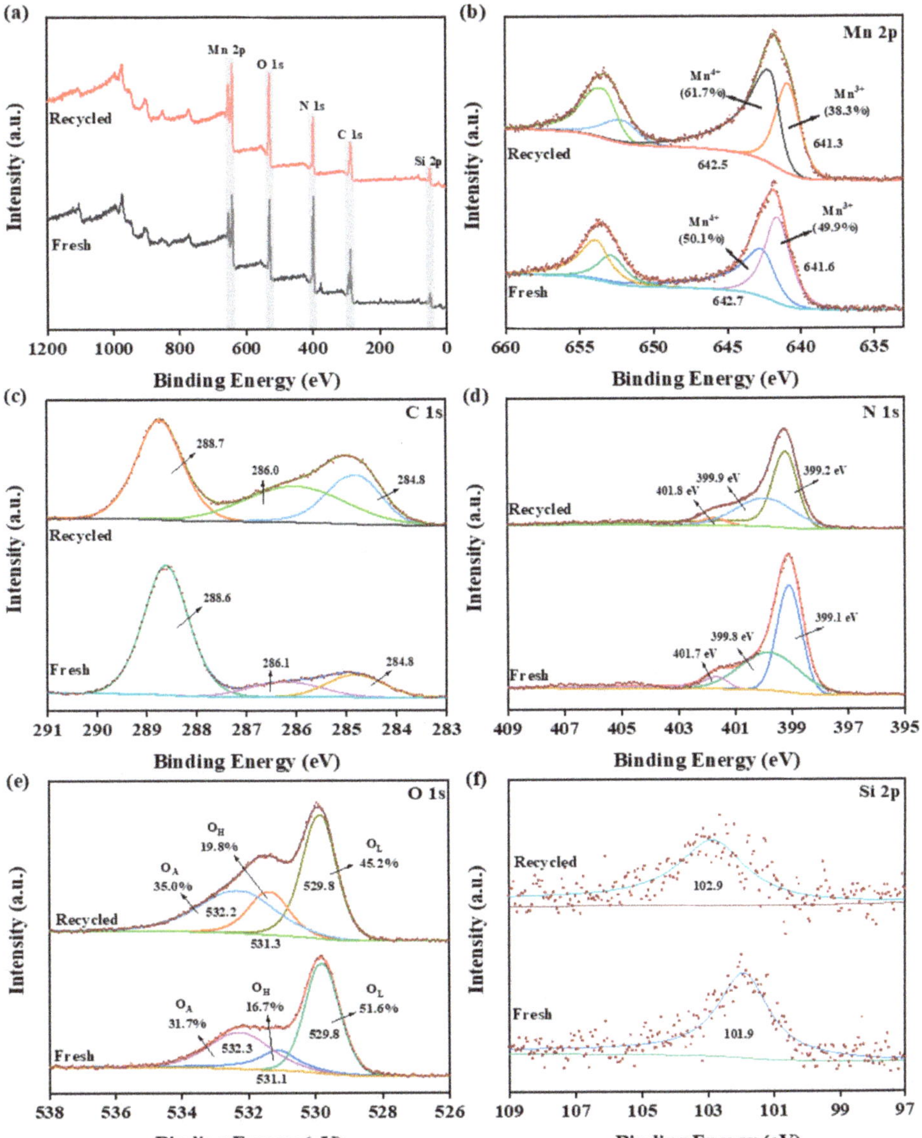

Figure 2. (**a**) XPS spectrum of survey, (**b**) Mn 2p, (**c**) C 1s, (**d**) N 1s, (**e**) O 1s, and (**f**) Si 2p of MCS.

Figure 3 demonstrates the SEM image, elemental mapping images, and energy dispersion spectrum of the MCS composites. As shown in Figure 3a, amorphous MnO_2 particles are distributed on the surface of the CN@SiO_2 substrate. The elemental mapping images (Figure 3b) reveal the distribution of the C, N, O, Si, and Mn elements across the substrate surface. Notably, the dispersion of Si is much denser than those of C and N, indicating the existence of a SiO_2-coated CN structure. Furthermore, the significant overlap of Mn, O, and Si elements indicates the successful preparation of the MCS composite.

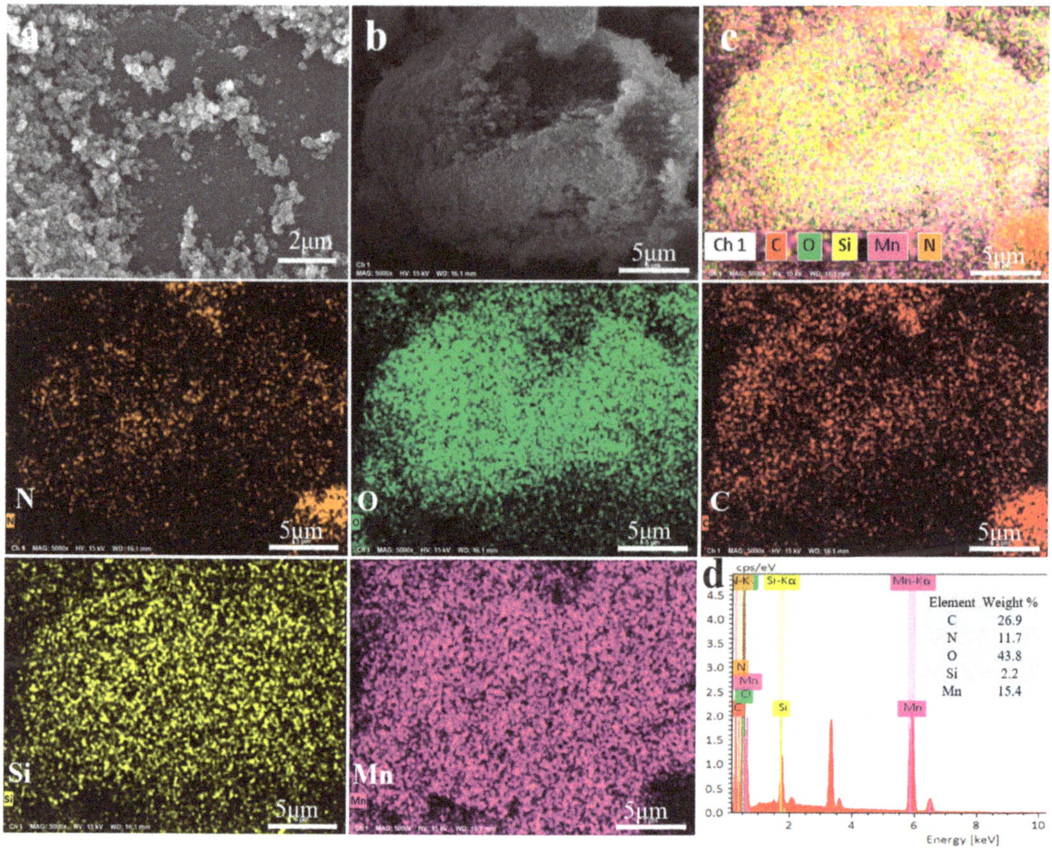

Figure 3. (**a**) SEM, (**b**,**c**) element mapping, and (**d**) EDS images of MCS.

2.2. *Effect of Catalyst Composition on LEV Degradation*

The degradation of LEV by different catalysts in the presence of PMS was investigated. As shown in Figure 4a, MnO_2/CN exhibited superior degradation performance compared to MnO_2/g–C_3N_4, which may be attributed to greater exposure of MnO_2 on CN than g–C_3N_4. Moreover, the binary CN/SiO_2 composite-supported catalyst demonstrated higher catalytic activity than the CN-supported catalyst. Figure 4b shows the effect of different MnO_2 contents on the LEV degradation efficiency. At a MnO_2 content of 50 wt%, the LEV degradation efficiency reached 80.8%, which is notably higher than that at 20 wt% (58.7%), 80 wt% (72.3%), and pure MnO_2 (67.4%).

The catalytic activities of the MnO_2/CN, MnO_2/SiO_2, and MCSs with various SiO_2/CN ratios for LEV degradation are presented in Figure 4c. The degradation efficiencies of MnO_2/CN, MCS-1 (SiO_2/CN = 1:4), MCS-2 (SiO_2/CN = 1:1), MCS-3 (SiO_2/CN = 4:1), and MnO_2/SiO_2 within 40 min are 68.2%, 80.8%, 77.0%, 76.9%, and 76.1%, respectively. As depicted in Figure 4d, the LEV degradation efficiencies with PMS alone, MnO_2 alone, and MCS alone are 1.9%, 7.8%, and 11.1%, respectively, indicating that neither PMS nor the catalysts can effectively degrade LEV. In contrast, MnO_2 can effectively activate PMS, leading to a LEV degradation efficiency of 67.4%. The MCS showed higher catalytic activity than the pure MnO_2 catalyst.

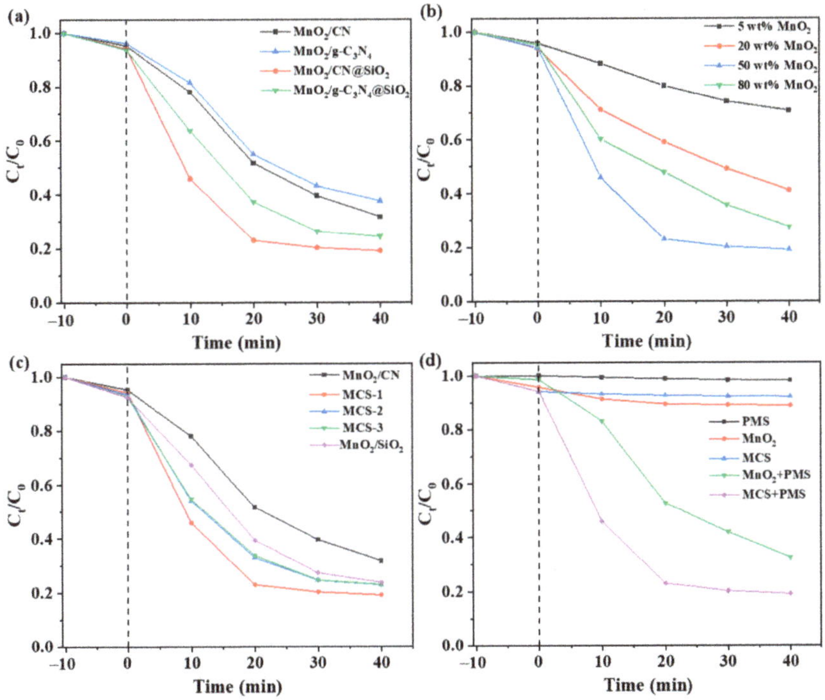

Figure 4. Effect of (**a**) different catalysts, (**b**) MnO_2 content, (**c**) SiO_2 content, and (**d**) reaction system on the degradation of LEV. Conditions: [LEV] = 50 mg/L; [PMS] = 0.3 g/L; [Catalyst] = 0.25 g/L; T = 20 °C.

2.3. Effect of Reaction Conditions on LEV Degradation

Figure 5 illustrates the effects of catalyst dosage, PMS concentration, initial LEV concentration, and reaction temperature on LEV degradation. As depicted in Figure 5a, the LEV degradation efficiency increased from 67.2% to 83.7%, with a corresponding increase in k_{app} from 0.0241 min^{-1} to 0.1354 min^{-1}, as the catalyst dosage was raised from 5 mg to 20 mg. This result is consistent with the expectation that an increase in catalyst dosage can provide more reactive sites, thus accelerating the PMS activation and improving the LEV degradation. As shown in Figure 5b, as the concentration of PMS increased from 0.5 mM to 1.0 mM, the degradation efficiency of LEV within 40 min increased from 75.0% to 80.1%, and the k_{app} of the reaction system increased from 0.0329 min^{-1} to 0.0945 min^{-1}. However, further increasing the PMS concentration resulted in insignificant enhancement in LEV degradation due to the limited number of active sites provided by fixed catalyst dosage. The effect of initial LEV concentration on LEV degradation is illustrated in Figure 5c. As expected, the degradation efficiency of LEV decreased from 84.5% to 80.8% with an increase in LEV concentration from 20 mg/L to 50 mg/L. This reduction can be attributed to the fixed catalyst dosage and PMS concentration, which result in a constant production of oxidative species throughout the reaction system. Consequently, under identical conditions, a higher LEV concentration leads to lower degradation efficiency. As displayed in Figure 5d, the degradation efficiency of LEV increased from 69.3% to 80.1% with an increase in reaction temperature from 10 °C to 40 °C. The elevated temperature favors the thermally driven activation of PMS, thereby promoting LEV degradation [43]. Based on the Arrhenius plot, the reaction activation energy was determined to be 18.176 kJ/mol.

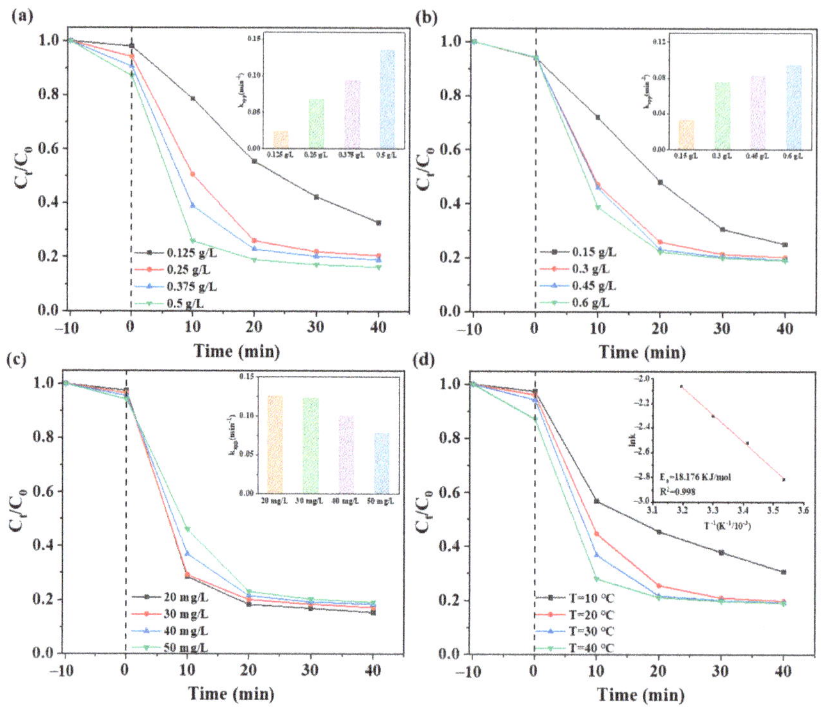

Figure 5. Effect of (**a**) catalyst dose, (**b**) PMS concentration, (**c**) initial LEV concentration, and (**d**) reaction temperature on the degradation efficiency of LEV. Conditions: [LEV] = 50 mg/L; [PMS] = 0.3 g/L; [Catalyst] = 0.25 g/L; T = 20 °C.

Figure 6 illustrates the effect of the initial solution pH on LEV degradation, PMS concentration, and potential Mn ion leaching. It is evident that pH significantly affects LEV degradation efficiency. An approximately 80.1% degradation efficiency was obtained in the pH range of 3–7. In contrast, only 24.8% and 17.7% LEV degradations were achieved at pH = 9 and pH = 11, respectively. Figure 6c presents the variation in PMS concentration with pH in LEV solutions without a catalyst. The PMS concentration was determined using the iodometric method after 10 min of stirring. PMS was stable under acidic and neutral conditions, but its concentration decreased in alkaline environments due to hydrolysis. This can explain the low LEV degradation efficiency at a high pH. At the same time, the effect of pH on catalyst stability was investigated by dispersing the catalyst in LEV solutions with varying pH levels for 10 min. Subsequently, PMS was added to initiate the reaction after filtering out the catalyst. As shown in Figure 6d, substantial LEV degradation occurred under acidic conditions (pH = 3–5), implying the happening of a homogeneous reaction due to leached Mn ions. According to previous studies, under acidic conditions, the catalyst enhances PMS activation via electrostatic attraction. However, as pH increases, the electrostatic effect diminishes or even transitions to electrostatic repulsion, significantly inhibiting PMS activation by the catalyst [44–46]. In conclusion, the pH of the solution not only impacts the stability of PMS and the catalyst but also interferes with catalytic PMS activation, thereby substantially influencing the system's degradation performance.

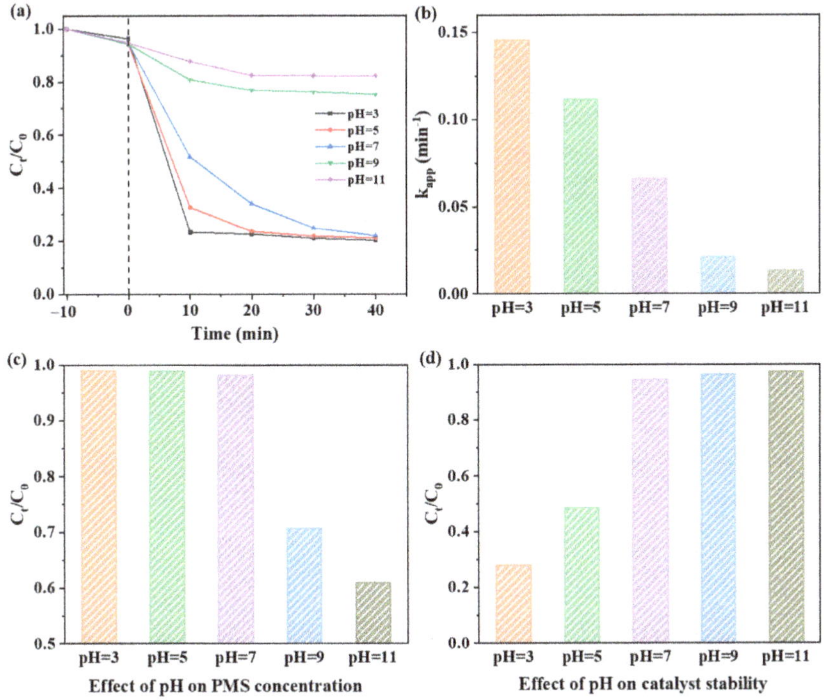

Figure 6. Effect of pH on (**a**) LEV degradation rate, (**b**) reaction system k_{app}, (**c**) PMS concentration, and (**d**) catalyst stability. Conditions: [LEV] = 50 mg/L; [PMS] = 0.3 g/L; [Catalyst] = 0.25 g/L; T = 20 °C.

Inorganic anions commonly found in natural aquatic environments, including NO_3^-, SO_4^{2-}, Cl^-, and CO_3^{2-}, can significantly influence the degradation of LEV. The effects of the four anions on the LEV degradation in the MCS-PMS system were investigated (Figure 7). The results show that Cl^- promotes LEV degradation, whereas the other three anions inhibit the reaction to varying degrees, with CO_3^{2-} being the most potent inhibitor. In solution, Cl^- reacts with $SO_4^{\bullet-}$ to produce additional active chlorine species (Cl_2^- and HOCl), which actively participate in the degradation process [47]. Therefore, increasing Cl^- concentration promotes LEV degradation efficiency [48]. Conversely, the inhibitory effects of NO_3^-, SO_4^{2-}, and CO_3^{2-} may arise from their reactions with free radicals and/or competition with HSO_5^- for the active site [49]. Notably, the pronounced inhibitory effect of CO_3^{2-} could also stem from its influence on solution pH. Upon the addition of CO_3^{2-}, the resulting alkaline conditions hinder PMS activation, thereby reducing LEV degradation efficiency (Figure 6).

Natural organic matter (NOM), ubiquitously present in natural aquatic environments, can inhibit the catalytic degradation process via reactive oxygen scavenging and competitive adsorption [50]. Humic acid (HA), a major component of NOM, interacts with functional groups on the catalyst surface, thereby hindering the binding of the catalyst to PMS and influencing the degradation reaction.

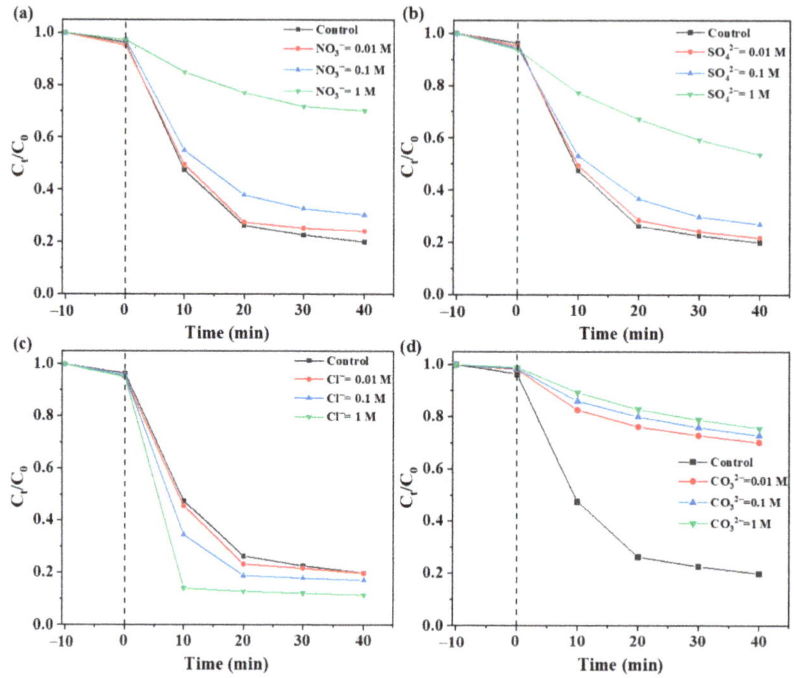

Figure 7. Effects of (**a**) NO_3^-, (**b**) SO_4^{2-}, (**c**) Cl^-, and (**d**) CO_3^{2-} concentration on the degradation efficiency of LEV. Conditions: [LEV] = 50 mg/L; [PMS] = 0.3 g/L; [Catalyst] = 0.25 g/L; T = 20 °C.

The effects of different HA concentrations on the degradation of LEV in the MnO_2/CN@SiO_2-PMS system were investigated, with the results presented in Figure 8a. When the HA concentration was less than 10 mg/L, its effect on LEV degradation was negligible. However, upon increasing the HA concentration to 20 mg/L, the degradation efficiency decreased significantly, from 80.8% to 72.2%.

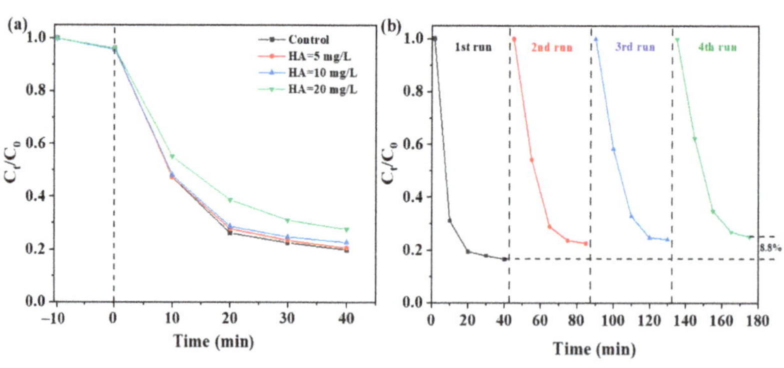

Figure 8. (**a**) Effect of HA concentration on the degradation efficiency of LEV and (**b**) catalyst stability test. Conditions: [LEV] = 50 mg/L; [PMS] = 0.3 g/L; [Catalyst] = 0.25 g/L; T = 20 °C.

The stability of MnO_2/CN@SiO_2 catalysts during degradation was investigated by cycling experiments, as shown in Figure 8b. Multiple replicate experiments were performed under the same conditions. After four cycles, the LEV degradation efficiency decreased from 83.5% to 74.7%. This reduction was likely attributed to the inevitable leaching of

Mn ions during the activation of PMS by MnO_2, which consequently reduced the MnO_2 content in the catalyst, thus diminishing its catalytic performance. Despite this, the cycling experiments confirmed the satisfactory stability of the catalysts, with only an 8.5% decline in activity after four cycles.

2.4. Identification of Active Species

Free radicals such as $\bullet SO_4^-$, $\bullet OH$, and $\bullet O_2^-$ are usually considered the main reactive substances in AOPs [51]. To identify the potential reactive substances in the MnO_2/CN@SiO_2-PMS system and their respective contributions to the degradation process, scavenging experiments were conducted. Tert-butyl alcohol (TBA), methanol (MeOH), p-benzoquinone (PBQ), and L-histidine (L-His) were used as scavengers because they can quench $\bullet OH$, $\bullet SO_4^-/\bullet OH$, $\bullet O_2^-$, and $^1O_2$, respectively.

Based on the reaction rates of the quencher and the radicals, different quencher concentrations were chosen to ensure that the same radicals were quenched in the same amount of time [52]. Figure 9 shows the effect of various quenchers on the degradation of LEV. Figure 9a shows the experimental results of tert-butanol quenching of $\bullet OH$ radicals in the MnO_2/CN@SiO_2-PMS system. Upon adding TBA at a concentration of 10 mM, the degradation efficiency of LEV decreased from 80.8% to 79.3%. Further increasing the TBA concentration to 50 mM led to a reduction in the LEV degradation efficiency. The experimental results of MeOH quenching of $\bullet SO_4^-$ and $\bullet OH$ radicals are shown in Figure 9b. When MeOH was added at a concentration of 10 mM, the LEV degradation rate decreased from 80.8% to 78.7%, and it further dropped to 74.6% at a concentration of 50 mM.

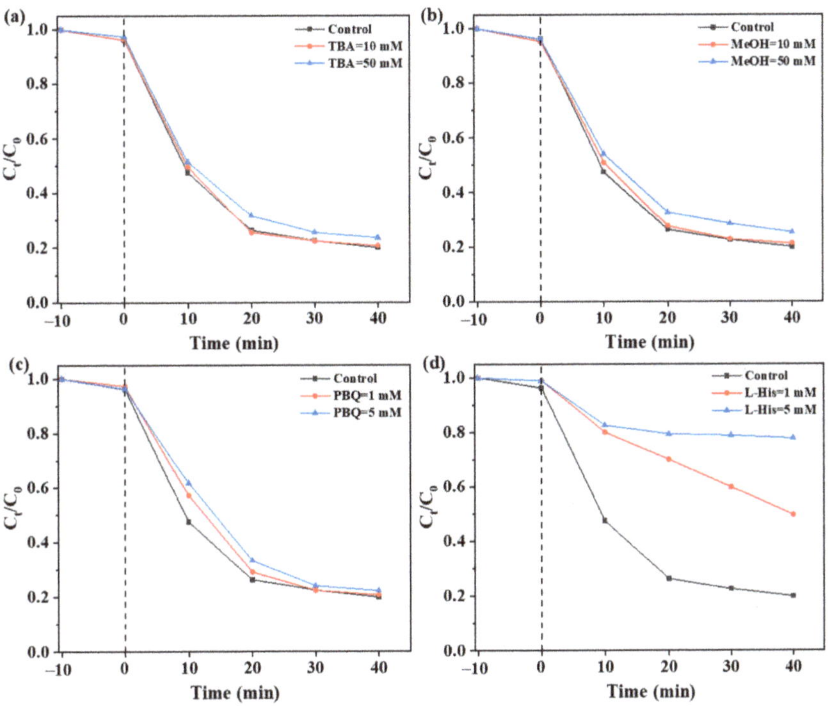

Figure 9. Effects of (**a**) TBA, (**b**) MeOH, (**c**) PBQ, and (**d**) L-His concentration on the degradation efficiency of LEV. Conditions: [LEV] = 50 mg/L; [PMS] = 0.3 g/L; [Catalyst] = 0.25 g/L; T = 20 °C.

Figure 9c illustrates the experimental results of the PBQ quenching system of $\bullet O_2^-$ radicals. At PBQ concentrations of 1 mM and 5 mM, the LEV degradation efficiencies were reduced to 79.2% and 77.6%, respectively. In summary, the quenching experiments targeting $\bullet SO_4^-$, $\bullet$OH, and $\bullet O_2^-$ showed that the free radical pathway is not the dominant pathway for LEV degradation. Consequently, it is hypothesized that the non-free radical pathway may play a more significant role in the degradation of LEV. The $^1O_2$ quenching experiment is shown in Figure 9d. L-His exerted a tremendous inhibitory effect on the degradation of LEV, with the degradation efficiency decreasing to 50.4% and 22.2% at concentrations of 1 mM and 5 mM, respectively. Previous studies have shown that furfural and L-His can not only quench $^1O_2$ but also directly react with PMS, thus inhibiting the reaction [53].

Figure 10a demonstrates the changes in PMS concentration after the addition of different scavengers. TBA, MeOH, PBQ, and L-His were added to the PMS solution with an initial concentration of 1 mM, and the concentration changes of PMS were determined using the iodometric method. The experimental results clearly indicate that the PMS concentration decreased markedly upon adding L-His, confirming its ability to quench PMS.

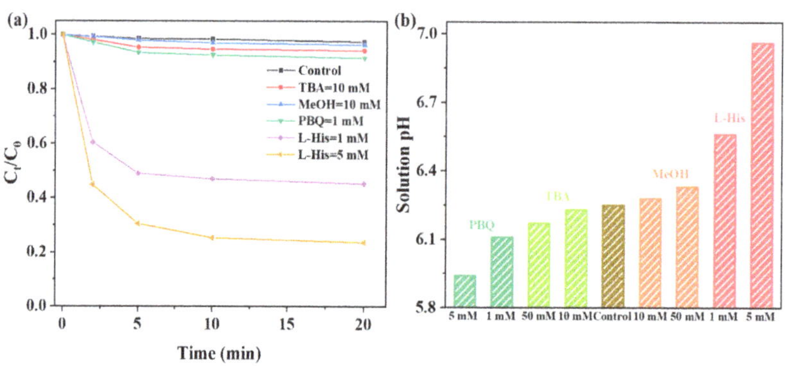

Figure 10. Effect of scavengers on (**a**) PMS concentration and (**b**) solution pH. Conditions: [LEV] = 50 mg/L; [PMS] = 0.3 g/L; [Catalyst] = 0.25 g/L; T = 20 °C.

As shown in Figure 10b, the effect of scavengers on solution pH was investigated. The presence of TBA and PBQ can decrease the solution pH, while the presence of MeOH and L-His can increase it. Slight changes in solution pH do not significantly affect the experimental results. Therefore, the inhibition of the reaction by the scavenger through changing the solution pH can be ignored.

To further investigate the role of $^1O_2$ in the degradation process, the production of $^1O_2$ during the reaction was examined by EPR spectroscopy. In this study, $^1O_2$ was captured using 2,2,6,6-tetramethylpiperidine (TEMP) as a spin-trapping agent, and the results are shown in Figure 11a. The MnO_2/CN@SiO_2-PMS system clearly detected the TEMP-$^1O_2$ triple peak signals with an approximate intensity ratio of 1:1:1, confirming the presence of $^1O_2$ in the reaction [54]. Conversely, no significant $^1O_2$ signal was detected in the MnO_2-PMS system, yet LEV degradation still occurred, suggesting that $^1O_2$-mediated degradation was not the sole pathway for LEV removal. As depicted in Figure 11b, PMS consumption was monitored at a reaction time of 10 min in the catalyst-free system, the MnO_2-PMS system, and the MCS-PMS system. Without the addition of a catalyst, the PMS concentration remained essentially constant. In contrast, in the MnO_2-PMS and MCS-PMS systems, the PMS concentrations decreased to 91.5% and 77.8% of the initial value, respectively. The PMS consumption in both systems was consistent with the $^1O_2$ signal intensities observed in the EPR spectra described above, indicating that the $MnO_2/CN@SiO_2$ catalyst could activate PMS more efficiently to degrade LEV.

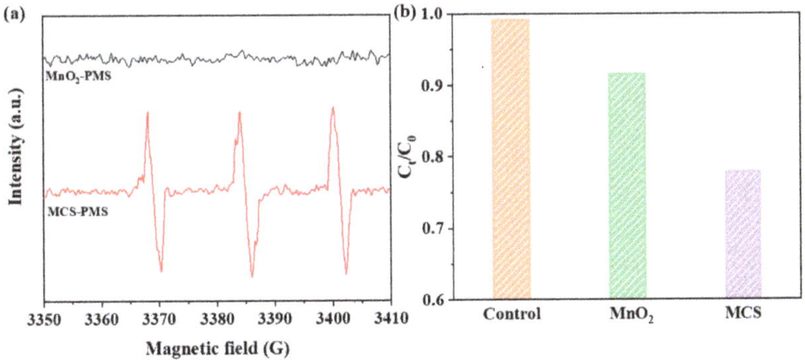

Figure 11. (**a**) EPR spectra of MnO_2-PMS and MCS-PMS and (**b**) changes in PMS concentration in different systems. Conditions: [LEV] = 50 mg/L; [PMS] = 0.3 g/L; [Catalyst] = 0.25 g/L; T = 20 °C.

2.5. Reaction Mechanism

Based on the above experimental and characterization results, a reaction mechanism for LEV degradation in the $MnO_2/CN@SiO_2$-PMS system was proposed. A non-radical pathway was proposed as the primary mechanism for LEV degradation. In the $MnO_2/CN@SiO_2$-PMS degradation system, MnO_2 serves as an activator for PMS, with Mn^{3+} acting as the main active center. The activation of PMS can be categorized into two pathways, which are elaborated as follows. Firstly, PMS binds to the active site of MnO_2 by substituting –OH groups, subsequently reacting with Mn^{3+} to form the complex Mn^{3+}–(O)OSO_3^- (Equation (1)) [55]. Electron transfer occurs between MnO_2 and PMS, enabling the complex to extract one electron from LEV, thereby oxidizing it. Simultaneously, Mn^{3+} loses one electron to form Mn^{4+}, while HSO_5^- gains two electrons to form SO_4^{2-} (Equation (2)). This pathway constitutes the dominant mechanism for LEV degradation in the system [56]. Secondly, HSO_5^- adsorbed onto the catalyst surface reacts directly with Mn^{3+} to form •OH, Mn^{4+}, and SO_4^{2-}, or reacts with Mn^{4+} to form Mn^{3+} and SO_5^- (Equations (3)–(6)). During the reaction, the reduction in Mn content leads to a decrease in lattice oxygen content (Figure 2e), resulting in the formation of oxygen vacancies and reactive oxygen species (O*). O* reacts with PMS to form $^1O_2$ (Equation (7)). In addition, $^1O_2$ is produced during the hydrolysis of $SO_5^{\bullet-}$ and the autolytic decomposition of PMS (Equations (8) and (9)). Ultimately, the ROS generated in the reaction contribute to the degradation of LEV [57].

$$\equiv Mn^{3+} - OH + HSO_5^- \rightarrow \equiv Mn^{3+} - (O)\ OSO_3^- + H_2O \quad (1)$$

$$\equiv Mn^{3+} - (O)\ OSO_3^- + e^- \rightarrow \equiv Mn^{4+} + SO_4^{2-} \quad (2)$$

$$Mn^{3+} + HSO_5^- \rightarrow Mn^{4+} + OH^- + SO_4^{\bullet-} \quad (3)$$

$$SO_4^{\bullet-} + H_2O \rightarrow \bullet OH + SO_4^{2-} + H^+ \quad (4)$$

$$SO_4^{\bullet-} + OH^- \rightarrow \bullet OH + SO_4^{2-} \quad (5)$$

$$Mn^{4+} + HSO_5^- \rightarrow Mn^{3+} + H^+ + SO_5^{\bullet-} \quad (6)$$

$$O* + HSO_5^- \rightarrow HSO_4^- + 1O_2 \quad (7)$$

$$SO_5^{\bullet-} + H_2O \rightarrow HSO_4^- + {}^1O_2 \tag{8}$$

$$HSO_5^- + SO_5^{2-} \rightarrow HSO_4^- + SO_4^{2-} + {}^1O_2 \tag{9}$$

3. Materials and Methods

3.1. Materials

Potassium permanganate (≥99.5% $KMnO_4$), sodium hydroxide (≥96.0% NaOH), manganese sulfate monohydrate (≥99.0% $MnSO_4 \cdot H_2O$), ammonium persulphate (≥98.0% $(NH_4)_2S_2O_8$), sodium chloride (≥99.5% NaCl), sodium sulfate (≥99.0% Na_2SO_4), sodium carbonate (≥99.0% Na_2CO_3), sodium nitrate (≥99.0% $NaNO_3$), ethanol (≥99.5% C_2H_5OH), methanol (≥99.5% CH_3OH), tert-butyl alcohol (≥99.0% TBA), para-benzoquinone (≥99.0% PBQ), sodium bicarbonate (≥99.5% $NaHCO_3$), and hydrochloric acid (36–38% HCl) were all purchased from Sinopharm Chemical Reagent Co., Ltd. (Beijing, China). Melamine (≥99.0% $C_3H_6N_6$), tetraethyl orthosilicate (≥99.0% TEOS), levofloxacin (≥99.0% LEV), potassium monopersulfate triple salt (≥42.0% $KHSO_5 \cdot 0.5KHSO_4 \cdot 0.5K_2SO_4$), 2,2,6,6-tetramethylpiperidine (≥98.0% TEMP), and humic acid (≥90.0% HA) were purchased from Shanghai Aladdin Biochemical Technology Co., Ltd. (Shanghai, China). Potassium iodide (≥99.5% KI), and L-histidine (≥99.0% L-His) were purchased from Shanghai Maclean's Biochemical Technology Co., Ltd. (Shanghai, China).

3.2. Preparation of MnO_2/CN@SiO_2

The MnO_2/CN@SiO_2 catalyst preparation consists of three steps, which are illustrated in Figure 12.

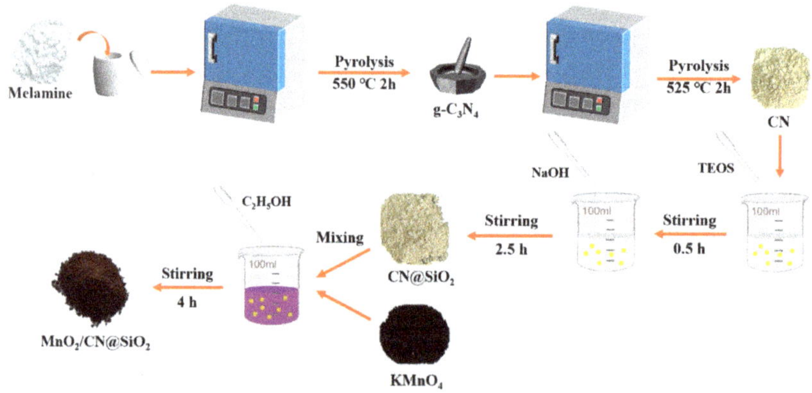

Figure 12. Synthesis route diagram of MnO_2/CN@SiO_2.

3.2.1. Synthesis of CN

A total of 5.0 g of melamine was added into a crucible with a lid and subsequently transferred to a muffle furnace for pyrolysis. The sample was heated to 550 °C at a rate of 5 °C/min and held at this temperature for 2 h. After natural cooling, yellow solid carbon nitride (g–C_3N_4) was successfully obtained. Thermally stripped carbon nitride (CN) was obtained by reheating g–C_3N_4 in a muffle furnace at 525 °C for 2 h.

3.2.2. Synthesis of CN@SiO_2

A total of 0.06 g of CN was added into a mixture of 22.5 mL of anhydrous ethanol and 7.5 mL of deionized water. Following ultrasonication for 10 min to disperse the CN, 0.045 mL of TEOS was added dropwise under stirring. After continuously stirring for

30 min, 0.2 mL of 6 wt% NaOH solution was added dropwise. Stirring was maintained for 2.5 h to ensure complete hydrolysis of TEOS, thereby obtaining CN@SiO_2. Finally, the product was centrifuged, washed three times with deionized water, and dispersed in 5 mL of deionized water.

3.2.3. Synthesis of MnO_2/CN@SiO_2

A total of 0.1125 g of $KMnO_4$ was dissolved in 30 mL of deionized water. Subsequently, the CN@SiO_2 was added to the $KMnO_4$ solution and stirred for 30 min. Then, 4.5 mL of anhydrous ethanol was added dropwise. After continuous stirring for 4 h, the resulting product was centrifuged, washed with deionized water and anhydrous ethanol, and dried at 70 °C to obtain the MnO_2/CN@SiO_2 catalyst.

Meanwhile, MnO_2/g–C_3N_4 and MnO_2/g–C_3N_4@SiO_2 composites were synthesized using g–C_3N_4 according to the above method, distinguishing between MnO_2/CN and MnO_2/CN@SiO_2 prepared with CN.

3.3. *Characterizations*

X-ray diffraction (XRD) was performed on a D/max-2500/PC model X-ray powder diffractometer (Rigaku, Akishima, Japan) to determine the crystal structure of the catalyst. X-ray photoelectron spectroscopy (XPS) analysis was performed on an ESCALAB XI+ photoelectron spectrometer (Thermo Fisher Scientific, Waltham, MA, USA), utilizing a Mg Kα source (hv = 1253.6 eV). The binding energy scale was calibrated by referencing the C 1s peak at 284.8 eV. The microscopic morphology of the samples was observed by scanning electron microscopy (SEM, TESCAN MIRA LMS, Brno, Czech Republic), with an energy-dispersive spectrometer (EDS, TESCAN MIRA LMS, Brno, Czech Republic) integrated for elemental analysis. Electron paramagnetic resonance (EPR) spectroscopy was performed using an EPR200-Plus spectrometer (CIQTEK, Hefei, China) to determine the generation of $^1O_2$ during the reaction. The test conditions were as follows: center field 3380 G, scanning range 60 G, microwave frequency 9.5 GHz, modulation frequency 100 kHz, and capture agent concentration 400 mM.

3.4. *Degradation Tests and Analytical Methods*

Degradation experiments were performed under dark conditions. A total of 10 mg of MnO_2/CN@SiO_2 catalyst was ultrasonically dispersed in 40 mL of 50 mg/L LEV solution. Subsequently, the solution was poured into a 50 mL quartz bottle with a lid, and the reaction temperature was controlled by a DC-0506 low-temperature constant temperature bath (Perfect Light, Beijing, China). After stirring for 10 min to reach adsorption equilibrium, 400 μL of 0.1 M PMS solution was added, resulting in a final 1 mM PMS concentration in the reaction mixture. The reaction solution was filtered several times through a 0.45 μm PTFE filter membrane. Then, samples were taken every 10 min, and the absorbance of each sample was measured at a wavelength of 287 nm.

The concentration of LEV was determined using a UV spectrophotometer (UV-8000, Metash, Shanghai, China) with a maximum absorption wavelength of 287 nm. The concentration of PMS in the solution was measured using the iodometric method. Specifically, 0.02 g of $NaHCO_3$ was dissolved in 2.8 mL of KI solution, followed by the addition of 0.2 mL of PMS. The absorbance of PMS was measured at 319 nm.

LEV degradation efficiency was calculated according to Equation (10). A proposed first-order kinetic model was used to simulate the LEV degradation reaction kinetics (Equation (11)). k_{app} is the apparent reaction constant, and C_0 and C_t represent the initial mass concentration of LEV and the mass concentration at a reaction time of t, respectively.

$$D = [1 - (C_t/C_0) \times 100\%] \quad (10)$$

$$k_{app} = -\ln(C_t/C_0)/t \quad (11)$$

4. Conclusions

In summary, $MnO_2/CN@SiO_2$ catalysts were successfully synthesized, and their catalytic performance for activating PMS to degrade LEV was systematically investigated. The catalyst, prepared with the mass ratio of 5:4:1 for MnO_2:CN:SiO_2, exhibited the highest LEV removal of 80.8%, which was higher than those of MnO_2 (67.4%), MnO_2/CN (68.2%), and MnO_2/SiO_2 (76.1%). The incorporation of SiO_2 significantly enhanced the catalytic activity, attributed to the rapid adsorption of PMS onto the catalyst surface via Si–OH groups, facilitating its reaction with MnO_2. The experimental results revealed that the $MnO_2/CN@SiO_2$ catalysts exhibited excellent tolerance to low concentrations of NO_3^-, SO_4^{2-}, Cl^-, and HA, and high stability in cycling experiments. Based on the free radical quenching experiment and EPR spectroscopy analysis, a non-radical-based pathway for LEV degradation was proposed. The primary degradation mechanism involved electron transfer from LEV to the Mn^{3+}–(O)OSO_3^- complex, which oxidizes LEV with high activity. The $^1O_2$-mediated reaction also contributes to LEV degradation. In conclusion, the $MnO_2/CN@SiO_2$ catalyst demonstrates superior performance in activating PMS and holds significant potential for application in treating antibiotic-contaminated wastewater.

Author Contributions: Methodology, software, formal analysis, and writing—original draft preparation, L.X.; conceptualization, supervision, resources, funding acquisition, and writing—review and editing, D.X.; validation and data curation, X.W.; data curation, J.L. All authors have read and agreed to the published version of the manuscript.

Funding: National Innovative Entrepreneurship Training Program for Undergraduates (202310426050, 202410426083).

Data Availability Statement: The data that support the findings of this study are available from the corresponding author upon reasonable request.

Acknowledgments: The authors acknowledge the Qingdao University of Science and Technology for providing resources and experimental facilities.

Conflicts of Interest: The authors declare no competing financial interests.

References

1. Sharma, K.; Thakur, I.S.; Kaushik, G. Occurrence and distribution of pharmaceutical compounds and their environmental impacts: A review. *Bioresour. Technol. Rep.* **2021**, *16*, 100841. [CrossRef]
2. Gao, F.Z.; Zou, H.Y.; Wu, D.L.; Chen, S.; He, L.Y.; Zhang, M.; Bai, H.; Ying, G.G. Swine farming elevated the proliferation of Acinetobacter with the prevalence of antibiotic resistance genes in the groundwater. *Environ. Int.* **2020**, *136*, 105484. [CrossRef] [PubMed]
3. Lyu, J.C.; Ge, M.; Hu, Z.; Guo, C.S. One-pot synthesis of magnetic $CuO/Fe_2O_3/CuFe_2O_4$ nanocomposite to activate persulfate for levofloxacin removal: Investigation of efficiency, mechanism and degradation route. *Chem. Eng. J.* **2020**, *389*, 124456. [CrossRef]
4. Jia, D.; Liu, Q.; Pan, Y.; Xu, S.D.; Li, H.X.; Tang, J.H. The Research Status, Potential Hazards and Toxicological Mechanisms of Fluoroquinolone Antibiotics in the Environment. *Antibiotics* **2023**, *12*, 1058. [CrossRef] [PubMed]
5. Sharma, V.K.; Johnson, N.; Cizmas, L.; McDonald, T.J.; Kim, H. A review of the influence of treatment strategies on antibiotic resistant bacteria and antibiotic resistance genes. *Chemosphere* **2016**, *150*, 702–714. [CrossRef]
6. Saya, L.; Malik, V.; Gautam, D.; Gambhir, G.; Balendra, W.; Singh, R.; Hooda, S. A comprehensive review on recent advances toward sequestration of levofloxacin antibiotic from wastewater. *Sci. Total Environ.* **2022**, *813*, 152529. [CrossRef]
7. Liu, J.Y.; Li, Z.L.; Wang, M.; Jin, C.Y.; Kang, J.; Tang, Y.W.; Li, S.Y. Eu_2O_3/Co_3O_4 nanosheets for levofloxacin removal via peroxymonosulfate activation: Performance, mechanism and degradation pathway. *Sep. Purif. Technol.* **2021**, *274*, 118666. [CrossRef]
8. Yang, W.Q.; Li, J.; Yao, Z.L.; Li, M. A review on the alternatives to antibiotics and the treatment of antibiotic pollution: Current development and future prospects. *Sci. Total Environ.* **2024**, *926*, 171757. [CrossRef]

9. Wang, N.N.; Zheng, T.; Zhang, G.Z.; Wang, P. A review on Fenton-like processes for organic wastewater treatment. *J. Environ. Chem. Eng.* **2016**, *4*, 762–787. [CrossRef]
10. Li, X.Y.; Liu, H.W.; Zhang, Y.S.; Mahlknecht, J.; Wang, C.Q. A review of metallurgical slags as catalysts in advanced oxidation processes for removal of refractory organic pollutants in wastewater. *J. Environ. Manag.* **2024**, *352*, 120051. [CrossRef]
11. Zhang, Q.; Zheng, D.; Bai, B.; Ma, Z.Y.; Zong, S.C. Insight into antibiotic removal by advanced oxidation processes (AOPs): Performance, mechanism, degradation pathways, and ecotoxicity assessment. *Chem. Eng. J.* **2024**, *500*, 157134. [CrossRef]
12. Li, S.; Yang, Y.L.; Zheng, H.S.; Zheng, Y.J.; Jing, T.; Ma, J.; Nan, J.; Leong, Y.K.; Chang, J.S. Advanced oxidation process based on hydroxyl and sulfate radicals to degrade refractory organic pollutants in landfill leachate. *Chemosphere* **2022**, *297*, 134214. [CrossRef]
13. Usman, M.; Cheema, S.A.; Farooq, M. Heterogeneous Fenton and persulfate oxidation for treatment of landfill leachate: A review supplement. *J. Clean. Prod.* **2020**, *256*, 120448. [CrossRef]
14. Chang, L.; Xue, X.J.; Deng, Q.C.; Xie, X.Y.; Zhang, X.D.; Cheng, C.; Chai, H.X.; Huang, Y. Modulating the electronic structure of Co center via MgO@C co-doping for PMS activation to remove levofloxacin. *Sep. Purif. Technol.* **2023**, *321*, 124151. [CrossRef]
15. Bai, Y.B.; Han, C.; Li, Z.M.; Zhang, H.M.; Deng, Q.Y.; Tong, H.N.; Zhang, C.L. Efficient degradation of levofloxacin using peroxymonosulfate activated by a novel magnetic catalyst derived from waste walnut shell biomass. *Colloids Surf. A Physicochem. Eng. Asp.* **2024**, *699*, 134668. [CrossRef]
16. Li, J.J.; Liang, Y.Q.; Jin, P.L.; Zhao, B.; Zhang, Z.H.; He, X.J.; Tan, Z.L.; Wang, L.; Cheng, X.W. Heterogeneous Metal-Activated Persulfate and Electrochemically Activated Persulfate: A Review. *Catalysts* **2022**, *12*, 1024. [CrossRef]
17. Deng, Q.C.; Zhang, X.D.; Chang, L.; Chai, H.X.; Huang, Y.M. The MOF/LDH derived heterostructured $Co_3O_4/MnCo_2O_4$ composite for enhanced degradation of levofloxacin by peroxymonosulfate activation. *Sep. Purif. Technol.* **2022**, *294*, 121182. [CrossRef]
18. Zheng, J.L.; Lin, Q.T.; Liu, Y.X.; Fan, X.D.; Xu, K.H.; Ma, Y.J.; He, J.; Fu, H.Y. Peroxymonosulfate activation by Mg-introduced Fe-N carbon nanotubes to accelerate sulfamethoxazole degradation: Singlet oxygen-dominated nonradical pathway. *Chem. Eng. J.* **2023**, *452*, 139233. [CrossRef]
19. Zheng, X.X.; Niu, X.J.; Zhang, D.Q.; Lv, M.Y.; Ye, X.Y.; Ma, J.L.; Lin, Z.; Fu, M.L. Metal-based catalysts for persulfate and peroxymonosulfate activation in heterogeneous ways: A review. *Chem. Eng. J.* **2022**, *429*, 132323. [CrossRef]
20. Huang, J.Z.; Dai, Y.F.; Singewald, K.; Liu, C.C.; Saxena, S.; Zhang, H.C. Effects of MnO_2 of different structures on activation of peroxymonosulfate for bisphenol A degradation under acidic conditions. *Chem. Eng. J.* **2019**, *370*, 906–915. [CrossRef]
21. Hatakeyama, T.; Okamoto, N.L.; Ichitsubo, T. Thermal stability of MnO_2 polymorphs. *J. Solid State Chem.* **2022**, *305*, 122683. [CrossRef]
22. Boyom-Tatchemo, F.W.; Devred, F.; Ndiffo-Yemeli, G.; Laminsi, S.; Gaigneaux, E.M. Plasma-induced redox reactions synthesis of nanosized α-, γ- and δ-MnO_2 catalysts for dye degradation. *Appl. Catal. B Environ.* **2020**, *260*, 118159. [CrossRef]
23. Zhou, Z.G.; Du, H.M.; Dai, Z.H.; Mu, Y.; Tong, L.L.; Xing, Q.J.; Liu, S.S.; Ao, Z.M.; Zou, J.P. Degradation of organic pollutants by peroxymonosulfate activated by MnO_2 with different crystalline structures: Catalytic performances and mechanisms. *Chem. Eng. J.* **2019**, *374*, 170–180. [CrossRef]
24. Yang, X.; Sun, X.S.; Yu, J.; Huang, Z.; Yu, J.; Deng, S.W.; Jiang, Y.Y.; Zhu, W.W. Insights into the potential applications of permanganate/peroxymonosulfate systems: Enhancement via amorphous MnO_2, effects of water matrices, and optimization using response surface methodology. *RSC Adv.* **2024**, *14*, 4116–4128. [CrossRef]
25. Edathil, A.A.; Kannan, P.; Haija, M.A.; Banat, F. Sulfide remediation from wastewater using hydrothermally synthesized δ-MnO_2/porous graphitic carbon as adsorbent. *Environ. Res.* **2021**, *196*, 110429. [CrossRef] [PubMed]
26. Li, M.; Zhang, H.; Liu, Z.L.; Su, Y.G.; Du, C.F. Oxygen vacancies and interfacial engineering of MnO/Mn_3O_4 heterojunction for peroxymonosulfate activation to promote Bisphenol A removal. *Appl. Surf. Sci.* **2023**, *628*, 157302. [CrossRef]
27. Wang, L.H.; Jiang, J.; Pang, S.Y.; Zhou, Y.; Li, J.; Sun, S.F.; Gao, Y.; Jiang, C.C. Oxidation of bisphenol A by nonradical activation of peroxymonosulfate in the presence of amorphous manganese dioxide. *Chem. Eng. J.* **2018**, *352*, 1004–1013. [CrossRef]
28. Oseghe, E.O.; Akpotu, S.O.; Mombeshora, E.T.; Oladipo, A.O.; Ombaka, L.M.; Maria, B.B.; Idris, A.O.; Mamba, G.; Ndlwana, L.; Ayanda, O.S.; et al. Multi-dimensional applications of graphitic carbon nitride nanomaterials–A review. *J. Mol. Liq.* **2021**, *344*, 117820. [CrossRef]
29. Ganesan, S.; Kokulnathan, T.; Sumathi, S.; Palaniappan, A. Efficient photocatalytic degradation of textile dye pollutants using thermally exfoliated graphitic carbon nitride (TE–g–C_3N_4). *Sci. Rep.* **2024**, *14*, 2284. [CrossRef]
30. Chen, H.M.; Fan, Y.Y.; Xu, H.Y.; Cui, D.F.; Xue, C.Y.; Zhang, W.D. A facile and green microwave hydrothermal method for fabricating g-C_3N_4 nanosheets with improved hydrogen evolution performance. *J. Alloys Compd.* **2021**, *863*, 158448. [CrossRef]
31. Khan, A.; Zhang, K.K.; Taraqqi-A-Kamal, A.; Wang, X.G.; Chen, Y.; Zhang, Y.R. Degradation of antibiotics in aqueous media using manganese nanocatalyst-activated peroxymonosulfate. *J. Colloid Interface Sci.* **2021**, *599*, 805–818. [CrossRef]
32. Sodipo, B.K.; Aziz, A.A. Recent advances in synthesis and surface modification of superparamagnetic iron oxide nanoparticles with silica. *J. Magn. Magn. Mater.* **2016**, *416*, 275–291. [CrossRef]

33. Ma, M.Y.; Li, H.W.; Xiong, Y.Z.; Dong, F.P. Rational design, synthesis, and application of silica/graphene-based nanocomposite: A review. *Mater. Des.* **2021**, *198*, 109367. [CrossRef]
34. Du, R.; Zhu, H.Y.; Zhao, H.Y.; Lu, H.; Dong, C.; Liu, M.T.; Yang, F.; Yang, J.; Pan, J. Coupling ultrafine plasmonic Co_3O_4 with thin-layer carbon over SiO_2 nanosphere for dual-functional PMS activation and solar interfacial water evaporation. *J. Alloys Compd.* **2023**, *940*, 168816. [CrossRef]
35. Yuan, D.S.; Li, Z.Y.; Chen, X.R.; Ding, J.; Wan, H.; Guan, G.F. Homodispersed B–CN/P–CN S-scheme homojunction for enhanced visible-light-driven hydrogen evolution. *Green Energy Environ.* **2022**, *7*, 1119–1127. [CrossRef]
36. Yang, Y.Y.; Zhang, P.P.; Hu, K.S.; Duan, X.G.; Ren, Y.X.; Sun, H.Q.; Wang, S.B. Sustainable redox processes induced by peroxymonosulfate and metal doping on amorphous manganese dioxide for nonradical degradation of water contaminants. *Appl. Catal. B Environ.* **2021**, *286*, 119903. [CrossRef]
37. Arkhipova, E.A.; Ivanov, A.S.; Maslakov, K.I.; Novotortsev, R.Y.; Savilov, S.V.; Xia, H.; Desyatov, A.V.; Aldoshin, S.M. Facile hydrothermal synthesis of α-MnO_2 and δ-MnO_2 for pseudocapacitor applications. *Ionics* **2022**, *28*, 3501–3509. [CrossRef]
38. Yang, W.; Su, Z.A.; Xu, Z.; Yang, W.; Peng, Y.; Li, J. Comparative study of α-, β-, γ- and δ-MnO_2 on toluene oxidation: Oxygen vacancies and reaction intermediates. *Appl. Catal. B Environ.* **2020**, *260*, 118150. [CrossRef]
39. Nguyen, M.D.; Nguyen, T.B.; Tran, L.H.; Nguyen, T.G.; Fatimah, I.; Kuncoro, E.P.; Doong, R.A. Z-scheme S, B co-doped g-C_3N_4 nanotube@MnO_2 heterojunction with visible-light-responsive for enhanced photodegradation of diclofenac by peroxymonosulfate activation. *Chem. Eng. J.* **2023**, *452*, 139249. [CrossRef]
40. Jing, L.Q.; Xie, M.; Xu, Y.G.; Tong, C.; Du, X.; Zhao, H.; Zhong, N.; Li, H.M.; Hu, J.G. Advanced oxidation via the synergy of C-defective/C–O band modified ultrathin porous g-C_3N_4 and PMS for efficient photothermal degradation of bisphenol pollutants and lignin derivatives. *Green Energy Environ.* **2024**, *9*, 1159–1170. [CrossRef]
41. Li, W.C.; Wen, X.Y.; Wang, X.J.; Li, J.; Ren, E.B.; Shi, Z.F.; Liu, C.M.; Mo, D.Q.; Mo, S.P. Oriented growth of δ-MnO_2 nanosheets over core-shell Mn_2O_3@δ-MnO_2 catalysts: An interface-engineered effects for enhanced low-temperature methanol oxidation. *Mol. Catal.* **2021**, *514*, 11847. [CrossRef]
42. Liu, M.T.; Ma, S.L.; Zhao, H.Y.; Lu, H.; Yang, J.; Tang, S.; Gao, S.Y.; Yang, F. Compositing ultrafine $CoFe_2O_4$ spinel with porous silica as catalyst for photothermal PMS activation and interfacial water evaporation. *J. Alloys Compd.* **2023**, *949*, 169901. [CrossRef]
43. Duan, X.G.; Indrawirawan, S.; Kang, J.; Tian, W.J.; Zhang, H.Y.; Sun, H.Q.; Wang, S.B. Temperature-dependent evolution of hydroxyl radicals from peroxymonosulfate activation over nitrogen-modified carbon nanotubes. *Sustain. Mater. Technol.* **2018**, *18*, e00082. [CrossRef]
44. Fan, J.H.; Yao, Z.L.; Wu, D.L.; Liu, X.; Ma, L.M. Pilot-scale study on an advanced Fe-Cu process for refractory wastewater pretreatment. *J. Hazard. Mater.* **2023**, *457*, 131756. [CrossRef] [PubMed]
45. He, P.; Gu, C.Y.; Tang, B.L.; Zhou, Y.H.; Gan, M.; Zhu, J.Y. Expeditious degradation of SMX by high-valent cobalt-oxo species derived from cobalt-doped C_3N_5-activated peroxymonosulfate with the assistance of visible light. *Sep. Purif. Technol.* **2022**, *301*, 122009. [CrossRef]
46. Qiao, L.L.; Zhang, X.Y.; Li, T.L.; Wang, H.T. Interfacial photothermal-enhanced FeCo@BC nanocomposites activating peroxymonosulfate for efficient tetracycline degradation. *Sep. Purif. Technol.* **2025**, *354*, 129391. [CrossRef]
47. Zhou, J.; Xiao, J.H.; Xiao, D.X.; Guo, Y.G.; Fang, C.L.; Lou, X.Y.; Wang, Z.H.; Liu, J.S. Transformations of chloro and nitro groups during the peroxymonosulfate-based oxidation of 4-chloro-2-nitrophenol. *Chemosphere* **2015**, *134*, 446–451. [CrossRef]
48. Cai, S.; Zhang, Q.; Wang, Z.Q.; Hua, S.; Ding, D.H.; Cai, T.M.; Zhang, R.H. Pyrrolic N-rich biochar without exogenous nitrogen doping as a functional material for bisphenol A removal: Performance and mechanism. *Appl. Catal. B Environ.* **2021**, *291*, 120093. [CrossRef]
49. Wang, Y.; Qiao, L.L.; Zhang, X.Y.; Liu, Z.L.; Li, T.L.; Wang, H.T. Green synthesis of FeCu@biochar nanocomposites through a mechanochemical method for enhanced tetracycline degradation via peroxymonosulfate activation. *Sep. Purif. Technol.* **2024**, *328*, 125077. [CrossRef]
50. Dong, J.R.; Li, P.; Ji, X.Y.; Kang, Y.; Yuan, X.; Tang, J.C.; Shen, B.X.; Dong, H.J.; Lyu, H.H. Electrons of d-orbital (Mn) and p-orbital (N) enhance the photocatalytic degradation of antibiotics by biochar while maintaining biocompatibility: A combined chemical and biological analysis. *J. Hazard. Mater.* **2023**, *451*, 131083. [CrossRef]
51. Coha, M.; Farinelli, G.; Tiraferri, A.; Minella, M.; Vione, D. Advanced oxidation processes in the removal of organic substances from produced water: Potential, configurations, and research needs. *Chem. Eng. J.* **2021**, *414*, 128668. [CrossRef]
52. Wang, L.L.; Lan, X.; Peng, W.Y.; Wang, Z.H. Uncertainty and misinterpretation over identification, quantification and transformation of reactive species generated in catalytic oxidation processes: A review. *J. Hazard. Mater.* **2021**, *408*, 124436. [CrossRef] [PubMed]
53. Wang, J.K.; Xie, T.P.; Han, G.K.; Zhu, Q.X.; Wang, Y.J.; Peng, Y.; Liu, S.L.; Yao, Z.P. SiO_2 mediated templating synthesis of γ-Fe_2O_3/MnO_2 as peroxymonosulfate activator for enhanced phenol degradation dominated by singlet oxygen. *Appl. Surf. Sci.* **2021**, *560*, 149984. [CrossRef]

54. Wang, Y.C.; Tong, Y.P.; Chen, D.Z.; Zhou, T.L.; Zhang, Q.Z.; Zou, J.P. Activation of peroxymonosulfate by g-C_3N_4/ε-MnO_2 microspheres for nonradical pathway degradation of organic pollutants in water: Catalytic mechanism and degradation path. *Chem. Eng. J.* **2023**, *459*, 141643. [CrossRef]
55. Zhao, Z.W.; Zhao, J.H.; Yang, C. Efficient removal of ciprofloxacin by peroxymonosulfate/Mn_3O_4-MnO_2 catalytic oxidation system. *Chem. Eng. J.* **2017**, *327*, 481–489. [CrossRef]
56. Yuan, R.X.; Jiang, Z.Q.; Wang, Z.H.; Gao, S.M.; Liu, Z.J.; Li, M.L.; Boczkaj, G. Hierarchical MnO_2 nanoflowers blooming on 3D nickel foam: A novel micro-macro catalyst for peroxymonosulfate activation. *J. Colloid Interface Sci.* **2020**, *571*, 142–154. [CrossRef]
57. Wang, Z.M.; Wang, Z.H.; Li, W.; Lan, Y.P.; Chen, C. Performance comparison and mechanism investigation of Co_3O_4-modified different crystallographic MnO_2 (α, β, γ, and δ) as an activator of peroxymonosulfate (PMS) for sulfisoxazole degradation. *Chem. Eng. J.* **2022**, *427*, 130888. [CrossRef]

catalysts

MDPI

Article

Natural Pyrolusite-Catalyzed Ozonation for Nanoplastics Degradation

Victor Mello [1,2,*], Julia Nieto-Sandoval [2], Márcia Dezotti [1] and Carmen Sans [2,*]

1 Chemical Engineering Program/COPPE, Federal University of Rio de Janeiro, Av. Horácio Macedo 2030, Bloco G, Cidade Universitária, Rio de Janeiro 21941-914, Brazil; mdezotti@peq.coppe.ufrj.br
2 Department of Chemical Engineering and Analytical Chemistry, Faculty of Chemistry, Universitat de Barcelona, C/Martí i Franqués 1, 08028 Barcelona, Spain; julia.nietosandoval@ub.edu
* Correspondence: victormello@peq.coppe.ufrj.br (V.M.); carmesans@ub.edu (C.S.); Tel.: +55-219-7362-3876 (V.M.)

Abstract: The increasing prevalence of polystyrene nanoplastics (PSNPs) in aquatic environments poses significant risks due to their persistence and potential toxicity. Conventional water treatment methods have proven ineffective in removing these emerging pollutants, highlighting the urgent need for sustainable and efficient treatment. This study investigates the application of catalytic ozonation using natural pyrolusite (n-MnO_2) and oxalic acid (OA) as a co-catalyst for the environmentally friendly degradation of PSNPs. Key operational parameters, including pH, applied ozone dose, pyrolusite dosage, and OA concentration, were systematically evaluated. Results demonstrate that the MnO_2 + OA + O_3 system enhances the generation of reactive oxygen species (ROS), leading to improved PSNP removal while maintaining the applied ozone dose compared to the single ozonation reaction. The highest TOC removal of 75% was achieved within 30 min of treatment under optimal conditions (pH = 4, [MnO_2] = 0.5 g L^{-1}, [OA] = 10 mg L^{-1}, and ozone dose of 37.5 mg min^{-1}), with significant turbidity reduction, indicating both chemical and physical degradation of PSNPs. Catalyst reusability after three consecutive cycles confirmed minimal loss in activity, reinforcing its potential as a sustainable catalytic system. These findings highlight natural MnO_2-driven catalytic ozonation as a green and effective strategy for nanoplastic removal in water treatment applications.

Keywords: catalytic ozonation; polystyrene nanoplastics; natural manganese dioxide; water treatment; advanced oxidation processes; green catalysis

1. Introduction

The widespread production and use of plastic materials have led to the increasing presence of microplastics (MPs) and nanoplastics (NPs) in aquatic environments. NPs, typically defined as plastic particles smaller than 1 μm, are particularly concerning due to their high surface-area-to-volume ratio, enhanced chemical reactivity, and ability to cross biological barriers, thereby posing significant environmental and health risks [1]. Unlike MPs, which are often removed to a large extent by conventional water treatment processes, NPs are more difficult to remove and have been detected in drinking water, freshwater bodies, and marine ecosystems [2]. Due to their small size, NPs can penetrate cellular membranes, cause oxidative stress, and disrupt metabolic processes in aquatic organisms [3]. Recognizing these risks, regulatory bodies like the European Union (EU) have implemented guidelines to limit and monitor plastic particles in water systems.

For instance, the EU Drinking Water Directive (Directive (EU) 2020/2184) mandates the monitoring of MPs in drinking water, and similar measures are being explored for NPs [4].

Among the various types of NPs, polystyrene nanoplastics (PSNPs) have garnered significant attention due to their extensive use in packaging, insulation, and disposable consumer products. Polystyrene (PS) is known for its chemical stability, hydrophobicity, and resistance to biodegradation, making PSNPs particularly persistent in aquatic environments [5]. Once released into aquatic systems, PSNPs undergo weathering and fragmentation, further increasing their abundance and bioavailability in water sources [6]. Their persistence and toxicity underscore the urgent need for efficient removal strategies. Conventional water treatment processes, such as filtration, coagulation, and sedimentation, are not sufficient to eliminate NPs completely due to their small size and colloidal nature [7]. As a result, advanced oxidation processes (AOPs) have emerged as promising alternatives for NPs removal.

Ozonation is one of the most widely studied AOPs, as ozone (O_3) is a strong oxidizing agent capable of decomposing a broad range of organic pollutants in water. However, the direct ozonation of PSNPs is limited due to the selective reactivity of molecular O_3. Ozone preferentially reacts with electron-rich sites, such as double bonds or amine groups, but it is less effective at breaking down stable aromatic structures like the benzene rings found in PS [8]. This limitation highlights the need for an approach that generates more reactive species capable of non-selective oxidation.

To overcome the limitations of direct ozonation, catalytic ozonation has been proposed as an enhanced AOP. Catalytic ozonation accelerates the decomposition of O_3 into reactive oxygen species (ROS) such as hydroxyl radicals ($^{\bullet}OH$) and superoxide radicals ($O_2^{\bullet-}$), which exhibit higher oxidation potential and non-selective reactivity [9,10]. These ROS are capable of breaking stable chemical bonds, including aromatic structures, thereby facilitating the degradation of PSNPs. Catalysts such as transition metal oxides have been widely investigated due to their ability to activate O_3. Among these, manganese dioxide (MnO_2) has gained attention for its low cost, availability, and environmental compatibility [11,12]. However, studies on the catalytic ozonation of NPs remain limited. To the best of our knowledge, only the work by Li et al. (2023) has addressed this issue in the literature by heterogenous catalytic ozonation, showing promising results on PSNPs molecular weight decreasing by ozonation in the presence of a CeOx@MnOx catalyst [13]. Nevertheless, no data about further mineralization were reported.

Natural pyrolusite, a naturally occurring mineral form of MnO_2, is particularly attractive as it provides a cost-effective and abundant source of MnO_2, thereby supporting the scalability of catalytic ozonation for water treatment [14,15]. The efficiency of catalytic ozonation can be further enhanced by introducing oxalic acid (OA) as a co-catalyst. OA forms stable Mn^{3+}–oxalate complexes that prevent the deactivation of the catalyst and maintain the continuous production of ROS. OA also acts as an additional radical promoter, promoting the formation of highly reactive $^{\bullet}OH$ and other oxygen species. This synergy between MnO_2, OA, and O_3 has improved the degradation of organic micropollutants in previous studies [16,17]. The mechanism underlying this enhancement is linked to the regeneration of redox sites on MnO_2, as the Mn^{3+}–oxalate complex facilitates the cycling of Mn from Mn^{3+} to Mn^{4+}, thereby sustaining catalytic activity.

Thus, this study aims to investigate the efficiency of heterogeneous catalytic ozonation for PSNPs removal using n-MnO_2 as a catalyst. In order to enhance PSNPs degradation, OA was incorporated into the system as a co-catalyst. A complete operating conditions study has been carried out to evaluate the effect of pH, catalyst dose, O_3 dose, and the presence and concentration of OA on PSNPs removal. The study also assesses catalyst reusability over consecutive cycles, offering insights into the long-term stability and practicality of catalytic ozonation for NPs removal in water treatment processes.

2. Results and Discussion

2.1. Pyrolusite Characterization

The characterization of n-MnO_2 was conducted using three analytical techniques: point of zero charge (pH_{PZC}), Fourier-transform infrared spectroscopy (FTIR), and X-ray photoelectron spectroscopy (XPS). Figure 1 shows the results obtained for each technique.

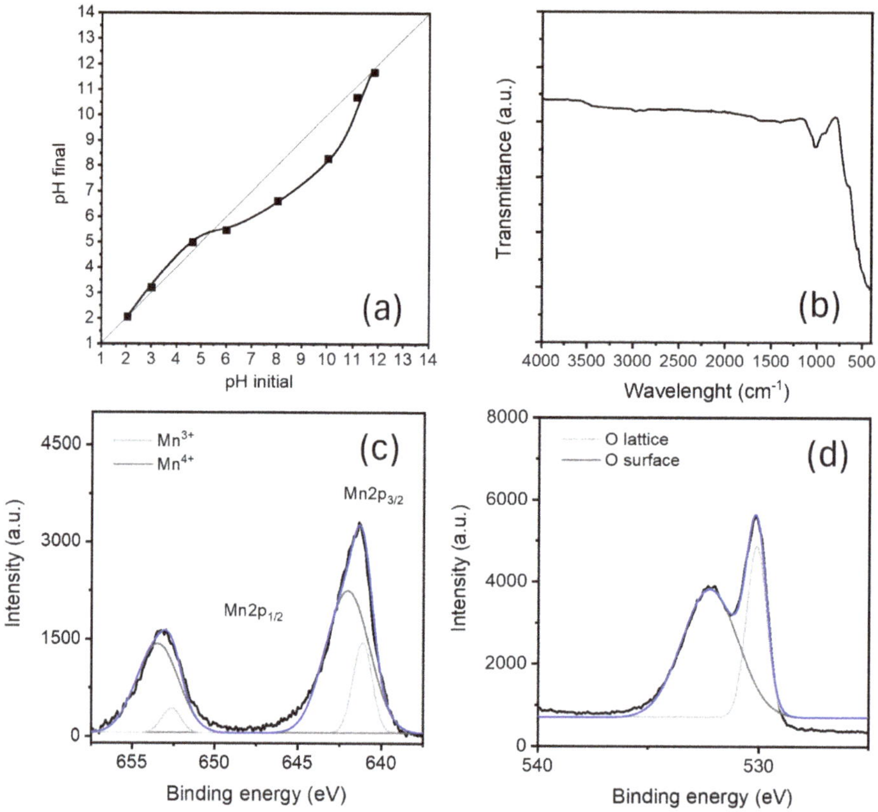

Figure 1. Pyrolusite characterization: pH_{PZC} (**a**), FTIR spectrum (**b**), Mn2p XPS spectra (**c**), and O1s XPS spectra of n-MnO_2 (**d**). In (**a**), black squares represent experimental pH values and the solid line corresponds to the linear fit used to determine the point of zero charge. In (**c**) and (**d**), the blue line is the experimental XPS spectrum, while the gray curves represent the deconvoluted component peaks.

The pH_{PZC} of pyrolusite was determined to be 5.5 (Figure 1a), indicating that below this pH, the surface of MnO_2 is positively charged, while above it, the surface becomes negatively charged. This charge variation significantly affects the adsorption of ionic species and, consequently, the formation of surface complexes. At pH values below the pH_{PZC}, the positively charged MnO_2 surface enhances the adsorption of negatively charged species, such as oxalate ions derived from the dissociation of OA. This electrostatic attraction facilitates the formation of surface complexes between MnO_2 and oxalate ions, which is crucial for catalytic processes [18,19]. The study by Omorogie et al. (2022) has demonstrated that the adsorption of anionic dyes onto MnO_2-modified biomass is more effective at pH levels below the pH_{PZC}, attributed to the increased positive surface charge promoting anion adsorption [20]. Conversely, at pH values above the pH_{PZC}, the MnO_2 surface acquires a negative charge, which can lead to repulsion of anionic species like oxalate ions, thereby reducing the efficiency of complex formation.

The FTIR spectrum (Figure 1b) reveals two characteristic peaks confirming MnO_2 presence in the pyrolusite sample. The peak at 1100 cm^{-1} is attributed to the coordination of manganese by hydroxyl groups (O-H), which are known to participate in surface-mediated catalytic reactions [21,22]. The second prominent peak, observed at approximately 500 cm^{-1}, corresponds to the stretching vibrations of Mn–O and Mn–O–Mn bonds, indicating the presence of octahedral MnO_6 structures characteristic of MnO_2 [23,24]. Similar FTIR signatures have been reported in other studies involving natural and synthetic manganese oxides used as ozonation catalysts [19,25].

The XPS analysis of the Mn2p region of n-MnO_2 (Figure 1c) revealed two primary peaks corresponding to $Mn2p_{3/2}$ and $Mn2p_{1/2}$, with a binding energy difference (ΔE) of 11.5 eV. This energy separation is characteristic of MnO_2 in the Mn^{4+} oxidation state [26,27]. The deconvolution of the $Mn2p_{3/2}$ peak showed contributions from two distinct components, indicating the presence of both Mn^{3+} and Mn^{4+} species. The calculated Mn^{3+}/Mn^{4+} ratio was 0.8, confirming that Mn^{4+} is the predominant oxidation state in the n-MnO_2 sample. The prevalence of Mn^{4+} is consistent with the mineralogical composition of pyrolusite and its catalytic activity in O_3 decomposition [28,29]. Mn^{4+} sites on the surface of pyrolusite facilitate the activation of ozone molecules, generating ROS such as $^{\bullet}OH$ and $O_2{\bullet}^-$, which contribute to the degradation of organic pollutants [30,31]. Moreover, the presence of Mn^{3+} indicates a redox-active surface, capable of cycling between Mn^{4+} and Mn^{3+} states during catalytic ozonation. This redox behavior enhances O_3 decomposition and maintains the catalytic activity of pyrolusite over multiple reaction cycles [19,31]. The presence of both oxidation states is essential for sustaining ROS generation and achieving efficient organic pollutants degradation. The O1s XPS spectrum of the n-MnO_2 catalyst, presented in Figure 1d, provides critical insights into the surface oxygen species involved in the catalytic ozonation process. The deconvolution of the spectrum reveals the presence of distinct peaks corresponding to different oxygen environments, which directly influence the catalyst's reactivity and its ability to generate ROS during ozonation. The O1s spectrum exhibits two primary peaks, which can be attributed to distinct oxygen species associated with the MnO_2 surface. The first peak (~529.5 eV, lower binding energy) corresponds to lattice oxygen (O_2^-) bound within the MnO_2 crystal structure [32]. Lattice oxygen is characteristic of metal oxides and is essential for the redox cycling of manganese species (Mn^{3+}/Mn^{4+}) that facilitates O_3 decomposition and ROS generation [32,33]. The dominance of this peak confirms the crystalline nature of the MnO_2 phase in pyrolusite, which serves as the active site for catalytic reactions. The second main peak (~531.5 eV, higher binding energy) is attributed to surface-adsorbed oxygen species, including hydroxyl groups (–OH) and chemisorbed oxygen (O^-) [34,35]. These species play a critical role in O_3 activation by providing sites for ozone adsorption and subsequent radical formation. Hydroxyl groups, in particular, act as proton donors and facilitate the formation of $^{\bullet}OH$, which are highly reactive toward organic pollutants like PSNPs [32,36]. The distribution of oxygen species observed in the O1s spectrum aligns with the previously determined point of zero charge (pH_{PZC} = 5.5). In acidic pH conditions (e.g., pH 4), the surface hydroxyl groups (O–H) are more likely to be protonated, promoting the adsorption of negatively charged ozone species (O_3^-) and oxalate ions (OA) [37]. The presence of chemisorbed oxygen species at ~531.5 eV supports this hypothesis, indicating a surface environment conducive to ROS production and organic pollutants degradation.

The surface oxygen groups detected via XPS are consistent with the FTIR results, which revealed a characteristic peak at 1100 cm^{-1}, associated with hydroxyl group coordination. This corroborates the presence of reactive –OH groups on the pyrolusite surface, essential for catalytic ozonation mechanisms [25].

2.2. Catalytic Ozonation of Polystyrene Nanoplastics

Previous studies have demonstrated that single ozonation effectively reduces the molecular weight of PSNPs, as evidenced mainly by gel permeation chromatography (GPC) analyses [38,39]. For instance, Li et al. (2022) observed a 99.9% reduction in molecular weight and a 42.7% mineralization of nano-sized PS after 240 min of ozonation under typical disinfection conditions [38]. Similarly, research by Li et al. (2023) reported that ozonation achieved a 32% decrease in PSNP molecular weight within the first 10 min, which increased to 51.67% in the presence of a MnOx catalyst, and further to 73.33% when employing a CeOx@MnOx catalyst [13]. However, complete removal of PSNPs is not guaranteed by ozonation alone. In fact, as demonstrated by Nieto-Sandoval (2024), single ozonation can lead to the formation of smaller NP particles, potentially increasing their environmental persistence and toxicity [39]. Therefore, to comprehensively assess NP removal, this study focuses on total organic carbon (TOC) reduction, providing deeper insights into the mineralization and effective elimination of PSNPs by heterogeneous catalytic ozonation.

In this study, the removal of PSNPs was evaluated under different systems, including single ozonation (O_3), catalytic ozonation (MnO_2 + O_3), ozonation with oxalic acid (OA + O_3), and catalytic ozonation with oxalic acid (MnO_2 + OA + O_3). Additionally, the individual effects of the catalysts, MnO_2 and OA, are presented to isolate their individual contributions to the removal process. Figure 2a shows TOC, and Figure 2b shows turbidity removal efficiencies for PSNPs under these systems. The initial TOC of the reactions containing 20 mg L^{-1} of PSNPs was 16.2 ± 1.7 mg L^{-1}, and the initial turbidity was 80.7 ± 3 NTU. The comparison provides insight into the relative feasibility of each treatment, highlighting the synergistic effects achieved when MnO_2 and OA are combined during catalytic ozonation. The subsequent analysis focuses on the key differences in TOC and turbidity removal efficiencies observed across these treatments, as well as the mechanistic implications of the results.

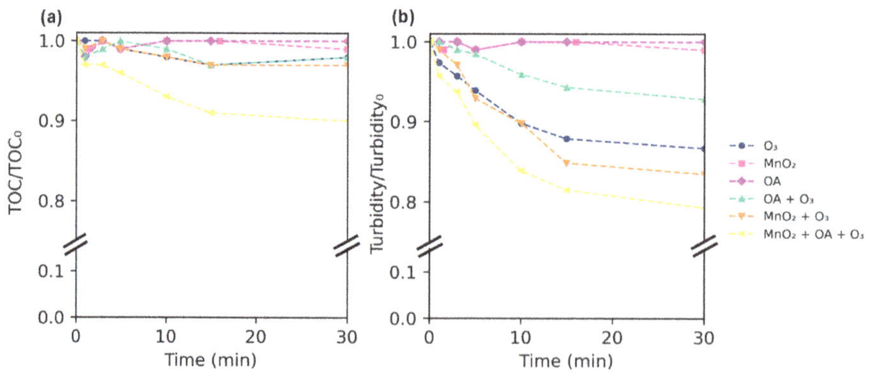

Figure 2. TOC (**a**) and turbidity (**b**) evolution upon PSNPs removal under different systems: O_3, MnO_2, OA, OA + O_3, MnO_2 + O_3, and MnO_2 + OA + O_3 ($[PSNPs]_0$ = 20 mg L^{-1}; O_3 dose= 7.5 mg min^{-1}; $[MnO_2]_0$ = 0.25 g L^{-1}, $[OA]_0$ = 2.5 mg L^{-1}; T = 20 °C; pH = 7).

Single ozonation showed limited efficacy, with TOC/TOC_0 decreasing by only 2% within 30 min, confirming ozone's inability to mineralize PSNPs alone. A similar trend was observed for turbidity, which decreased from 1.00 to 0.87, suggesting minor physical disaggregation of PSNP aggregates. This limited performance is attributed to the selective nature of molecular O_3, which reacts with electron-rich moieties but struggles to break down the stable aromatic structure of PS [8]. Similar findings were reported by Li et al. (2022), who demonstrated that significant PSNP degradation by ozonation requires extended contact times (up to 240 min) to achieve molecular weight reductions [38].

The individual application of MnO_2 and OA also resulted in negligible TOC and turbidity reduction (TOC/TOC_0 = 0.99 and 1.00; $turbidity/turbidity_0$ = 0.99 and 1.00, respectively). In the absence of O_3, MnO_2 lacks an oxidizing mechanism, and OA alone does not promote PSNP oxidation. Even the OA + O_3 combination yielded no improvement (TOC/TOC_0 = 0.99; $turbidity/turbidity_0$ = 0.90), indicating that OA alone is not sufficient to enhance ozone-based degradation.

A slight improvement was observed with the MnO_2 + O_3 system, reducing TOC/TOC_0 to 0.97 and turbidity to 0.84. MnO_2 serves as a heterogeneous catalyst that decomposes O_3 into reactive oxygen species (ROS) [18,40], including $^{\bullet}OH$ and $O_2^{\bullet-}$, which exhibit higher oxidation potential and broader reactivity than ozone alone. The redox cycling between Mn^{3+} and Mn^{4+} on the catalyst surface supports the generation of these ROS, as shown in Equation (1):

$$MnO_2 + O_3 \rightarrow MnO_2^{*} + O_2 + ROS \quad (1)$$

where MnO_2^{*} refers to Mn^{3+} on the catalyst surface. The reactivity of these radicals toward PSNPs leads to oxidative cleavage of polymeric chains, promoting fragmentation and partial mineralization.

The most significant enhancement occurred in the MnO_2 + OA + O_3 system, which achieved a TOC/TOC_0 reduction of 0.90 and $turbidity/turbidity_0$ = 0.81 after 30 min. Notably, the system did not reach a constant value, indicating that TOC/TOC_0 continued to decrease throughout the reaction time. This system benefits from the synergistic interaction between MnO_2 and OA. Oxalic acid contributes by forming Mn^{3+}–oxalate complexes (Equation (2)) that maintain the redox cycling and prevent catalyst deactivation [19]:

$$Mn^{3+} + C_2O_4^{2-} \rightarrow Mn^{3+}(oxalate)_{complex}. \quad (2)$$

Additionally, OA acts as a radical promoter under acidic conditions, facilitating further ROS formation. Studies by Andreozzi et al. (1998) and Xiao et al. (2008) have shown that low-molecular-weight organic acids like OA enhance ozone decomposition by donating electrons and stabilizing redox intermediates [9,41]. This synergy promotes the sustained generation of ROS, leading to both polymer chain cleavage and partial mineralization of PSNPs.

In line with these findings, the formation of ROS is strongly supported by the catalytic conditions employed and the previous literature involving Mn-based systems. Under acidic pH, MnO_2 facilitates the decomposition of O_3 into $\bullet OH$ and $O_2\bullet^-$ via redox cycling between Mn^{4+} and Mn^{3+}, particularly when oxalate ions are present to stabilize Mn^{3+} intermediates. Prior studies have shown that OA promotes additional ROS formation by accelerating ozone decomposition and enabling Mn^{3+} regeneration through complexation [9,19,40]. While the exact contribution of each radical species was not quantified, the observed improvements in TOC and turbidity removal under catalytic conditions, together with mechanistic consistency reported in similar systems, strongly suggest that both $\bullet OH$ and $O_2\bullet^-$ were involved in PSNP degradation. Future work should include targeted scavenger assays and ESR analysis to confirm the dominant oxidative pathways.

Recent research has emphasized the importance of surface interactions in these systems. Liu et al. (2024) demonstrated that MnO_2–OA surface complexes (particularly β-MnO_2) improve ROS generation through electron transfer and oxygen vacancy formation [19]. Similarly, Wang et al. (2023) observed that MnO_2–oxalate interactions in δ-MnO_2 systems support continuous ROS cycling, optimizing pollutant degradation [42].

2.3. *Effects of pH and Ozone Dosage*

Although the combination of MnO_2 and OA exhibited a higher removal efficiency compared to other systems, the TOC removal achieved was just 10% under the tested experimental conditions. This underscores the necessity of optimizing operational parameters to enhance overall removal efficiency.

The efficiency of catalytic ozonation is strongly influenced by pH and O_3 dosage, as these parameters dictate the formation, stability, and reactivity of ROS. pH affects both ozone decomposition kinetics and the surface charge of the MnO_2 catalyst, while O_3 dosage determines oxidant availability and mass transfer efficiency [43,44].

To evaluate these effects, experiments were conducted at different initial pH values (3, 4, 7, 9, 11, and 12) and different O_3 doses ranging between 7.5 and 37.5 mg min^{-1}. The catalytic performance was assessed by monitoring TOC removal and turbidity reduction as indicators of PSNP degradation. Figure 3 shows the influence of pH on the degradation of PSNPs based on the TOC/TOC_0 ratios observed after 30 min of reaction using the MnO_2 + OA + O_3 system.

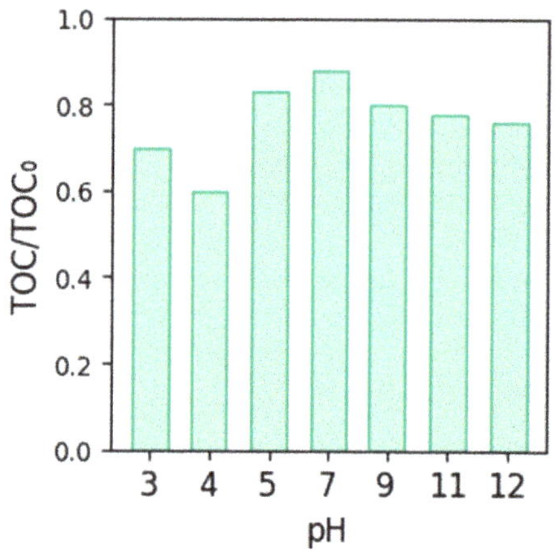

Figure 3. Effect of pH on TOC removal through catalytic ozonation ($[PSNP]_0$ = 20 mg L^{-1}, $[MnO_2]$ = 0.25 g L^{-1}, [OA] = 2.5 mg L^{-1}, O_3 dose= 7.5 mg min^{-1}, t = 30 min).

The results reveal that both pH and O_3 dosage play critical roles in the effectiveness of catalytic ozonation. At pH_0 = 4 and 7.5 mg min^{-1} of ozone dose, TOC removal reached 40% after 30 min, demonstrating the highest catalytic ozonation efficiency across all tested conditions (Figure 3). This improvement can be attributed to enhanced O_3 stability at acidic pH, reducing its self-decomposition into oxygen; strong adsorption of negatively charged ozone intermediates (O_3^-) and oxalate species ($C_2O_4^{2-}$) onto the positively charged MnO_2 surface; and the formation of Mn^{3+}–oxalate complexes, which facilitate redox cycling and sustained ROS generation [40]. These factors collectively improve the efficiency of catalytic ozonation by increasing ROS availability for PSNP oxidation. Similar effects were reported in previous studies, where carboxylate ligands, such as oxalate, stabilized Mn^{3+} and enhanced radical formation under acidic conditions [19]. Under near-neutral conditions (pH $\approx$ pH_{PZC}), TOC removal was not as effective when compared to the results found under acidic conditions. At pH_0 = 7, TOC removal dropped significantly (TOC/TOC_0 = 0.88 at 7.5 mg min^{-1} ozone dose), indicating that near-neutral conditions were suboptimal for catalytic ozonation. This decrease in efficiency can be attributed to limited adsorption of O_3 and oxalate species, as the MnO_2 surface charge is close to neutral, and lower production of ROS [43]. These findings align with reports from Li et al. (2023), who observed a similar decline in catalytic ozonation performance for CeOx@MnOx catalysts at neutral pH [13]. At alkaline conditions, pH_0 between 9 and 12 ($pH_0 > pH_{PZC}$), TOC removal improved compared to neutral conditions (TOC/TOC_0 = 0.78 at pH_0 =11). This is consistent with the higher rate of O_3 self-decomposition into $^{\bullet}OH$ in alkaline media, indicating that radical-based oxidation plays a key role in alkaline conditions [40].

Figure 4 illustrates the effect of increasing ozone dose (7.5–37.5 mg min^{-1}) on the removal of TOC (Figure 4a) and turbidity (Figure 4b) through single ozonation and catalytic ozonation systems after 30 min of reaction. In terms of TOC removal, catalytic ozonation consistently outperformed single ozonation across all tested doses. At the highest dose of 37.5 mg min^{-1}, TOC/TOC_0 reached 0.48 for catalytic ozonation, compared to 0.65 for single ozonation. These results indicate that the catalytic system responds more effectively to increased ozone input, enhancing pollutant degradation through intensified ROS generation. Regarding turbidity removal, catalytic ozonation also demonstrated superior performance. The turbidity/$turbidity_0$ ratio decreased from 0.63 to 0.56 for catalytic ozonation across the tested ozone doses, while single ozonation only reached 0.84 at its lowest. This indicates not only greater chemical degradation but also enhanced disaggregation of PSNP aggregates through catalytic ozonation, supporting the role of ROS in disrupting particle cohesion [8].

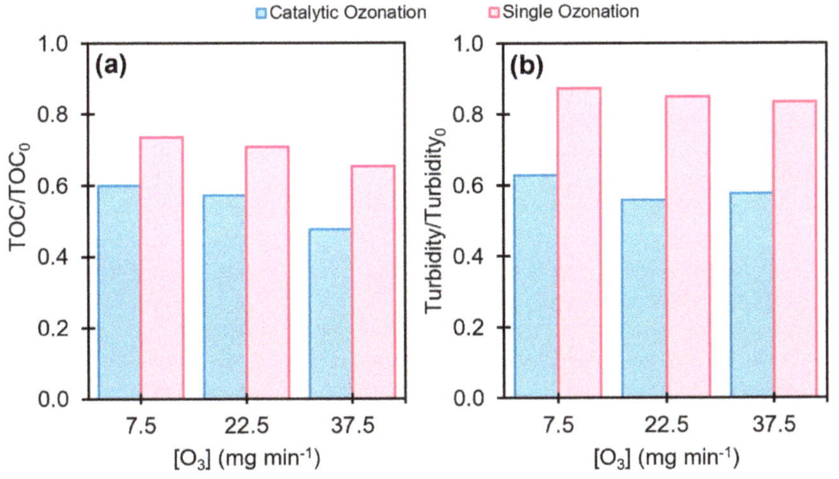

Figure 4. Effect of O_3 dose on TOC (**a**) and turbidity (**b**) removal through single ozonation and catalytic ozonation after 30 min ($[PSNP]_0$ = 20 mg L^{-1}, $[MnO_2]$ = 0.25 g L^{-1}, [OA] = 2.5 mg L^{-1}, pH_0 = 4).

2.4. *Effect of Pyrolusite and Oxalic Acid Doses*

The efficiency of catalytic ozonation can depend significantly on the concentration of both MnO_2 and OA, as these components work synergistically to promote the generation of ROS and maintain catalyst activity. Figure 5 provides insight into the synergistic interactions between MnO_2 and OA by identifying the optimal operating conditions for PSNP removal.

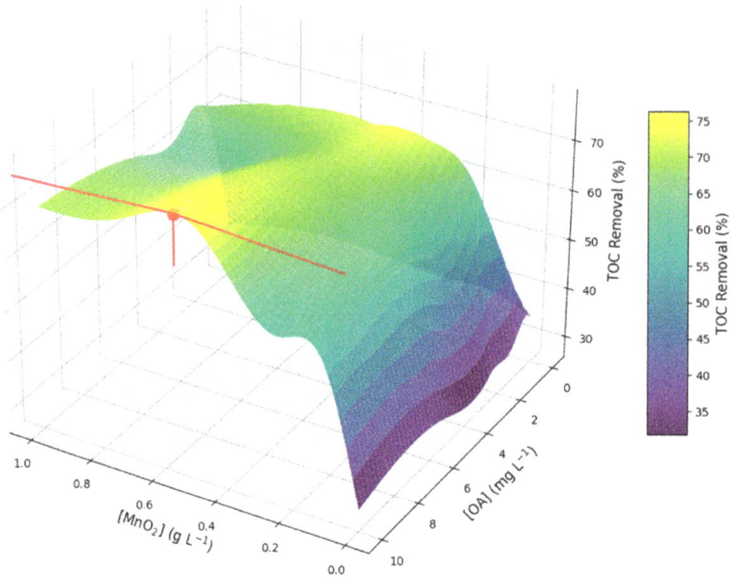

Figure 5. Surface response plot of the effects of MnO_2 and OA doses on TOC removal upon catalytic ozonation ($[PSNP]_0$ = 20 mg L^{-1}, O_3 dose= 37.5 mg min^{-1}, t = 30 min, pH_0 = 4). The red dot and line indicate the maximum TOC removal point identified in the response surface.

As shown in Figure 5, increasing MnO_2 concentration from 0 to 0.5 g L^{-1} led to a progressive enhancement in TOC removal, achieving a maximum efficiency of 75% at 0.5 g L^{-1} when combined with 10 mg L^{-1} of OA. This increased removal efficiency is attributed to the greater availability of active Mn sites, which facilitate O_3 decomposition into ROS, in agreement with previous observations regarding the role of transition metal oxides in catalytic ozonation. However, at concentrations exceeding 0.5 g L^{-1}, the efficiency plateaued and exhibited a slight decline, suggesting that an excessive MnO_2 load may promote ROS scavenging and contribute to catalyst agglomeration, limiting the availability of active sites. Similar trends have been reported for Mn-based ozonation catalysts, where excessive catalyst loading resulted in non-productive $^{\bullet}OH$ consumption and reduced pollutant degradation efficiency [11,34].

The presence of OA also played a crucial role in sustaining catalyst activity and maintaining continuous ROS generation, as indicated by the increase in TOC removal with OA concentration up to 10 mg L^{-1}. This effect is primarily due to the formation of Mn^{3+}–oxalate complexes, which prevent the accumulation of inactive Mn species and promote continuous redox cycling, ensuring the sustained production of ROS. However, when OA concentrations exceeded 10 mg L^{-1}, a reduction in PSNP degradation was observed, likely due to excessive radical scavenging, as organic acids can react with $^{\bullet}OH$, $O_2^{\bullet-}$, and O_3, reducing their availability for polymer oxidation. This observation is consistent with previous reports on the dual role of OA as both a radical promoter and a scavenger depending on its concentration [45,46].

A strong synergistic interaction between MnO_2 and OA was observed, as neither component alone achieved comparable TOC removal efficiencies. In the absence of OA, TOC removal remained below 40%, reinforcing the necessity of OA in preventing catalyst deactivation and enhancing ROS generation. This cooperative effect is further supported by previous studies [9,16,19] demonstrating that MnO_2–oxalate complexes enhance O_3

decomposition by stabilizing Mn^{3+} intermediates and increasing oxygen vacancy formation on the catalyst surface.

The highest PSNPs degradation efficiency was achieved at 0.5 g L^{-1} of MnO_2 and 10 mg L^{-1} of OA, reinforcing the importance of optimizing their ratio to maximize ROS formation while minimizing scavenging effects.

To assess the structural integrity of the PSNPs and confirm the occurrence of chemical degradation during ozonation, FTIR spectra were obtained for three samples: (i) the concentrated stock solution (50,000 mg L^{-1}), (ii) the diluted PSNP solution used in the experiments (20 mg L^{-1}), and (iii) the treated sample after 30 min of ozonation. Results are shown in Figure 6.

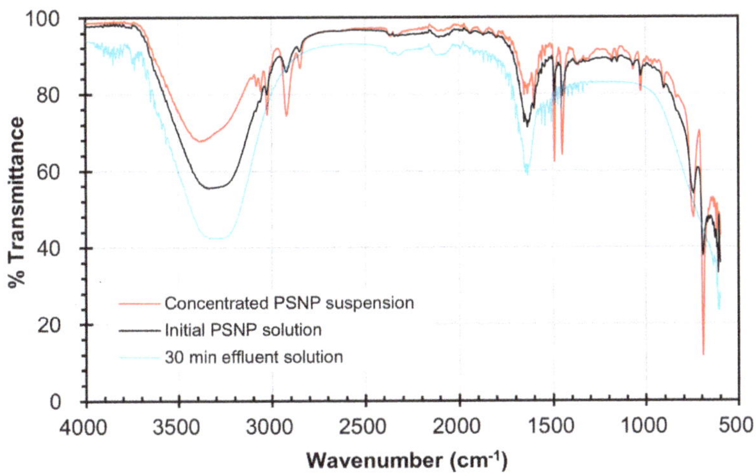

Figure 6. FTIR spectra of PSNPs: concentrated reference solution (red), initial solution of catalytic ozonation reaction (black), and effluent after 30 min reaction (blue) ($[PSNP]_0$ = 20 mg L^{-1}, O_3 dose = 37.5 mg min^{-1}, pH_0 = 4, $[MnO_2]$ = 0.5 g L^{-1}, [OA] = 10 mg L^{-1}).

In Figure 6, the initial spectrum of the PSNPs revealed characteristic absorption bands of polystyrene, including peaks at ~3025 cm^{-1} (C–H stretching of the aromatic ring), ~2920 and 2850 cm^{-1} (aliphatic C-H stretching), ~1600 cm^{-1} and ~1490 cm^{-1} (C=C stretching of the benzene ring), ~1452 cm^{-1} (CH_2 bending), and ~758 cm^{-1} and 698 cm^{-1} (out-of-plane bending of aromatic C–H) at both the concentrated PSNP suspension and the initial PSNP solution [47]. These assignments are consistent with the studies from di Luca et al. (2024) on PSNPs characterization [48]. After 30 min of ozonation, a notable decrease in intensity was observed for the bands associated with aromatic structures, particularly in the region from 1600 to 700 cm^{-1}, indicating that the ozonation process led to the cleavage of the benzene rings and progressive breakdown of the polymer chains. This behavior is consistent with the known reactivity of ozone, which initiates electrophilic attack on electron-rich aromatic rings, forming hydroxylated intermediates and ultimately yielding smaller oxygenated species and CO_2. The observed spectral changes, in combination with TOC and turbidity reductions, reinforce the evidence of PSNP degradation under the applied catalytic ozonation conditions.

Figure 7 illustrates the reduction in TOC over 120 min reaction time during the removal of PSNPs by catalytic ozonation (MnO_2 + OA + O_3) and single ozonation (O_3 only). The results clearly demonstrate the superior performance of catalytic ozonation, with TOC decreasing significantly faster and achieving a notable final reduction (~80%) compared to single ozonation (~37%). It must be noted that, during the early phase of the reaction (0–30 min), the difference in performance between catalytic and single ozonation is most pronounced. By the 30 min mark, the TOC/TOC_0 ratio for catalytic ozonation had decreased to 0.2768, while single ozonation only reached 0.6528, indicating that the catalytic system achieves a removal rate that is more than twice as effective as ozone alone. As the reaction progresses, the TOC reduction rate slows down, which is common in ozonation processes due to the accumulation of oxidation by-products (e.g., carboxylic acids, aldehydes) that are more resistant to oxidation. This plateau effect is more noticeable in the single ozonation process, where the lack of continuous ROS production limits further degradation. In contrast, catalytic ozonation maintains higher efficiency due to the sustained generation of ROS on the MnO_2 surface, highlighting the benefits of the combined MnO_2 and OA catalytic system.

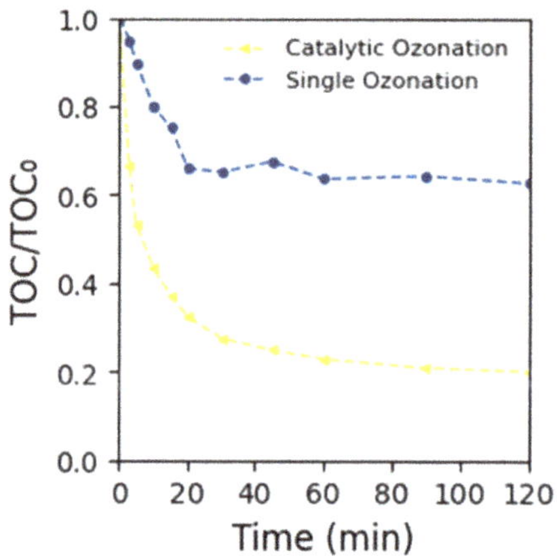

Figure 7. TOC reduction evolution upon catalytic ozonation and single ozonation of PSNPs ($[PSNP]_0$ = 20 mg L^{-1}, O_3 dose = 37.5 mg min^{-1}, pH_0 = 4, $[MnO_2]$ = 0.5 g L^{-1}, [OA] = 10 mg L^{-1}).

2.5. Catalyst Reusability

While catalytic ozonation demonstrated a faster and more substantial TOC decrease, its practical implementation also requires an understanding of catalyst durability. To address this, a study on catalyst reusability was conducted over three consecutive reuse cycles, and the results are shown in Figure 8. The catalyst (n-MnO_2) was dried at 80 °C overnight between cycles.

A progressive decline in removal efficiency was observed. In the first cycle, the system achieved a removal efficiency of 75%, which decreased to 73% in the second cycle and further to 66% in the third cycle. This gradual decrease in efficiency highlights the gradual deactivation of the MnO_2 catalyst, a phenomenon commonly observed in catalytic ozonation systems [19,49].

One of the primary causes of catalyst deactivation could be surface fouling by oxidation by-products. As PSNPs undergo oxidation, intermediate by-products, such as carboxylic acids and aldehydes, are generated. These intermediates can adsorb onto the surface of MnO_2, blocking active sites and reducing the availability of sites for O_3 decomposition. The accumulation of these by-products increases with each successive cycle, explaining the observed reduction in removal efficiency [43,50].

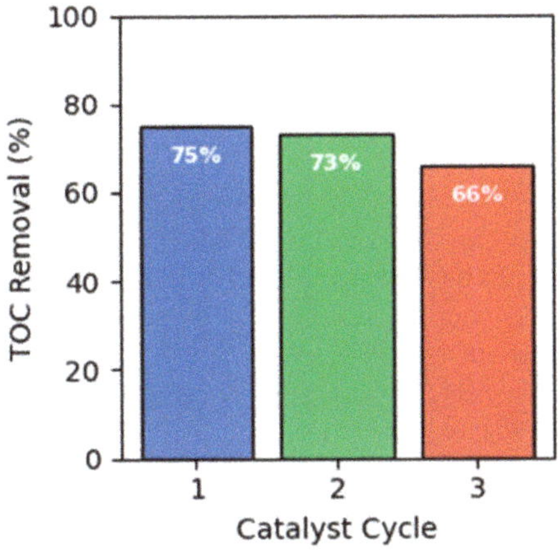

Figure 8. TOC removal efficiency of PSNPs after 30 min during three consecutive cycles ($[PSNP]_0$ = 20 mg L^{-1}, pH = 4, O_3 dose = 37.5 mg min^{-1}, t = 30 min, $[MnO_2]$ = 0.5 g L^{-1}, [OA] = 10 mg L^{-1}). Bar colors are used solely to differentiate catalyst cycles and do not represent additional variables.

Catalyst leaching is another factor that may contribute to the decline in performance. Under acidic conditions, manganese may dissolve into the solution, releasing Mn^{2+} ions into the reaction medium. This process reduces the active catalyst mass available for ROS production. Manganese leaching was measured during the study, and no leaching was detected, indicating that the MnO_2 catalyst remained stable throughout the reaction cycles. This result contrasts with findings from other studies on MnO_2-based catalytic ozonation systems, where leaching has been identified as a potential deactivation mechanism [51,52].

Despite these deactivation mechanisms, it is important to note that the efficiency decreased by only 12% after three consecutive cycles, indicating that the MnO_2-based system maintains a notable activity. This result suggests that, although some loss of catalytic activity occurs, the system remains viable for extended use. To improve catalyst durability, several regeneration strategies could be considered, such as thermal reactivation, oxidative washing, or chemical cleaning to remove fouling agents [50,53].

3. Materials and Methods

3.1. Materials and Reagents

The materials used in this study included PSNPs, catalysts, co-catalysts, ozone gas, and various reagents essential for experimental procedures and analytical measurements. The PSNPs were commercially available (microParticles GmbH, Berlin, Germany) and characterized by a nominal mean diameter of 140 nm. Detailed physicochemical characterization of this commercial PSNPs suspension was previously reported in recent studies [39,48,54], including transmission electron microscopy (TEM), dynamic light scattering (DLS), and nanoparticle tracking analysis (NTA). These analyses confirmed the nanoscale size and spherical morphology of the particles, with a DLS hydrodynamic diameter of 149.7 nm (PDI = 0.06), modal size of 135.9 nm (NTA), and mean diameter of 137.4 nm observed by TEM. Optical and chemical characteristics were also confirmed by UV–Vis spectroscopy (λ = 223 nm) and FTIR (700–760 cm^{-1}), consistent with polystyrene signatures. These particles were used without further modification. The commercial PSNPs solution was 50,000 mg L^{-1}. Natural MnO_2 in its pyrolusite form was used as the primary catalyst due to its availability, low cost, and high redox potential. This naturally occurring mineral was used as a heterogeneous catalyst without prior modification or pretreatment, with a particle size distribution ranging from 25 μm to 70 μm. The natural manganese dioxide (MnO_2) used in this study was obtained from Suministros Anper S.L. (Barcelona, Spain), under the reference OX841027. The chemical composition of the natural pyrolusite (MnO_2) used in this study, as reported by the supplier, is as follows: 78.4% MnO_2, 5.39% Al_2O_3, 3.76% SiO_2, 3.78% Fe_2O_3, and 0.2% P_2O_5. The material was used without further purification. An additional ICP-MS analysis (Thermo Scientific iCAP RQ, Waltham, MA, USA) confirmed Mn as the predominant element (~73.5 wt.%), with minor Fe content (~3.2 wt.%), consistent with the supplier's data. The co-catalyst used in this study was analytical-grade oxalic acid (purity $\geq$ 99%), which was sourced from Panreac, AppliChem, Barcelona, Spain. Hydrochloric acid (HCl, 37%) and sodium hydroxide (NaOH, $\geq$97%) were both obtained from Sigma-Aldrich (Merck KGaA, Darmstadt, Germany) and used for pH adjustments throughout the experimental procedures.

3.2. *Experimental Setup and Procedure*

The experimental setup consisted of a semi-batch reactor equipped with a continuous O_3 inlet. The borosilicate glass vessel reactor had a volume of 1.5 L and was continuously stirred at 800 rpm to ensure homogeneity of the reaction medium. The temperature was maintained at 25 $\pm$ 2 °C (room temperature) throughout all experiments.

Ozone (O_3) gas was generated in situ using an ozone generator (Model 301.19, Sander Elektrogerätebau GmbH, Uetze, Germany) and was introduced at a constant flow rate of 0.75 L min^{-1} and a concentration of 10 mg L^{-1}, unless otherwise indicated, through a porous diffuser located at the base of the reactor. A gas trap containing an aqueous potassium iodide (KI) solution was used to capture unreacted O_3, preventing its release into the laboratory environment.

The PSNPs suspension was prepared by diluting the commercial solution of PSNPs to a target concentration of 20 mg L^{-1} in ultrapure deionized water (resistivity $\geq$ 18.2 MΩ·cm). The pH of the solution was adjusted to the desired initial pH (3–11) using 0.1 M HCl or 0.1 M NaOH. Then, MnO_2 and OA were introduced in this order. The initial concentration of both was investigated in the range from 0.125 to 1 g L^{-1} and from 1.25 to 10 mg L^{-1}, respectively. Moreover, O_3 dose was evaluated between 7.5 and 37.5 mg min^{-1}.

Control experiments were conducted to assess the effects of single ozonation (without MnO_2 or OA) and to evaluate the performance of catalytic ozonation under different conditions. Blank tests (without PSNPs) were also performed to check for background interference. All experiments were performed in triplicate, and the experimental data reported represent the average values of these replicates to ensure consistency and reproducibility.

3.3. *Analytical Methods*

Samples were periodically withdrawn from the reactor and filtered when necessary to separate catalyst particles and analyzed using fiberglass membranes with an average pore diameter below 1 μm (Sartorius Glass Microfiber Filters, 13400-47-Q, Sartorius AG, Göttingen, Germany).

Turbidity was measured using a turbidimeter (Model 2100P, Hach Company, Düsseldorf, Germany), following the ISO 7027 method [55]. Turbidity was recorded at regular intervals during each experiment to monitor the physical disaggregation of PSNPs. Changes in turbidity provided indirect evidence of PSNP breakdown [39].

TOC analysis was performed using a TOC analyzer (multi N/C 3100 Series, Analytik Jena GmbH, Jena, Germany), which quantifies the total concentration of organic carbon in the aqueous phase. TOC reflects the chemical degradation of PSNPs, as it captures the conversion of polymeric carbon into dissolved intermediates or complete mineralization into CO_2. TOC samples were filtered to remove catalyst particles before analysis. All measurements were conducted in triplicate, and the results were reported as mean values.

Manganese leaching was assessed to evaluate the stability of the MnO_2 catalyst during the ozonation process. The elemental composition of the MnO_2 catalyst was evaluated using inductively coupled plasma mass spectrometry (ICP-MS, iCAP RQ, Thermo Scientific Inc., Waltham, MA, USA), providing semiquantitative data on Mn and Fe content in the solid material. This analysis confirmed that manganese release was not significant under the tested conditions, ensuring the stability and reusability of the catalyst.

Fourier-transform infrared (FTIR) spectra were acquired using a spectrometer (Cary 630 FTIR, Agilent Technologies Inc., Santa Clara, CA, USA) equipped with a single-reflection diamond ATR (attenuated total reflectance) accessory, to evaluate the chemical structure of PSNPs before and after ozonation. The measurements were carried out in the spectral range of 4000–500 cm^{-1}, with a resolution of 4 cm^{-1} and 32 scans per sample. The ATR crystal is composed of type IIa diamond, providing high chemical resistance and an optimal IR signal for polymeric materials. Prior to each analysis, the sample press was used to ensure proper contact between the PSNP powders and the crystal surface.

3.4. Catalyst Characterization

The characterization of the n-MnO_2 catalyst was conducted using three complementary techniques: Fourier-transform infrared spectroscopy (FTIR), pH at the point of zero charge (pH_{PZC}) determination, and X-ray photoelectron spectroscopy (XPS). These techniques provided insights into the surface functional groups, chemical bonding, and oxidation states of manganese species, which are critical for understanding the catalytic behavior of pyrolusite during the ozonation of PSNPs.

The FTIR spectra of the catalyst were obtained using a PerkinElmer Spectrum Two™ FTIR spectrometer (PerkinElmer Inc., Shelton, CT, USA). The analyses were performed by recording multiple scans across a wavenumber range of 450 to 4000 cm^{-1} to identify the vibrational modes of surface functional groups. This spectral range encompasses characteristic peaks associated with manganese–oxygen bonding, as well as the presence of hydroxyl groups that can influence catalytic ozonation performance.

The FTIR technique was employed to elucidate the chemical interactions occurring on the surface of pyrolusite, particularly the role of hydroxyl groups in the adsorption and decomposition of O_3. The data collected were subsequently analyzed to correlate the structural characteristics of the catalyst with its catalytic efficiency in PSNP degradation.

The elemental composition and oxidation states of manganese within the pyrolusite catalyst were investigated using a PHI VersaProbe 4 Scanning XPS Microprobe (ULVAC-PHI Inc., Chanhassen, MN, USA). This surface-sensitive technique enabled the identification of manganese species and their respective oxidation states, which are directly related to the catalytic activity of the material.

The XPS measurements were performed by irradiating the sample with a monochromatic Al Kα X-ray source (1486.6 eV) and recording the photoelectron emission from the Mn2p and O1s regions. The high-resolution spectra were deconvoluted to distinguish between different manganese oxidation states, particularly Mn^{3+} and Mn^{4+}, which play a key role in the generation of ROS during ozonation.

Additionally, the pH_{PZC} of the pyrolusite catalyst was determined using the batch equilibrium method. This analysis provides insights into the surface charge properties of the material, which influence its interaction with O_3 and organic pollutants. The pH_{PZC} was obtained by suspending the catalyst in deionized water at different initial pH values (adjusted using 0.1 M HCl or 0.1 M NaOH), followed by equilibration for 24 h. The final pH was then measured, and the pH_{PZC} was identified as the pH at which no net surface charge was observed on the catalyst surface. This parameter is essential for understanding the adsorption behavior of reactive species during the catalytic ozonation process.

4. Conclusions

The growing presence of PSNPs in aquatic environments poses significant environmental and health risks due to their chemical stability, small size, and bioavailability. This study demonstrates the potential of catalytic ozonation as an effective method for PSNP removal, particularly when enhanced by the synergistic action of natural pyrolusite (n-MnO_2) and OA. Operational parameters were evaluated, including pH, applied O_3 dose, OA, and catalyst concentration, providing a comprehensive understanding of the effects of each operational parameter on PSNP degradation.

The results reveal that the combination of MnO_2 and OA significantly enhances the generation of ROS compared to single ozonation or the use of MnO_2 alone. The optimal conditions for PSNP removal were observed at pH 4, with an O_3 dose of 37.5 mg min^{-1}, 0.5 g L^{-1} of MnO_2, and 10 mg L^{-1} of OA. Under these conditions, the system achieved a maximum TOC removal of 75% after 30 min ozonation, with a notable improvement in turbidity reduction, reflecting both the chemical degradation and physical disaggregation of PSNP aggregates. These findings underscore the importance of catalyst–co-catalyst interactions in promoting ROS generation and enhancing oxidation efficiency.

Beyond removal efficiency, the reusability of the MnO_2 catalyst was evaluated across consecutive reaction cycles. Although a slight decline in removal efficiency was observed (from 74.69% to 65.73% after three cycles), the catalyst retained most of its activity without detecting manganese leaching, indicating its potential for repeated use in water treatment systems. This durability positions MnO_2 as a promising candidate for scalable applications, especially when paired with low-cost natural pyrolusite, a readily available source of MnO_2.

Author Contributions: Conceptualization, V.M. and J.N.-S.; methodology, V.M. and J.N.-S.; validation, V.M., J.N.-S. and C.S.; formal analysis, V.M.; investigation, V.M. and J.N.-S.; resources, M.D. and C.S.; data curation, V.M.; writing—original draft preparation, V.M.; writing—review and editing, J.N.-S., M.D. and C.S.; visualization, V.M. and J.N.-S.; supervision, M.D. and C.S.; project administration, C.S.; funding acquisition, M.D. and C.S. All authors have read and agreed to the published version of the manuscript.

Funding: This work was financially supported through the projects TED2021-131569B-I00 by MCIN/AEI/10.13039/501100011033, Next Generation EU/PRTR; PID2023-146338OB-I00 by MCIU/AEI/10.13039/501100011033/ FEDER,UE; as well as by the Coordenação de Aperfeiçoamento de Pessoal de Nível Superior—Brasil (CAPES), Finance Code 001. Additional financial support was provided by Fundação Carlos Chagas Filho de Amparo à Pesquisa do Estado do Rio de Janeiro (FAPERJ), under projects E-26/204.177/2024 and E-26/200.979/2021, and by the Conselho Nacional de Desenvolvimento Científico e Tecnológico (CNPq), project 304175/2020-0.

Data Availability Statement: The original contributions presented in this study are included in the article. Further inquiries can be directed to the corresponding authors.

Acknowledgments: The authors thank the Surface Analysis Laboratory of the CCiTUB for carrying out the XPS measurements. Victor Mello thanks CAPES for the PRINT scholarship through the research projects 88887.311613/2018-00 (UFRJ) and 88887.310122/2018-00 (UFRJ).

Conflicts of Interest: The authors declare no conflicts of interest. The funders had no role in the design of the study; in the collection, analyses, or interpretation of data; in the writing of the manuscript; or in the decision to publish the results.

Abbreviations

The following abbreviations are used in this manuscript:

AOP	Advanced oxidation process
FTIR	Fourier-transform infrared spectroscopy
GPC	Gel permeation chromatography
MnO_2	Manganese dioxide
MP	Microplastic
NP	Nanoplastic
n-MnO_2	Natural manganese dioxide (pyrolusite)
OA	Oxalic acid
O_3	Ozone
$^{\bullet}OH$	Hydroxyl radical
$O_2^{\bullet-}$	Superoxide radical
pH_{PZC}	Point of zero charge
PS	Polystyrene
PSNPs	Polystyrene nanoplastics
ROS	Reactive oxygen species
TOC	Total organic carbon
XPS	X-ray photoelectron spectroscopy

References

1. Koelmans, A.A.; Besseling, E.; Foekema, E.; Kooi, M.; Mintenig, S.; Ossendorp, B.C.; Redondo-Hasselerharm, P.E.; Verschoor, A.; Van Wezel, A.P.; Scheffer, M. Risks of Plastic Debris: Unravelling Fact, Opinion, Perception, and Belief. *Environ. Sci. Technol.* **2017**, *51*, 11513–11519. [CrossRef] [PubMed]
2. Wright, S.L.; Kelly, F.J. Plastic and Human Health: A Micro Issue? *Environ. Sci. Technol.* **2017**, *51*, 6634–6647. [CrossRef] [PubMed]
3. Gigault, J.; ter Halle, A.; Baudrimont, M.; Pascal, P.Y.; Gauffre, F.; Phi, T.L.; El Hadri, H.; Grassl, B.; Reynaud, S. Current Opinion: What Is a Nanoplastic? *Environ. Pollut.* **2018**, *235*, 1030–1034. [CrossRef] [PubMed]
4. European Commission Directive (EU). 2020/2184 on the Quality of Water Intended for Human Consumption. *Off. J. Eur. Union* **2020**, *L435*, 1–62.
5. Andrady, A.L. Microplastics in the Marine Environment. *Mar. Pollut. Bull.* **2011**, *62*, 1596–1605. [CrossRef]
6. Song, Y.K.; Hong, S.H.; Jang, M.; Kang, J.H.; Kwon, O.Y.; Han, G.M.; Shim, W.J. Large Accumulation of Micro-Sized Synthetic Polymer Particles in the Sea Surface Microlayer. *Environ. Sci. Technol.* **2014**, *48*, 9014–9021. [CrossRef]
7. Wagner, M.; Lambert, S. Freshwater Microplastics: Emerging Environmental Contaminants? In *The Handbook of Environmental Chemistry*; Barceló, D., Kostianoy, A.G., Eds.; Springer International Publishing: Cham, Switzerland, 2018; Volume 58, ISBN 978-3-319-61614-8. [CrossRef]
8. von Gunten, U. Ozonation of Drinking Water: Part I. Oxidation Kinetics and Product Formation. *Water Res.* **2003**, *37*, 1443–1467. [CrossRef]
9. Xiao, H.; Liu, R.; Zhao, X.; Qu, J. Enhanced Degradation of 2,4-Dinitrotoluene by Ozonation in the Presence of Manganese(II) and Oxalic Acid. *J. Mol. Catal. A Chem.* **2008**, *286*, 149–155. [CrossRef]
10. Issaka, E.; AMU-Darko, J.N.O.; Yakubu, S.; Fapohunda, F.O.; Ali, N.; Bilal, M. Advanced Catalytic Ozonation for Degradation of Pharmaceutical Pollutants—A Review. *Chemosphere* **2022**, *289*, 133208. [CrossRef]
11. Rekhate, C.V.; Srivastava, J.K. Recent Advances in Ozone-Based Advanced Oxidation Processes for Treatment of Wastewater—A Review. *Chem. Eng. J. Adv.* **2020**, *3*, 100031. [CrossRef]
12. Issaka, E.; Baffoe, J.; Adams, M. Exploring Heterogeneous Catalytic Ozonation: Catalyst Types, Reaction Mechanisms, Applications, Challenges, and Future Outlook. *Sustain. Chem. Environ.* **2024**, *8*, 100185. [CrossRef]
13. Li, Y.; Zhang, C.; Shen, C.; Jiang, G.; Guan, B. Enhanced Ozonation of Polystyrene Nanoplastics in Water with CeOx@MnOx Catalyst. *Environ. Res.* **2023**, *220*, 115220. [CrossRef] [PubMed]
14. Isibor, P.O.; Devi, G.; Enuneku, A.A. (Eds.) *Environmental Nanotoxicology Combatting the Minute Contaminants*; Springer: Berlin/Heidelberg, Germany, 2024; ISBN 978-3-031-54153-7.
15. Huang, J.; Dai, Y.; Singewald, K.; Liu, C.C.; Saxena, S.; Zhang, H. Effects of MnO_2 of Different Structures on Activation of Peroxymonosulfate for Bisphenol A Degradation under Acidic Conditions. *Chem. Eng. J.* **2019**, *370*, 906–915. [CrossRef]
16. Andreozzi, R.; Insola, A.; Caprio, V.; D'Amore, M.G. The Kinetics of Mn(II)-Catalysed Ozonation of Oxalic Acid in Aqueous Solution. *Water Res.* **1992**, *26*, 917–921. [CrossRef]

17. Andreozzi, R.; Caprio, V.; Giovanna D'amore, M.; Insola, A. Manganese Catalysis in Water Pollutants Abatement by Ozone. *Environ. Technol.* **1995**, *16*, 885–891. [CrossRef]
18. Nawrocki, J.; Kasprzyk-Hordern, B. The Efficiency and Mechanisms of Catalytic Ozonation. *Appl. Catal. B Environ.* **2010**, *99*, 27–42. [CrossRef]
19. Liu, J.; Yuan, X.; Sans, C. Insights into the Role of β-MnO_2 and Oxalic Acid Complex Expediting Ozonation: Structural Properties and Mechanism. *Sep. Purif. Technol.* **2024**, *341*, 126904. [CrossRef]
20. Omorogie, M.O.; Agbadaola, M.T.; Olatunde, A.M.; Helmreich, B.; Babalola, J.O. Surface Equilibrium and Dynamics for the Adsorption of Anionic Dyes onto MnO2/Biomass Micro-Composite. *Green Chem. Lett. Rev.* **2022**, *15*, 49–58. [CrossRef]
21. Dubal, D.P.; Kim, W.B.; Lokhande, C.D. Galvanostatically Deposited Fe: MnO2 Electrodes for Supercapacitor Application. *J. Phys. Chem. Solids* **2012**, *73*, 18–24. [CrossRef]
22. Yu, Y.; Li, X.; Zhang, R.; Guo, W. Insights into the Morphology-Dependent Adsorption of Aged Polystyrene Nanoplastics on Manganese Oxides. *Colloids Surfaces A Physicochem. Eng. Asp.* **2023**, *679*, 132565. [CrossRef]
23. Pang, M.; Long, G.; Jiang, S.; Ji, Y.; Han, W.; Wang, B.; Liu, X.; Xi, Y. Rapid Synthesis of Graphene/Amorphous α-MnO2 Composite with Enhanced Electrochemical Performance for Electrochemical Capacitor. *Mater. Sci. Eng. B* **2015**, *194*, 41–47. [CrossRef]
24. Kumar, Y.; Chopra, S.; Gupta, A.; Kumar, Y.; Uke, S.J.; Mardikar, S.P. Low Temperature Synthesis of MnO_2 Nanostructures for Supercapacitor Application. *Mater. Sci. Energy Technol.* **2020**, *3*, 566–574. [CrossRef]
25. Malvestiti, J.A.; Cavalcante, R.P.; Tornisielo, V.L.; Dantas, R.F. Metals as Catalysts for Ozonation. In *Heavy Metals–Recent Advances*; Almayyahi, B., Ed.; IntechOpen: London, UK, 2023; ISBN 978-1-83768-515-8.
26. Chen, S.; Shu, X.; Wang, H.; Zhang, J. Thermally Driven Phase Transition of Manganese Oxide on Carbon Cloth for Enhancing the Performance of Flexible All-Solid-State Zinc-Air Batteries. *J. Mater. Chem. A* **2019**, *7*, 19719–19727. [CrossRef]
27. Jiang, Y.; Ba, D.; Li, Y.; Liu, J. Noninterference Revealing of "Layered to Layered" Zinc Storage Mechanism of δ-MnO_2 toward Neutral Zn–Mn Batteries with Superior Performance. *Adv. Sci.* **2020**, *7*, 1902795. [CrossRef]
28. De la Cruz, I.J.; Rodríguez, S.J.L.; Fuentes, I.; Tiznado, H.; Vazquez-Arce, J.L.; Romero-Ibarra, I.; Guzmán, C.J.I.; Gutiérrez, H.M. Effect of Crystalline Phase of MnO2 on the Degradation of Bisphenol A by Catalytic Ozonation. *J. Environ. Chem. Eng.* **2023**, *11*, 110753. [CrossRef]
29. Zhang, W.; Hao, M.; Wang, Y.; Sun, P.; Zeng, D.; Wang, X.; Liang, P. Effect of PH on Microstructure and Catalytic Oxidation of Formaldehyde in MnO2 Catalyst. *Catalysts* **2023**, *13*, 490. [CrossRef]
30. Zhu, G.; Zhu, J.; Jiang, W.; Zhang, Z.; Wang, J.; Zhu, Y.; Zhang, Q. Surface Oxygen Vacancy Induced A-MnO2nanofiber for Highly Efficient Ozone Elimination. *Appl. Catal. B Environ.* **2017**, *209*, 729–737. [CrossRef]
31. Wang, Y.; Chen, L.; Cao, H.; Chi, Z.; Chen, C.; Duan, X.; Xie, Y.; Qi, F.; Song, W.; Liu, J.; et al. Role of Oxygen Vacancies and Mn Sites in Hierarchical Mn2O3/LaMnO3-Δ Perovskite Composites for Aqueous Organic Pollutants Decontamination. *Appl. Catal. B Environ.* **2019**, *245*, 546–554. [CrossRef]
32. Xia, L.; Liang, W.; Chen, G.; Li, W.; Gao, M. Catalytic Ozonation of Quinoline Utilizing Manganese-Based Catalyst with Abundant Oxygen Vacancies. *Catal. Letters* **2022**, *152*, 1669–1677. [CrossRef]
33. Radinger, H.; Connor, P.; Stark, R.; Jaegermann, W.; Kaiser, B. Manganese Oxide as an Inorganic Catalyst for the Oxygen Evolution Reaction Studied by X-Ray Photoelectron and Operando Raman Spectroscopy. *ChemCatChem* **2021**, *13*, 1175–1185. [CrossRef]
34. Luo, K.; Zhao, S.X.; Wang, Y.F.; Zhao, S.J.; Zhang, X.H. Synthesis of Petal-like δ-MnO_2 and Its Catalytic Ozonation Performance. *New J. Chem.* **2018**, *42*, 6770–6777. [CrossRef]
35. He, Y.; Wang, L.; Chen, Z.; Shen, B.; Wei, J.; Zeng, P.; Wen, X. Catalytic Ozonation for Metoprolol and Ibuprofen Removal over Different MnO2 Nanocrystals: Efficiency, Transformation and Mechanism. *Sci. Total Environ.* **2021**, *785*, 147328. [CrossRef]
36. Tang, W.; Wu, X.; Li, D.; Wang, Z.; Liu, G.; Liu, H.; Chen, Y. Oxalate Route for Promoting Activity of Manganese Oxide Catalysts in Total VOCs' Oxidation: Effect of Calcination Temperature and Preparation Method. *J. Mater. Chem. A* **2014**, *2*, 2544–2554. [CrossRef]
37. Kasprzyk-Hordern, B.; Ziółek, M.; Nawrocki, J. Catalytic Ozonation and Methods of Enhancing Molecular Ozone Reactions in Water Treatment. *Appl. Catal. B Environ.* **2003**, *46*, 639–669. [CrossRef]
38. Li, Y.; Li, J.; Ding, J.; Song, Z.; Yang, B.; Zhang, C.; Guan, B. Degradation of Nano-Sized Polystyrene Plastics by Ozonation or Chlorination in Drinking Water Disinfection Processes. *Chem. Eng. J.* **2022**, *427*, 131690. [CrossRef]
39. Nieto-Sandoval, J.; Ammar, R.; Sans, C. Enhancing Nanoplastics Removal by Metal Ion-Catalyzed Ozonation. *Chem. Eng. J. Adv.* **2024**, *19*, 100621. [CrossRef]
40. Rodríguez, A.; Rosal, R.; Perdigón-Melón, J.A.; Mezcua, M.; Agüera, A.; Hernando, M.D.; Letón, P.; Fernández-Alba, A.R.; García-Calvo, E. Ozone-Based Technologies in Water and Wastewater Treatment. *Handb. Environ. Chem. Vol. 5 Water Pollut.* **2008**, *5* (Suppl. S2), 127–175. [CrossRef]
41. Andreozzi, R.; Caprio, V.; Insola, A.; Marotta, R.; Tufano, V. The Ozonation of Pyruvic Acid in Aqueous Solutions Catalyzed by Suspended and Dissolved Manganese. *Water Res.* **1998**, *32*, 1492–1496. [CrossRef]

42. Wang, J.; Yue, W.; Teng, Y.; Zhai, Y.; Zhu, H. Degradation Kinetics and Transformation Pathway of Methyl Parathion by δ-MnO_2/Oxalic Acid Reaction System. *Chemosphere* **2023**, *320*, 138054. [CrossRef]
43. Psaltou, S.; Mitrakas, M.; Zouboulis, A. Heterogeneous Catalytic Ozonation: Solution PH and Initial Concentration of Pollutants as Two Important Factors for the Removal of Micropollutants from Water. *Separations* **2022**, *9*, 413. [CrossRef]
44. Fallah, N.; Bloise, E.; Santoro, D.; Mele, G. State of Art and Perspectives in Catalytic Ozonation for Removal of Organic Pollutants in Water: Influence of Process and Operational Parameters. *Catalysts* **2023**, *13*, 324. [CrossRef]
45. Wu, Z.; Abramova, A.; Nikonov, R.; Cravotto, G. Sonozonation (Sonication/Ozonation) for the Degradation of Organic Contaminants—A Review. *Ultrason. Sonochem.* **2020**, *68*, 105195. [CrossRef]
46. Zhao, Y.; An, H.; Dong, G.; Feng, J.; Ren, Y.; Wei, T. Elevated Removal of Di-n-Butyl Phthalate by Catalytic Ozonation over Magnetic Mn-Doped Ferrospinel ZnFe2O4 Materials: Efficiency and Mechanism. *Appl. Surf. Sci.* **2020**, *505*, 144476. [CrossRef]
47. Smith, B.C. The Infrared Spectra of Polymers V: Epoxies. *Spectroscopy* **2022**, *37*, 17–19. [CrossRef]
48. di Luca, C.; Garcia, J.; Munoz, M.; Hernando-Pérez, M.; de Pedro, Z.M.; Casas, J.A. Strategies for the Quantification and Characterization of Nanoplastics in AOPs Research. *Chem. Eng. J.* **2024**, *493*, 152490. [CrossRef]
49. Li, X.; Fu, L.; Chen, F.; Zhao, S.; Zhu, J.; Yin, C. Application of Heterogeneous Catalytic Ozonation in Wastewater Treatment: An Overview. *Catalysts* **2023**, *13*, 342. [CrossRef]
50. Cao, H.; Xie, Y.; Wang, Y.; Xiao, J. (Eds.) *Advanced Ozonation Processes for Water and Wastewater Treatment: Active Catalysts and Combined Technologies*; Royal Society of Chemistry: Cambridge, UK, 2022; ISBN 978-1-83916-389-0.
51. Nawaz, F.; Xie, Y.; Cao, H.; Xiao, J.; Wang, Y.; Zhang, X.; Li, M.; Duan, F. Catalytic Ozonation of 4-Nitrophenol over an Mesoporous α-MnO2 with Resistance to Leaching. *Catal. Today* **2015**, *258*, 595–601. [CrossRef]
52. Yang, K.; Yu, J.; Guo, Q.; Wang, C.; Yang, M.; Zhang, Y.; Xia, P.; Zhang, D.; Yu, Z. Comparison of Micropollutants' Removal Performance between Pre-Ozonation and Post-Ozonation Using a Pilot Study. *Water Res.* **2017**, *111*, 147–153. [CrossRef]
53. Saputra, E.; Muhammad, S.; Sun, H.; Ang, H.M.; Tadé, M.O.; Wang, S. Manganese Oxides at Different Oxidation States for Heterogeneous Activation of Peroxymonosulfate for Phenol Degradation in Aqueous Solutions. *Appl. Catal. B Environ.* **2013**, *142–143*, 729–735. [CrossRef]
54. Ortiz, D.; Munoz, M.; Nieto-Sandoval, J.; Romera-Castillo, C.; de Pedro, Z.M.; Casas, J.A. Insights into the Degradation of Microplastics by Fenton Oxidation: From Surface Modification to Mineralization. *Chemosphere* **2022**, *309*, 136809. [CrossRef]
55. ISO. *ISO 7027:2016*; Water Quality—Determination of Turbidity. International Organization for Standardization: Geneva, Switzerland, 2016.

catalysts

MDPI

Article

Green Innovation: Multifunctional Zinc Oxide Nanoparticles Synthesized Using *Quercus robur* for Photocatalytic Performance, Environmental, and Antimicrobial Applications

Selma Redjili [1], Houria Ghodbane [1,*], Hichem Tahraoui [2,3], Lokmane Abdelouahed [4], Derradji Chebli [5], Mohammad Shamsul Ola [6], Amine Aymen Assadi [7], Mohammed Kebir [8], Jie Zhang [9], Abdeltif Amrane [3,*] and Sabrina Lekmine [10,*]

1 Laboratory of Physics for Matter and Radiation, Faculty of Science and Technology, Souk Ahras University, P.O. Box 1553, Souk-Ahras 41000, Algeria; s.redjili@univ-soukahras.dz
2 Laboratory of Biomaterials and Transport Phenomena, University of Médéa, Médéa 26000, Algeria; hichemm.tahraouii@gmail.com
3 Ecole Nationale Supérieure de Chimie de Rennes, CNRS, ISCR—UMR6226, University Rennes, 35000 Rennes, France
4 Laboratoire de Sécurité des Procédés Chimiques—LSPC EA 4704, Normandie Univ, INSA Rouen, UNIROUEN, 76000 Rouen, France; lokmane.abdelouahed@insa-rouen.fr
5 Laboratory of Chemical Process Engineering, Department of Process Engineering, Faculty of Technology, University of Ferhat Abbas Setif-1, Setif 19000, Algeria; derradji_chebli@yahoo.fr
6 Department of Biochemistry, College of Science, King Saud University, Riyadh 11451, Saudi Arabia; mola@ksu.edu.sa
7 College of Engineering, Imam Mohammad Ibn Saud Islamic University, IMSIU, Riyadh 11432, Saudi Arabia
8 Research Scientific and Technical Center on Physico-Chemical Analysis (CRAPC), BP 384, Industrial Zone Bou-Ismail, Tipaza 42000, Algeria; medkebir@yahoo.fr
9 School of Engineering, Merz Court, Newcastle University, Newcastle upon Tyne NE1 7RU, UK; jie.zhang@newcastle.ac.uk
10 Biotechnology, Water, Environment and Health Laboratory, Abbes Laghrour University, Khenchela 40000, Algeria
* Correspondence: h.ghodbane@univ-soukahras.dz (H.G.); abdeltif.amrane@univ-rennes1.fr (A.A.); lekmine.sabrina@univ-khenchela.dz (S.L.)

Abstract: This study investigates the green synthesis of zinc oxide nanoparticles (ZnO NPs) using leaf extract as a natural reducing agent, evaluating their antimicrobial and photocatalytic properties. The nanoparticles were annealed at 320 °C and 500 °C, and the effects of leaf extract concentration and annealing temperature on their structural, morphological, and electronic properties were systematically explored. X-ray diffraction (XRD) analysis confirmed the hexagonal wurtzite structure of ZnO, with crystallite size and defect density being influenced by the concentration of the extract. Scanning electron microscopy (SEM) revealed the formation of smaller, spherical particles, with increased aggregation observed at higher extract concentrations. Fourier-transform infrared spectroscopy (FTIR) identified key functional groups, such as hydroxyl groups, C–O bonds, and metal–oxygen vibrations. UV–Vis spectroscopy showed a reduction in band gap energy and an increase in Urbach energy as the extract concentration and annealing temperature were increased. The antimicrobial activity of the ZnO NPs was evaluated against Gram-positive and Gram-negative bacteria as well as Candida albicans, demonstrating significant antibacterial efficacy. Photocatalytic degradation studies of methylene blue dye revealed a superior efficiency of up to 74% for the annealed samples, particularly at 500 °C. This research highlights the potential of green-synthesized ZnO NPs for a wide range of applications, including antimicrobial agents, water purification, and environmental catalysis. It contributes to the advancement of sustainable nanotechnology, offering promising solutions for both technological and ecological challenges.

Keywords: degradation; green synthesis; zinc oxide nanoparticles; *Quercus robur*; optical properties; antimicrobial activity

1. Introduction

The growing population, combined with rising industrialization and economic demands, has necessitated the extensive use of synthetic chemicals and organic dyes, particularly in industries that produce by-products during their processes [1,2]. Despite the economic benefits provided by sectors such as textiles and chemicals, the untreated discharge of industrial wastewater leads to significant challenges, including health hazards, ecosystem imbalances, and reduced biodiversity due to its toxicity and persistence [3–5]. To address these issues, various treatment methods have been developed to eliminate contaminants before they are released into the environment [6,7]. These approaches include biological [8], chemical [9–12], physicochemical [13,14], and hybrid techniques for water purification [15,16].

Nanoremediation is an advanced remediation strategy that leverages nanomaterials to break down or remove pollutants efficiently [17]. This technique is recognized for its superior effectiveness in managing environmental contamination and enabling pollutant detection, monitoring, and treatment [18]. Nanoremediation methods are cost-efficient and significantly reduce the time needed to degrade pollutants, often lowering their concentration to near-zero levels [19,20]. This technology has been successfully applied to address various pollutants in water, such as chlorinated compounds, heavy metals, organic chemicals, hydrocarbons, pesticides, and inorganic ions [18,21,22].

A broad spectrum of nanomaterials, including zeolites, metal oxides, carbon nanostructures, and bimetallic nanoparticles, is employed in nanoremediation to transform or degrade environmental contaminants [23–25]. Conventional wastewater treatment techniques, such as filtration, coagulation, reverse osmosis, sedimentation, flotation, oxidation, precipitation, evaporation, and adsorption, remain widely used [26,27]. Among these, advanced oxidation processes (AOPs) are particularly effective for treating industrial wastewater [28,29]. Photocatalysis, a subset of AOPs, is especially advantageous because it does not generate secondary pollutants during the treatment process [29,30].

Heterogeneous photocatalysis is considered a highly promising method for breaking down organic pollutants due to its simplicity, high efficiency, and environmentally friendly characteristics. This technique is further valued for its non-toxic nature, low energy consumption, ease of use, and reproducibility [31,32]. It involves generating reactive oxygen species (ROS), such as hydroxyl radicals (OH•), superoxide ions ($•O_2^-$), and hydrogen peroxide (H_2O_2), to oxidize a wide range of organic contaminants. Semiconductor materials play a central role in photocatalysis, with important attributes including optimal band gap values, electronic structure, light absorption efficiency, charge transport, stability, and reusability [33,34]. For decades, metal oxides such as TiO_2, ZnO, SnO_2, and CeO_2 have been extensively utilized as photocatalysts due to their desirable properties [35–40]. More specifically, they have played a crucial role as heterogeneous photocatalysts in various applications [41].

Among these, ZnO is a promising alternative to TiO_2 due to its affordability, accessibility, and environmental compatibility. ZnO nanoparticles exhibit superior photocatalytic efficiency because of their high surface reactivity, which promotes the formation of numerous defect sites due to oxygen non-stoichiometry, making them more effective than many other metal oxides [42]. Additionally, ZnO nanoparticles have shown potential for enhancing the performance of perovskite-based photodetectors [43,44].

Several nanoparticle synthesis methods have been developed, including hydrothermal synthesis, sol–gel techniques, gas-phase processes, thermolysis, and hydrolysis. However, these approaches often face challenges such as high costs, environmental risks, and difficulties in controlling nanoparticle size and surface characteristics [45]. To overcome these issues, green synthesis methods using biological systems, such as plants and fungi, have gained traction [46,47]. These environmentally friendly methods are increasingly being explored as alternatives to conventional physical and chemical techniques [48].

Plants are particularly suitable for large-scale nanoparticle production due to their accessibility, safety, and ability to function as both reducing and stabilizing agents. They facilitate the synthesis of stable nanoparticles with diverse sizes and shapes [47,49–52]. Several plants, including Hibiscus sabdariffa [7], Carica papaya [53], Aloe barbadensis [54], Solanum nigrum [55], Agathosma betulina [56], Ruellia tuberosa [57], Foeniculum vulgare [58], have been employed to synthesize ZnO nanoparticles for various applications.

In this work, a green synthesis method was developed to produce zinc oxide nanoparticles (ZnO NPs) using an extract from *Quercus robur* leaves. This extract, rich in bioactive compounds such as flavonoids and polyphenols, was utilized as both a reducing and stabilizing agent to replace conventional synthesis methods, which are often costly and environmentally damaging. The synthesis parameters were carefully optimized to ensure precise control over the properties of the nanoparticles. The samples were dried at 25 °C and then calcined at 320 °C and 500 °C. The concentrations of the leaf extract varied between 0.03 g/L and 10 g/L. These adjustments allowed for the study of the impact of these parameters on the physical, chemical, and functional characteristics of the ZnO NPs.

Characterization techniques were employed to analyze the synthesized nanoparticles. X-ray diffraction (XRD) was used to determine the crystalline structure and crystallite size while scanning electron microscopy (SEM) provided information on the morphology, size, and distribution of the particles. Fourier transform infrared spectroscopy (FTIR) revealed the functional groups present on the nanoparticles, indicating interactions between the extract compounds and the ZnO NPs. Additionally, UV–Vis spectroscopy was used to study the optical properties of the nanoparticles, including their band gap energy and Urbach energy. The photocatalytic activity of the ZnO NPs was evaluated using methylene blue (MB) as a model organic pollutant. Tests were conducted under UV irradiation to track the degradation kinetics of the dye. This evaluation allowed for an assessment of the nanoparticle efficiency as a function of synthesis conditions, such as extract concentration and calcination temperature. The antimicrobial properties of the ZnO NPs were tested against several microbial strains, including Gram-positive and Gram-negative bacteria, as well as a fungal strain. The minimum inhibitory concentration (MIC) method was used to determine the lowest concentration of nanoparticles required to inhibit microbial growth. The tests evaluated the efficacy of the ZnO NPs in relation to calcination conditions and extract concentrations. Finally, this study explored the potential applications of these nanoparticles in environmental and biomedical fields, highlighting their potential as photocatalysts for pollutant degradation and as antimicrobial agents. This work represents a significant contribution to the development of sustainable and environmentally friendly nanomaterials.

The novelty of this work lies in the use of an extract from *Quercus robur* leaves as a reducing and stabilizing agent for the green synthesis of zinc oxide nanoparticles (ZnO NPs), offering an environmentally friendly and innovative approach to producing multifunctional nanomaterials. Unlike conventional methods that require costly and sometimes toxic chemicals, this approach harnesses natural bioactive compounds, such as flavonoids and polyphenols, to simultaneously control the size, crystalline structure, and optical properties of the ZnO NPs. Furthermore, this study introduces a systematic optimization of synthesis parameters, including calcination temperature and extract concentration, to tailor the properties of the ZnO NPs. This research stands out by combining evaluations of the photocatalytic and antimicrobial performance of the nanoparticles, paving the way for dual applications in environmental remediation and bacterial infection control. Finally, this research makes a significant contribution by integrating green synthesis strategies with thorough characterizations, positioning ZnO NPs as sustainable and versatile materials suited to address contemporary environmental and biomedical challenges.

2. Results

2.1. Characterisation

2.1.1. X-Ray Diffraction

Figure 1 illustrates the XRD patterns of both the as-synthesized and calcined ZnO nanoparticles, which were synthesized using varying concentrations of leaf extract (c = 0, 0.03, 0.1, 2, 10 g/L). The observed diffraction peaks correspond to the hexagonal wurtzite crystal structure of ZnO, in alignment with the data from the JCPDS card No. 36-1451 [59]. These peaks are present at specific angular positions (2θ) of 31.77°, 34.42°, 36.26°, 47.53°, 56.58°, 62.84°, 66.34°, 67.92°, 69.06°, 72.52°, and 76.92°. These positions are attributed to the (100), (002), (101), (102), (110), (103), (200), (112), (201), (004), and (202) Bragg planes, respectively. The diffraction patterns do not display any additional peaks that would suggest the presence of other phases, thereby confirming the high purity of the synthesized ZnO nanoparticles. The crystallite size of the nanoparticles was estimated using Scherrer Equation (1), providing insights into the material's structural properties [42].

$$D = \frac{k\lambda}{\beta \cos\theta} \quad (1)$$

where D is the crystallite size in Å; k is the correction factor that accounts for the shape of the particles ($k = 0.94$); λ is the wavelength of the XRD beam in nm ($\lambda = 0.15406$ nm); β is the full-width at half maximum (FWHM) of the peak in radians, and θ is half of the Bragg's diffraction angle [59].

The microstrain ($\varepsilon\%$) and the density of the dislocations (δ) were computed based on relations (2) (3), respectively [60]:

$$\varepsilon\% = \frac{\beta}{4tan\theta} \quad (2)$$

$$\delta = \frac{1}{D^2} \quad (3)$$

- D denotes the mean crystallite size;
- β represents the full width at half-maximum (FWHM) of the XRD peak, quantified in radians;
- θ represents the Bragg angle, defined as half the angle formed between the incident X-ray beam and the dispersed X-ray beam.

Figure 2 reveals that for both the as-synthesized and the annealed green-synthesized ZnO nanoparticles, increases in lattice parameters and microstrain were observed at the nanometric scale when high concentrations of plant extract were used. This increase in lattice spacing can be attributed to the higher presence of plant-derived molecules that interact with the ZnO structure, potentially causing a modification of its crystallographic arrangement. At the macroscopic scale, however, the crystallite size was found to decrease, which, in turn, led to the creation of structural defects. These defects were reflected in a noticeable increase in dislocation density, which indicates the presence of imperfections in the crystal structure.

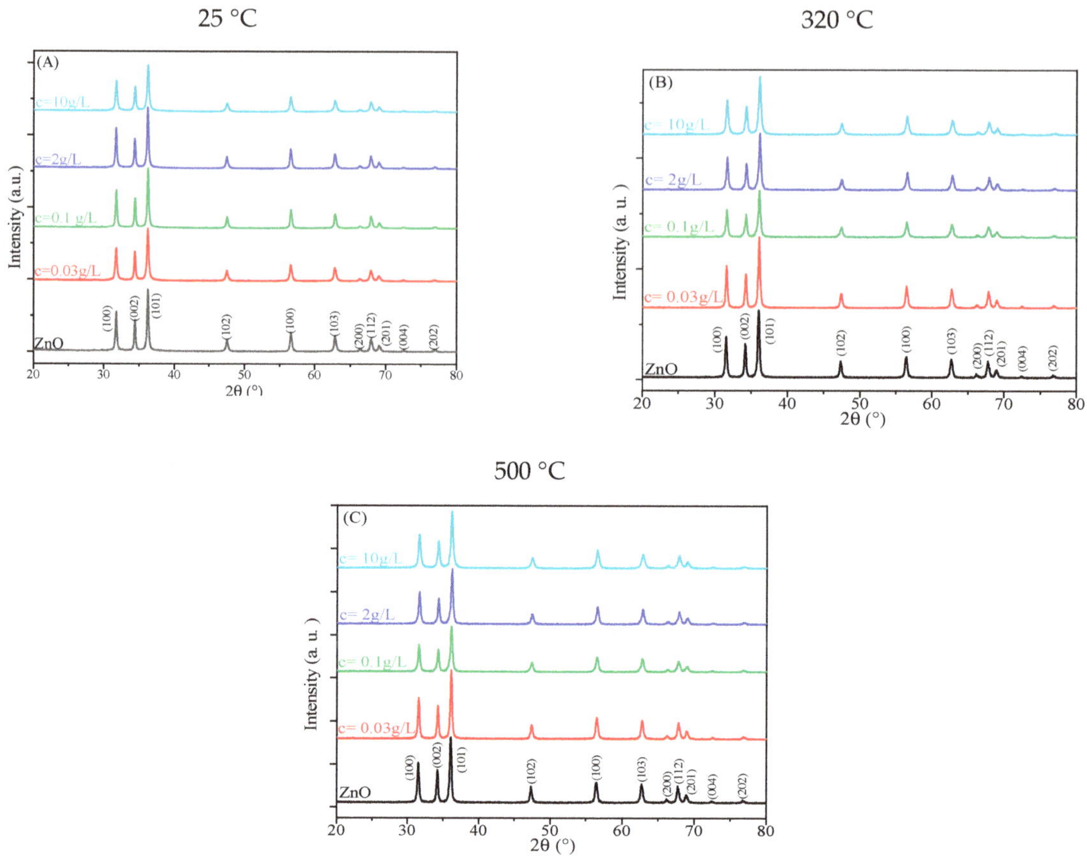

Figure 1. (**A–C**) XRD patterns of ZnO at different calcination temperatures vs. extract concentration.

The results suggest that as the concentration of plant extract increased, the size of the crystalline domains of the green-synthesized ZnO nanoparticles decreased. The decrease in the size of the crystalline domains of the green-synthesized ZnO nanoparticles with an increase in the concentration of plant extract is primarily influenced by the interaction between the plant extract and the ZnO precursor during the synthesis process. As the concentration of the plant extract increases, the functional groups in the extract can better control the nucleation and growth of ZnO nanoparticles, leading to smaller crystalline domains. While thermal energy during the calcination process may also contribute, particularly in influencing crystallization and particle size, it is the concentration of plant extract that more directly drives the observed trend. The reduction in the size of the crystalline domains was primarily attributed to the increased concentration of plant extract, with thermal energy playing a secondary role during the calcination process [61].

These findings emphasize the importance of both the quantity of plant extract and the annealing temperature, as they have a significant impact on the crystallinity, structural defects, and the overall size of the ZnO nanoparticles. This highlights the delicate balance required to optimize the properties of the nanoparticles, ensuring their effectiveness for various applications [62,63].

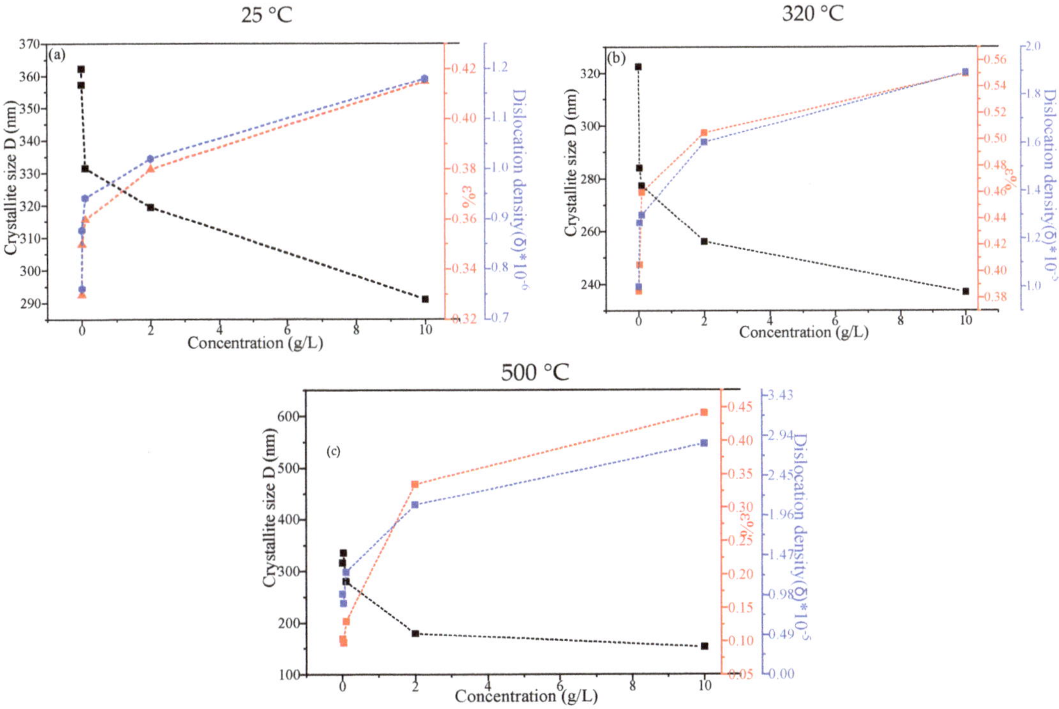

Figure 2. (**a–c**) Crystallite size (*D*), microstrain (ε%), and dislocation density (*δ*) vs. extract concentration.

2.1.2. Scanning Electron Microscopy (SEM)

Figure 3 presents the SEM images of ZnO nanoparticles synthesized via a green method under different annealing temperatures. These images offer valuable insights into the morphology, distribution, and degree of aggregation of the nanoparticles, which predominantly exhibit a spherical shape. Additionally, the particle size distribution of all samples was determined using SEM images and quantitatively analyzed with ImageJ software 1.51p, providing a comprehensive assessment of nanoparticle dimensions [64].

The results reveal a spherical shape and that both the leaf extract concentration and the annealing temperature significantly influence the nanoparticle size. Specifically, in the absence of leaf extract, the average nanoparticle size ranges from 10 to 35 nm. However, increasing the leaf extract concentration results in a noticeable reduction in particle size, albeit irregularly, as shown in Figure 3. This suggests that the leaf extract acts as a capping and stabilizing agent, crucial in controlling nanoparticle size by affecting nucleation and growth processes.

Furthermore, higher concentrations of leaf extract combined with elevated annealing temperatures reduce nanoparticle aggregation and lead to new surface morphologies. These changes, visible in Figure 3, reveal more distinct surface features than nanoparticles synthesized under lower extract concentrations and annealing temperatures. This enhancement in morphology can be attributed to the stabilizing effect of the leaf extract and the increased thermal energy during annealing, which prevent excessive aggregation and promote controlled growth.

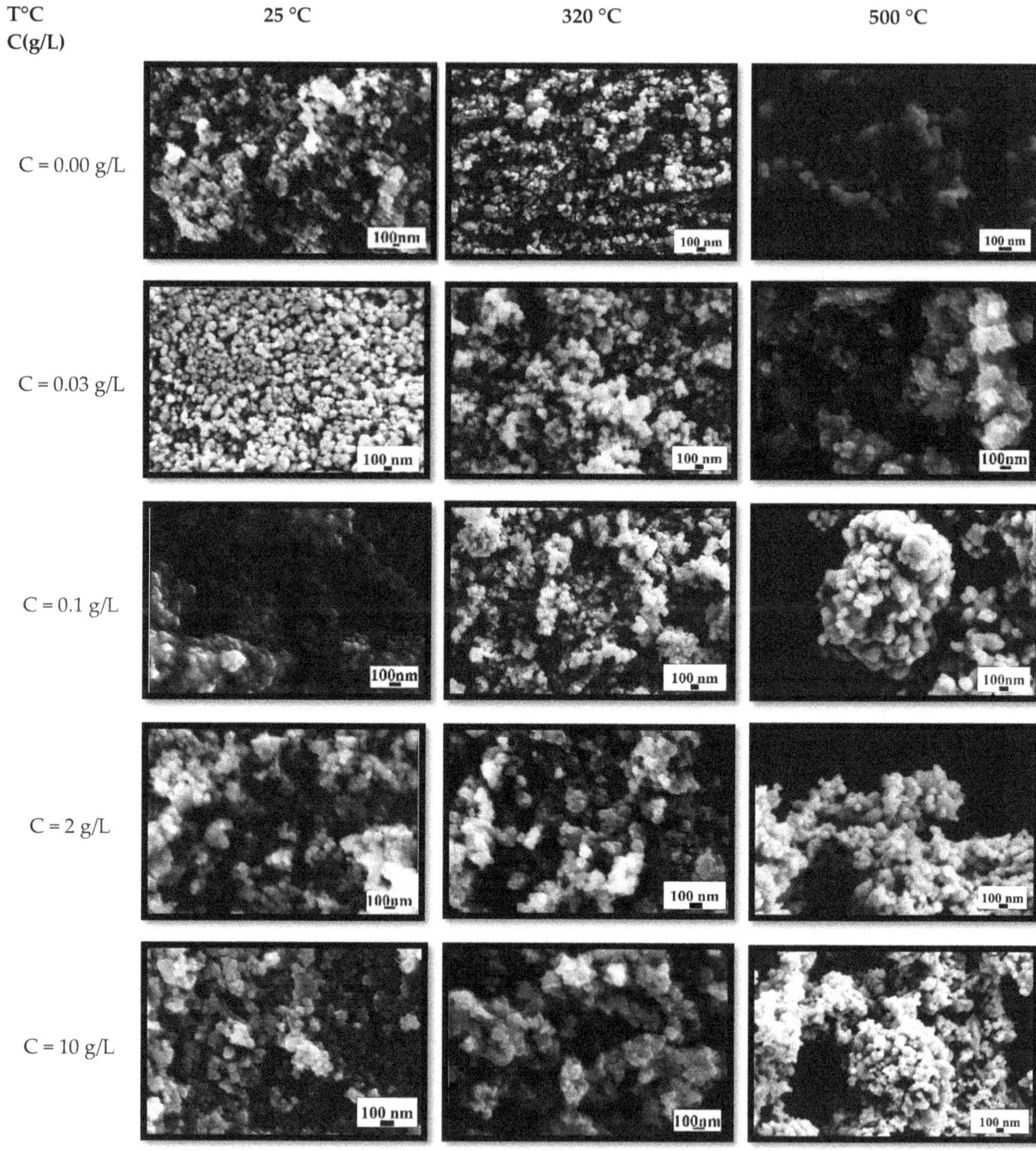

Figure 3. SEM images of green synthesized ZnO NPs. Top images 100 nm.

In summary, the amount of plant extract used in the synthesis of ZnO nanoparticles (ZnO-NPs) plays a crucial role in controlling the nucleation, growth, and stabilization of these particles, ultimately influencing their size and morphology. As the concentration of the extract increases, the abundance of active phytochemical compounds, such as phenolics and flavonoid groups, also rises, enhancing their role as reducing and capping agents. This increased availability of reducing agents promotes faster nucleation of ZnO and inhibits particle growth by preventing agglomeration and uncontrolled crystallization. Studies have shown that when the extract volume is increased, the particle size decreases significantly. For example, the use of higher volumes of Azadirachta indica extract reduced ZnO-NP size from 25–60 nm (at 20 mL) to 8.2–11.9 nm (at 40 mL) due to more uniform nucleation and stabilization [65]. Similarly, varying extract concentrations with Moringa leaf and Dysphania ambrosioides demonstrated a clear trend: higher extract amounts yielded smaller, more stable particles with controlled morphology. The organic functional groups from the extract bind to Zn ions and modulate growth by capping the particles, which inhibits further growth and leads to uniform nanoscale structures [65]. However, insufficient extract levels may result in larger particles and agglomeration due to incomplete capping of ZnO surfaces. Therefore, optimizing the extract amount is key to controlling particle size, which is critical for applications where surface area and reactivity are important [64].

2.1.3. Fourier Transformation Infrared Spectroscopy (FTIR)

Spectroscopic analysis serves as a powerful tool for tracking the chemical and structural transformations that occur during heat treatment processes. Figure 4 presents the FTIR spectra of both as-synthesized and annealed green-synthesized ZnO nanoparticles, providing insight into their molecular characteristics. The Zn–O stretching vibration, which is indicative of the Zn–O bond, is confirmed by a distinct absorption band between 400–600 cm^{-1} [66]. This region is associated with the primary lattice vibration in ZnO, which is an essential feature of its crystal structure. The band observed between 800 and 900 cm^{-1} is attributed to the presence of Zn–OH bonds, suggesting the existence of hydroxyl groups that might be part of the nanoparticle surface or the surrounding environment [67].

Furthermore, vibrations in the 1500–1300 cm^{-1} region are likely due to the presence of hydroxyl (O–H) or carboxylate groups that are bound to the ZnO surface, confirming the interaction between the nanoparticles and these functional groups [68]. The range of 1600–1500 cm^{-1} is particularly significant as it corresponds to the carbonyl (C=O) stretch and other functional groups, such as C=C and −CO, originating from organic residues that are adsorbed onto the surface of the ZnO nanoparticles during synthesis [69]. Additionally, the broad absorption band between 3600–3200 cm^{-1} indicates the stretching vibrations of (OH) groups, which could either be due to adsorbed water molecules or hydroxyl groups that are chemically bound to the surface of the nanoparticles [70].

When ZnO nanoparticles are annealed at 320 °C and 500 °C, a significant reduction in the intensity of the infrared bands is observed. This reduction is likely caused by structural modifications occurring during the annealing process, such as the formation of ZnO-based composites or changes to the nanoparticle surface. These structural changes can disrupt or weaken the interactions between the functional groups and the nanoparticles, leading to the observed decline in band intensity [62].

The decrease in infrared intensity could also suggest a reduction in the number of surface-bound organic residues or hydroxyl groups, further supporting the idea that thermal treatment is influencing nanoparticle surface chemistry.

The FTIR spectrum of the leaf extract powder, which was used in the green synthesis of ZnO nanoparticles, also reveals distinct absorption bands corresponding to various functional groups in the plant material. One such peak, located at 1732 cm^{-1}, is indicative of the carbonyl (C=O) groups present in the leaf extract, which are likely involved in the reduction and stabilization of the nanoparticles [71]. Another prominent peak around 1600 cm^{-1}, within the 1600–1670 cm^{-1} range, reflects the stretching vibrations of carbon–carbon double bonds (C=C), which are common in aromatic compounds and may play a role in stabilizing the nanoparticles [71]. The peak at 1365 cm^{-1} is attributed to the in-plane bending vibration of O–H groups, further confirming the presence of hydroxyl groups in the leaf extract that contribute to the reduction in metal ions during synthesis [72]. Finally, the absorption feature at 1176 cm^{-1} is related to the symmetric stretching of C–O bonds, which could be involved in the coordination between the organic molecules and the ZnO surface [73]. These findings not only provide valuable information about the chemical composition and interactions within the green-synthesized ZnO nanoparticles but also underscore the important role that both the leaf extract and the annealing temperature play in determining the final properties of the nanoparticles. The FTIR analysis offers a comprehensive understanding of how the functional groups in the plant extract interact with the metal oxide, helping to stabilize the nanoparticles and influence their structural characteristics.

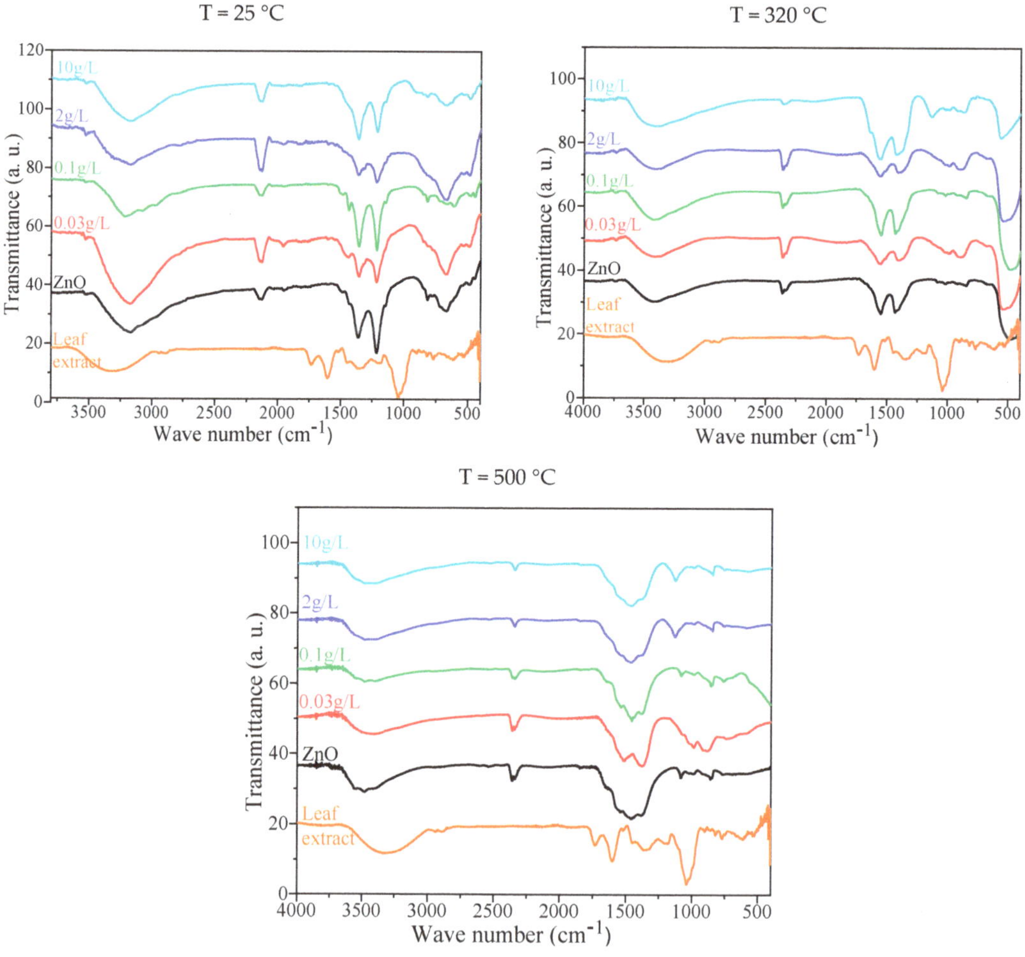

Figure 4. FTIR spectra of green synthesised ZnO NPs.

2.1.4. UV–Vis Spectroscopy

UV–Vis–DRS (Ultraviolet–Visible Diffuse Reflectance Spectroscopy) is a highly effective technique for analyzing the electronic properties of materials, particularly their band gap and Urbach energy, which are crucial for understanding their optical characteristics. Figure 5 presents the diffuse reflectance spectra of ZnO nanoparticles synthesized through a green method. The analysis of these spectra allows for the estimation of the band gap

("E_g") and Urbach energy ("E_u") of the nanoparticles. To quantify these properties, the absorption coefficients, denoted by $F(R)$, were derived from the reflectance data using the Kubelka–Munk Equation (4) [74]:

$$F(R) = \frac{(1-R)^2}{2R} \tag{4}$$

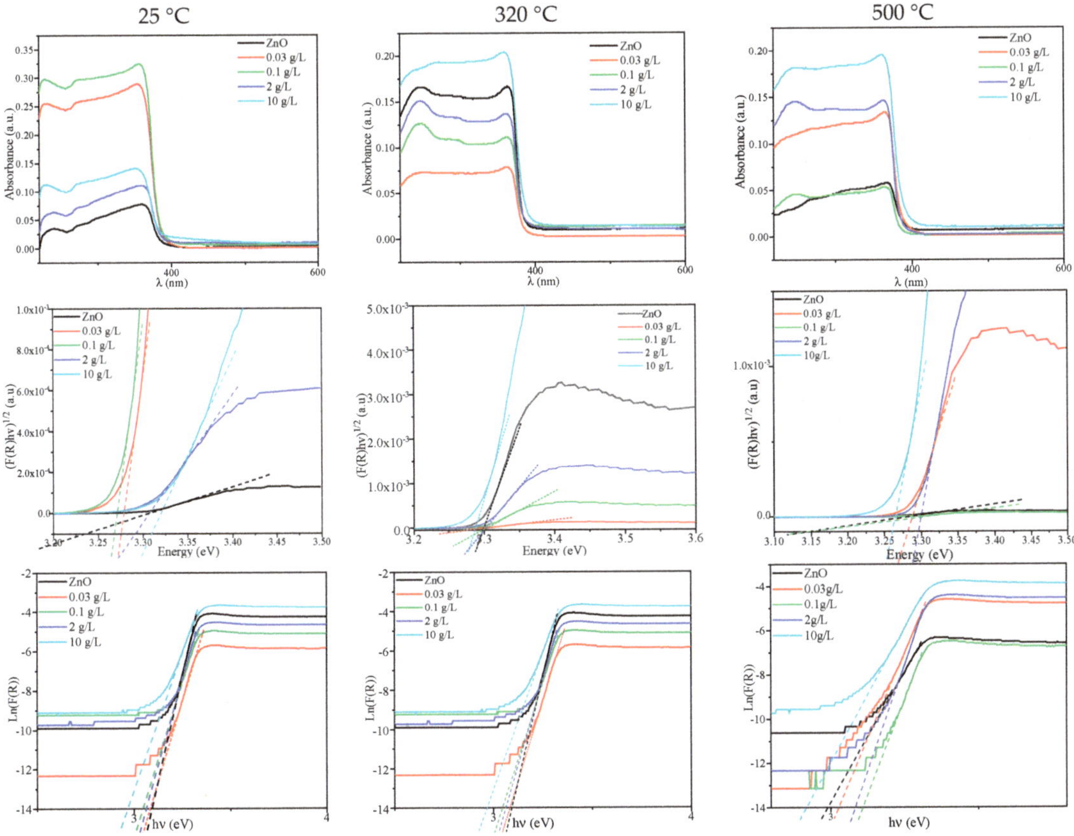

Figure 5. UV–visible spectra of green synthesized ZnO NPs. The inset figure shows a variation in the band gap, e.g., according to the Kubelka–Munk method, and a variation in Urbach energy. All spectra are normalized to one by the maximum absorbance.

In Figure 5 (insert), the variations in both the Urbach energy ("E_u") and the band gap ("E_g") are illustrated according to the Kubelka–Munk method. The band gap ("E_g") is computed using the Tauc plot, which is a graphical method for determining the optical band gap of semiconductors. In the case of ZnO, the Tauc plot's linear portion of the Equation (5) is as follows:

$$(\alpha h\nu)^n \ \alpha h\nu - E_g \tag{5}$$

For direct band gap semiconductors, such as ZnO, the parameter 'n' is taken as 2. The photon energy, $h\nu$, is used, and the corresponding energy values are extracted accordingly.

The results of the band gap measurements are presented in Table 1, where the estimated band gap values for each sample are listed.

Table 1. Direct band gap energy and Urbach energy of green synthesized ZnO NPs at 25 °C (c = 0, 0.03, 1, 2, and 10 g/L).

Concentration (g/L)	ZnO	0.03	0.1	2	10
Direct gap energy (eV)	3.29	3.29	3.29	3.28	3.29
Urbach energy (eV)	2.94	2.90	3.01	2.97	2.90

The Urbach energy ("E_u") is an important parameter that gives insight into the tailing of the density of states near the conduction band edge. It is calculated using the following Equation (6), where E_0 and α_0 are constants that depend on the specific material being studied [74]:

$$\alpha = \alpha^0 exp\frac{(hv - E^0)}{Eu} \quad (6)$$

The Urbach energy ("E_u") can be extracted from the slope of the linear portion of the plot of ln[$F(R)$] vs. hv, where hv is the photon energy. The Urbach energy reflects the degree of disorder in the material's electronic structure, which is influenced by various defects such as vacancies, interstitial sites, lattice strain, and dislocations. These structural features contribute to the broadening of the band tail and affect the material's optical properties.

UV–Vis–DRS spectroscopy, by directly analyzing the absorption and reflection spectra, provides valuable experimental data for determining key optical properties like the band gap and Urbach energy. Figure 3 showcases the spectra of ZnO nanoparticles that were synthesized using green methods at room temperature (25 °C), as well as those that were annealed at higher temperatures of 320 °C and 500 °C.

At the ambient temperature of 25 °C, the values for E_c (conduction band energy) and E_g (band gap energy) remain constant across all leaf extract concentrations (show Table 1). However, when the nanoparticles are annealed at 320 °C, the E_g values remain relatively uniform for all concentrations of leaf extract, while the Urbach energy decreases as the concentration of leaf extract increases (show Table 2). This suggests that higher concentrations of leaf extract may promote the formation of a more ordered crystal structure with fewer defects, thereby lowering the Urbach energy.

Table 2. Direct band gap energy and Urbach energy of green synthesized ZnO NPs annealed at 320 °C (c = 0, 0.03, 1, 2, and 10 g/L).

Concentration (g/L)	ZnO	0.03	0.1	2	10
Direct gap energy (eV)	3.28	3.27	3.25	3.29	3.28
Urbach energy (eV)	3.05	3.03	2.99	3.01	2.90

At an annealing temperature of 500 °C, both the E_c and E_g values exhibit an increase with higher concentrations of leaf extract (show Table 3). This trend is indicative of changes in the crystallinity and electronic structure of the nanoparticles as the temperature and extract concentration are varied. By comparing the calcined ZnO nanoparticles with those that were synthesized without annealing, a significant reduction in the band gap (E_g) is observed for the annealed nanoparticles. This decrease in the band gap is associated with an increase in the crystallite size of the nanoparticles, which tends to occur at higher calcination temperatures. As the crystallite size increases, the energy levels in the material become more defined, leading to a narrowing of the band gap.

Table 3. Direct band gap energy and Urbach energy of green synthesized ZnO NPs annealed at 500 °C (c = 0, 0.03, 0.1, 2 and 10 g/L).

Concentration (g/L)	ZnO	0.03	0.1	2	10
Direct gap energy (eV)	2.98	3.01	3.15	3.27	3.26
Urbach energy (eV)	2.89	3.01	3.10	3.07	2.90

Essentially, the increase in calcination temperature promotes the growth of ZnO nanoparticles, which, in turn, leads to the reduction in the band gap. This phenomenon is consistent with the general principle that larger particles typically exhibit a smaller band gap due to the quantum size effects becoming less pronounced as the particle size increases. Therefore, a narrower band gap results when the nanoparticles grow larger as the calcination temperature is increased [42]. This observation highlights the critical role of both the leaf extract concentration and the annealing temperature in modulating the size, electronic properties, and optical behavior of ZnO nanoparticles.

2.2. Photocatalysis

2.2.1. Photocatalytic Activity

A series of ZnO nanoparticles were synthesized using a green synthesis method with varying calcination temperatures to examine the influence of these temperatures on the photocatalytic performance of the ZnO semiconductors. The synthesis utilized different concentrations of leaf extract (c = 0, 0.03, 0.2, 2, and 10 g/L) to test the ability of the ZnO nanoparticles to degrade MB under UV light irradiation. The degradation process was monitored using UV–visible spectroscopy, and the decay of the normalized concentration (C(t)/C0) of MB was recorded at a wavelength of $\lambda = 664$ nm. The kinetic data were then fitted to an exponential decay model described by Equation (7):

$$\frac{C(t)}{C0} = A_0 + B_0\, e^{-k_{ap}t} \quad (7)$$

In this equation, A_0 represents the integrated baseline, and $A_0 + B_0$ corresponds to the initial concentration of MB at t = 0 (after adsorption). The term k_{ap} denotes the apparent photocatalytic rate constant, which reflects the rate at which the ZnO nanoparticles break down the MB under UV light exposure.

The findings showed that the photocatalytic degradation of MB was more efficient for ZnO nanoparticles calcined at 320 °C and 500 °C compared to those that were not calcined. This improvement in photocatalytic efficiency can be attributed to several factors related to changes in the morphology and optical properties of the nanoparticles during calcination. Higher calcination temperatures likely increased the surface area of the ZnO nanoparticles, which is beneficial for photocatalytic reactions, as a larger surface area provides more active sites for the adsorption and degradation of the dye molecules. Additionally, the calcination process may have improved the crystallinity of the ZnO nanoparticles, making them more efficient in promoting photocatalytic reactions [75,76]. Using the pseudo-first-order reaction rate equation, the following equation is used:

$$\mathrm{Ln}\left(\frac{C0}{Ct}\right) = kt \quad (8)$$

This equation allows for the determination of the rate constant *kt*, where C0 is the initial concentration of MB, and Ct is the concentration of MB at time t. By plotting ln(C0/Ct) against reaction time, the rate constant can be extracted, providing insight into the efficiency of the photocatalytic degradation process. Tables 4–6 show the calculated rate constants for each experimental condition.

Table 4. Elimination efficiency and apparent rate constants of the photocatalysis (25 °C).

Concentration (g/L)	0	0.03	0.1	2	10
E%	50	44	38	35	26
k (10^{-3} min^{-1})	4.90	3.85	3.14	2.66	1.90

Table 5. Elimination efficiency and apparent rate constants of the photocatalysis (320 °C).

Concentration (g/L)	0	0.03	0.1	2	10
E%	62	74	55	53	47
k (10^{-3} min^{-1})	6.90	9.09	5.54	5.05	3.89

Table 6. Elimination efficiency and apparent rate constants of the photocatalysis (500 °C).

Concentration (g/L)	0	0.03	0.1	2	10
E%	56	74	62	60	50
k (10^{-3} min^{-1})	4.52	6.41	5.49	4.93	3.70

The results indicated that both the concentration of leaf extract used in the synthesis of the ZnO nanoparticles and the calcination temperature significantly impacted the degradation rates. The calcination temperature plays a pivotal role in determining the photocatalytic performance of materials, particularly in the context of materials calcined at 320 °C and 500 °C. Research suggests that this temperature range often leads to an optimal balance between crystallite size, surface area, and the material's optical properties, which are crucial for photocatalysis. The calcining ZnO materials at 500 °C results in an increased crystallinity and surface area, facilitating the efficient separation of photogenerated electron–hole pairs, a critical factor in photocatalytic degradation. Additionally, the calcination process at these intermediate temperatures is ideal for maintaining a high surface-to-volume ratio while preventing excessive sintering, which would otherwise hinder the material's photocatalytic capabilities. Moreover, the calcination temperature of 320 °C plays an important role in the green synthesis of ZnO nanoparticles, especially in terms of their photocatalytic performance. In the context of green synthesis, this temperature range is typically used to avoid high-energy consumption while ensuring the formation of crystalline and pure ZnO. At 320 °C, the precursor material undergoes decomposition, which promotes the development of ZnO with optimal crystallinity while preserving the eco-friendly aspects of the synthesis method. Research indicates that calcination at this moderate temperature helps achieve the right balance between particle size and surface area, both of which are crucial for photocatalytic efficiency. For instance, studies have demonstrated that ZnO nanoparticles calcined at lower temperatures (around 320 °C) tend to have a high surface area, which is beneficial for photocatalytic degradation due to an increased number of active sites available for reactions. Additionally, green synthesis methods, such as using plant extracts, are employed to minimize harmful chemical residues, contributing to both environmental sustainability and enhanced photocatalytic properties. The low-temperature calcination ensures the stability of these materials, which is vital for their application in long-term photocatalytic processes. Thus, the 320 °C calcination temperature is often chosen for its ability to optimize the physical properties of ZnO while maintaining an energy-efficient and environmentally friendly process [42].

Furthermore, the impact of the plant extract on the ZnO NPs and calcination temperature deserves attention. The plant extract reduces the particle size of ZnO, increasing the specific surface area accessible for catalytic reactions. This expanded surface area exposes more active sites, allowing for better adsorption of MB dye molecules and more effective catalytic turnover. The plant extract may add catalytic sites to the surface of ZnO NPs, increasing their activity. Adding more active sites to the photocatalyst improves its overall efficiency by providing extra catalyst centers [74].

The calcination temperature plays a crucial role in enhancing the photocatalytic performance of ZnO nanoparticles. We studied the photodegradation of BM dye using green-synthesized ZnO NPs (c = 0, 0.03, 0.5, 1, 2 g/L) at different calcination temperatures, as illustrated in Figure 6. Higher temperatures result in larger particles with better crystallinity, which are more effective at degrading pollutants like MB. Furthermore, the concentration of leaf extract used during synthesis also affects the photocatalytic efficiency, with higher concentrations generally leading to improved degradation performance. These findings highlight the importance of optimizing both the synthesis conditions and calcination parameters to maximize the photocatalytic potential of ZnO nanoparticles for environmental applications. In fact, the key factors influencing photocatalytic performance, such as crystallinity, surface area, and the presence of active sites, may stabilize at moderate calcination temperatures. Temperatures like 320 °C and 500 °C provide sufficient enhancement without introducing detrimental effects. ZnO synthesized through green methods has shown that both higher and lower calcination temperatures could lead to similar photocatalytic outcomes when specific material properties, such as particle size and surface defects, are optimized. For instance, ZnO calcined at 500 °C has been found to perform similarly to materials calcined at 320 °C in terms of photocatalytic efficiency despite differences in crystallinity. This suggests that the optimal temperature for photocatalytic performance may depend more on the specific properties of the synthesized material rather than solely on a higher calcination temperature. The recommendation to reconsider the assertion that the material calcined at 500 °C is the best is, therefore, well-supported, as both temperatures offer favorable outcomes for photocatalytic degradation, particularly under UV irradiation.

2.2.2. Photocatalytic Cycling Test for Methylene Blue Degradation Under UV Light

Photostability plays a vital role in determining the long-term functionality and practicality of a photocatalyst. To comprehensively evaluate the structural integrity and catalytic performance of the ZnO nanocomposite, a series of three consecutive photocatalytic degradation tests of MB were carried out. This approach not only provides insight into the material's catalytic efficiency but also examines how well it can sustain its activity over multiple cycles of degradation.

Each photocatalytic test involved the degradation of MB under UV light, followed by a precise recycling process. After the completion of each test, the nanocomposite was separated from the reaction mixture through centrifugation, a key step to isolate the photocatalyst. To ensure the removal of any residual contaminants and maintain the material's catalytic potential, the nanocomposite underwent a thorough three-step washing procedure. This involved washing with distilled water, followed by a mixture of distilled water and ethanol, and finally, a wash with double-distilled water. These steps helped eliminate any adsorbed organic molecules or by-products that could affect the photocatalyst's performance in subsequent cycles. After the washing procedure, the ZnO material was carefully collected and dried at 60 °C for 24 h. The drying step was essential to ensure that the photocatalyst was completely free of any solvents and ready for reuse in the next cycle.

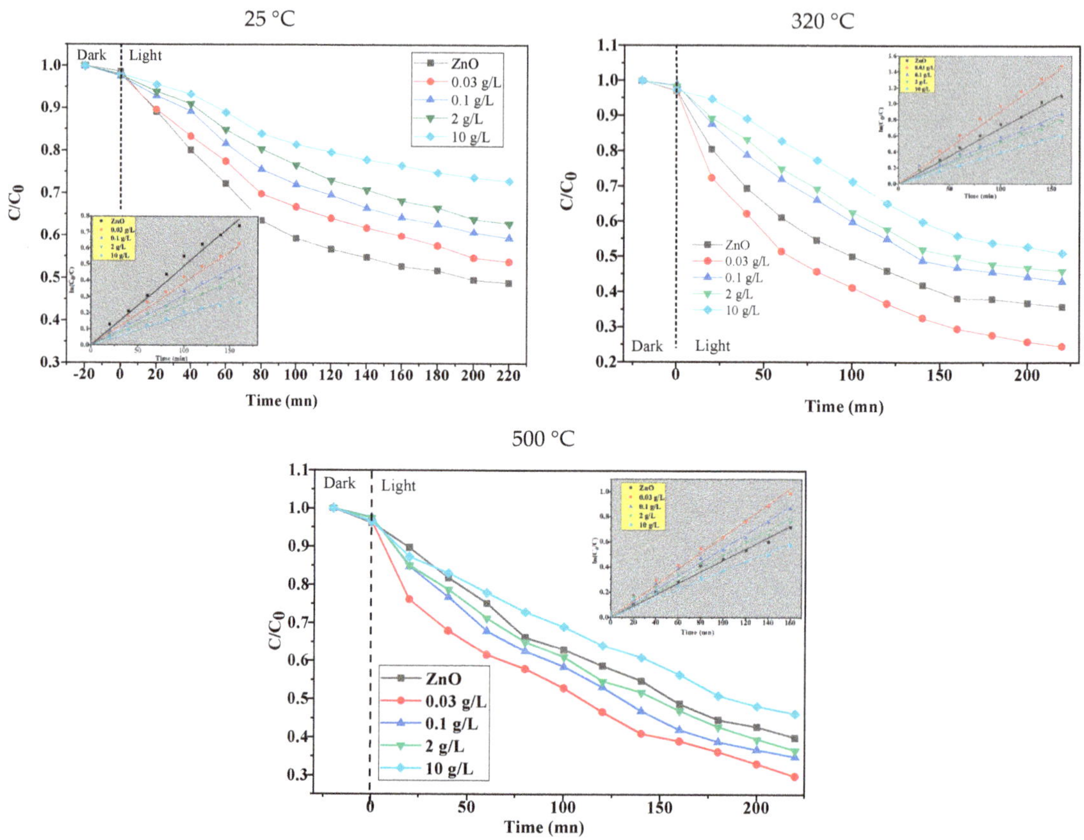

Figure 6. Photodegradation of BM dye using green synthesized ZnO NPs (c = 0, 0.03, 0.5, 1, 2 g/L) at different calcination temperatures.

Figure 7 presents the results of the three consecutive photocatalytic degradation cycles, with the data showing consistent photocatalytic activity across all cycles. This stability indicates that the ZnO nanocomposite retained its efficiency in degrading MB under UV light despite undergoing multiple testing cycles. The results highlight the intrinsic stability and resilience of the pre-synthesized and annealed ZnO nanocomposite. This suggests that the material has a robust structure capable of maintaining its catalytic properties over extended use, a crucial factor for any practical application of photocatalysts in environmental remediation processes.

1. Surface Degradation: Prolonged exposure to reactive species or harsh conditions may lead to surface degradation of ZnO, such as oxidation, reduction, or the formation of defects that diminish its active sites;
2. Structural Changes: High temperatures or repeated cycles of catalysis can induce morphological or structural changes in ZnO, such as sintering or particle agglomeration, reducing the surface area available for catalysis;
3. Contamination or Fouling: The active surface of ZnO may become contaminated by impurities, byproducts, or undesired species, leading to a reduction in catalytic efficiency over time;

4. Leaching of Active Sites: In liquid-phase catalysis, ZnO may experience leaching, where catalytic components are gradually dissolved into the reaction medium, leading to performance decay;
5. Formation of Inactive Phases: Chemical interactions between ZnO and the reaction medium could result in the formation of secondary, less active, or inactive phases, altering the material's catalytic behavior;
6. Photo-Induced Changes: For photocatalytic applications, prolonged exposure to UV or visible light can induce photocorrosion or electron–hole recombination, which can degrade the material over time.

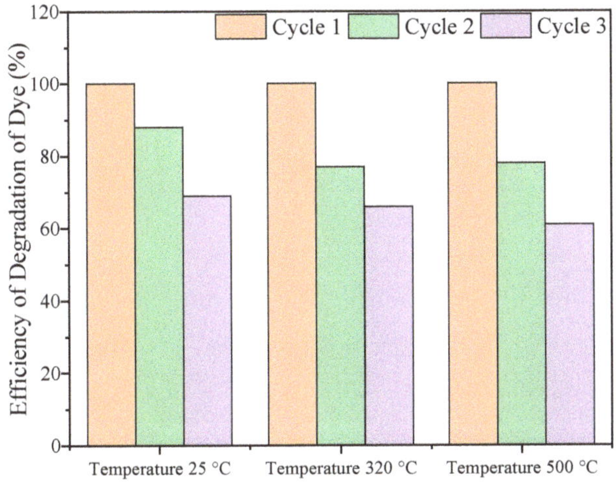

Figure 7. Comparative Performance of Recycled ZnO Photocatalyst.

The decline in photocatalytic performance of ZnO over successive reuse cycles is generally attributed to several factors. First, the surface deactivation of ZnO can occur due to the accumulation of by-products or contaminants from the reactions, which block the active sites necessary for photocatalysis. Additionally, the crystallinity of ZnO nanoparticles may degrade over time, especially under harsh conditions during photocatalytic reactions (e.g., UV light exposure). This degradation can lead to a reduction in the available surface area, further diminishing the material's efficiency. Another key factor is the leaching of ZnO components during the photocatalytic process, which can result in a loss of catalytic activity with repeated cycles. Finally, structural changes, such as particle agglomeration or morphological changes, could also hinder the photocatalytic process by reducing the material's surface-to-volume ratio and available active sites.

2.2.3. Effect of Catalyst Dosage

Figure 8 illustrates the relationship between catalyst dose and photocatalytic activity, measured by the C/C_0 ratio over time, for varying catalyst concentrations of 0.25, 0.5, 1, and 2 g/L. At lower concentrations of 0.25 g/L and 0.5 g/L, the degradation of MB is moderate, with efficiency levels that are significantly lower compared to the 1 g/L concentration Table 7. This suggests that lower concentrations do not provide enough active surface area for efficient photocatalytic reactions, leading to incomplete degradation of the MB dye.

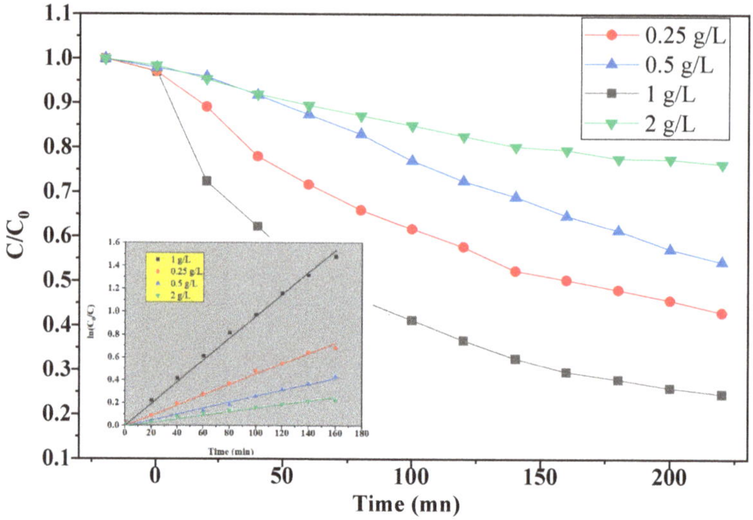

Figure 8. Effect of catalyst dose on photodegradation of MB by green synthesized ZnO (neutral pH, [MB] = 10 mg/L).

Table 7. Rate constant K of degradation of MB with different dosages of ZnO.

Concentration (g/L)	0.25	0.5	1	2
E%	66	44	74	22
k (10^{-3} min^{-1})	4.54	2.61	9.57	1.88

When the concentration is increased to 1 g/L, the degradation efficiency reaches an optimal value of 74%. This improvement can be attributed to a sufficient specific surface area and effective light transmission, which minimizes shading or saturation effects. At this concentration, the catalyst particles are dispersed enough to allow light to reach all active sites, promoting enhanced photon absorption. This, in turn, leads to the generation of more electron–hole (e^-, h^+) pairs, which are crucial for the photocatalytic process, and increases the production of reactive radicals that facilitate the breakdown of organic contaminants such as MB [77].

However, at the highest concentration of 2 g/L, the photocatalytic activity drops dramatically, with only 22% degradation observed. This decrease in efficiency is primarily due to two factors: shading and saturation. As the concentration increases beyond 1 g/L, the catalyst particles tend to agglomerate or overlap, blocking light from reaching the active sites on the particles. This shading effect prevents efficient photon absorption and hampers the photocatalytic process. Furthermore, the saturation effect occurs when the number of active sites becomes limited, meaning that even though the catalyst concentration is higher, there are not enough available sites to accommodate the increased amount of particles. This leads to a decrease in the overall catalytic performance.

The results from this study underline the importance of selecting the optimal catalyst concentration for photocatalytic applications. Too low concentrations result in inefficient use of the available surface area, while too high concentrations lead to inefficiencies due to shading and saturation effects. The 1 g/L concentration strikes a balance, offering the best combination of surface area and light transmission, ensuring maximum degradation of MB. This emphasizes that for efficient photocatalytic degradation, careful control over the catalyst concentration is necessary to avoid diminishing returns. Thus, 1 g/L is identified as the optimal concentration, where the photocatalytic activity is maximized, demonstrating the critical role of catalyst dosage in photocatalytic processes.

2.2.4. Effect of Reaction pH

Figure 9 provides a detailed overview of the effect of pH on the photocatalytic degradation of MB, with results measured by the C/C_0 ratio as a function of time across a range of pH values (4, natural pH, 8, and 10). The data reveals that the natural pH condition (pH 7) is optimal for photocatalytic activity, achieving a degradation rate of 74%. This suggests that the natural pH maintains the stability of the ZnO photocatalyst surface and allows for efficient interaction between the photocatalyst and the pollutant, promoting the generation of active species required for degradation.

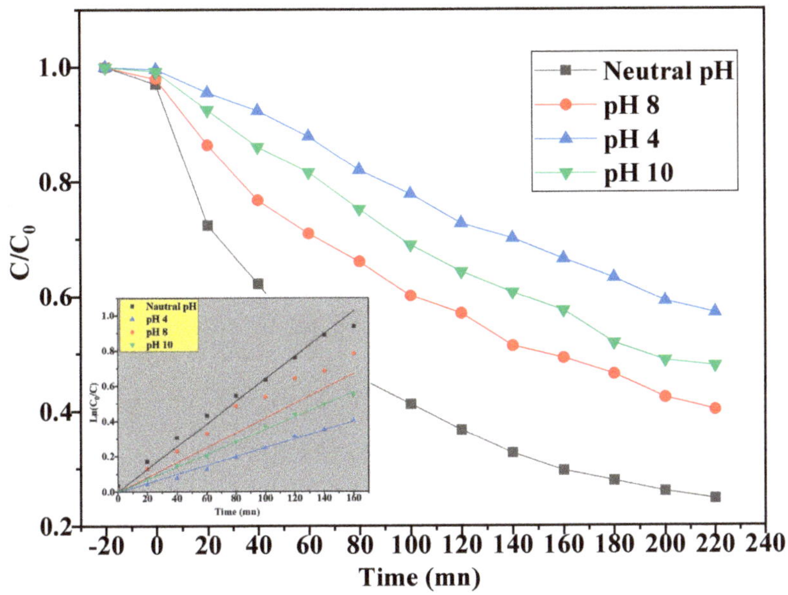

Figure 9. Effect of reaction pH on photodegradation of MB by ZnO (catalyst one g/L, [MB] = 10 mg/L).

At pH 8, a slight decrease in degradation efficiency was observed, with a reduction of 58%. This result still indicates considerable photocatalytic activity, although it is not as efficient as at natural pH. The slightly basic environment may slightly alter the surface charge or the ability of the photocatalyst to interact effectively with the pollutant molecules, thus reducing the overall efficiency of the degradation process.

In contrast, the lowest pH value tested, pH 4, resulted in only 42% degradation, while at pH 10, the degradation was 51% (Table 8). These findings suggest that both acidic and basic environments are less ideal for photocatalytic activity. At acidic pH, the surface of the photocatalyst may develop a highly positive charge due to protonation, which could reduce its interaction with certain negatively charged pollutant ions. Additionally, ZnO is known to be sensitive to weak acids like acetic acid, leading to corrosion of the material and a subsequent decrease in catalytic performance. This degradation effect might explain the observed reduction in activity. Studies on the functionalization of ZnO powders for applications such as gas sensors may provide insights into strategies for improving its stability under acidic conditions.

Table 8. Rate constant K of degradation of MB with different pH reactions.

pH	4	neutral	8	10
E%	42	74	58	51
k (10^{-3} min^{-1})	2.52	6.41	5.24	3.56

This study underscores the importance of maintaining a neutral or slightly acidic pH range for optimal photocatalytic performance. The results suggest that pH significantly influences the photocatalytic degradation of MB, with neutral to slightly acidic conditions (pH 7–8) being the most favorable for achieving the highest degradation efficiency. On the other hand, very acidic (pH 4) or basic (pH 10) conditions diminish the photocatalytic efficiency, likely due to destabilization of the catalyst's surface or unfavorable interactions with reactive species.

2.2.5. Effect of Initial Concentrations of MB

To investigate the impact of initial dye concentration on the photocatalytic degradation of MB, a series of experiments were conducted by varying the concentration of MB from 10 to 30 mg/L. The photocatalytic degradation process was monitored for a reaction time of 220 min, and the dye removal efficiency was calculated for each concentration. The results, shown in Figure 10, indicated that the removal efficiency decreased as the dye concentration increased. Specifically, at 10 mg/L MB, a high removal rate of 74% was achieved, with a corresponding rate constant (k) of 9.57×10^{-3} min^{-1}. In contrast, for 20 mg/L, the removal efficiency dropped to 48% (k = 3.05×10^{-3} min^{-1}), and at the highest concentration of 30 mg/L, the removal efficiency further decreased to 29% (k = 1.68×10^{-3} min^{-1}) as it shown in Table 9.

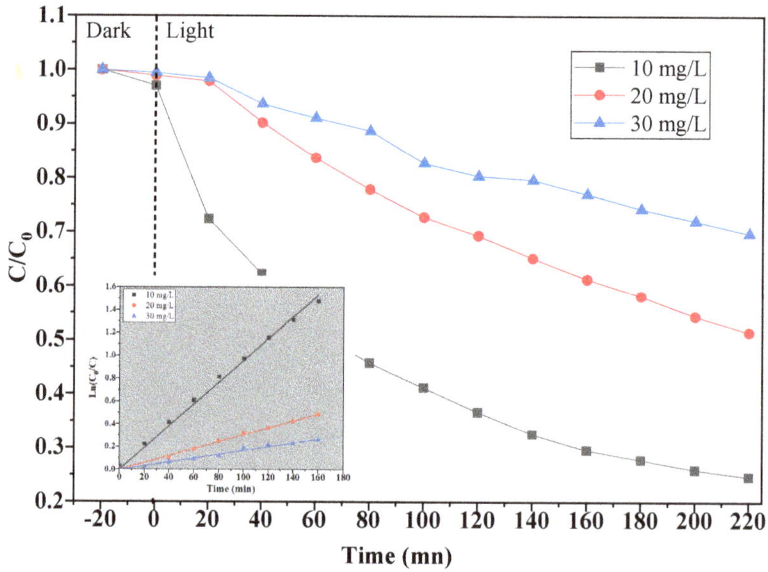

Figure 10. Effect of the initial concentration of methylene blue on photodegradation performance of ZnO NPs.

This decline in removal efficiency with increasing MB concentration can be attributed to several factors. As the dye concentration increases, more MB molecules adsorb onto the surface of the photocatalyst, occupying the active sites required for the adsorption of reactive species such as oxygen and hydroxide ions. This reduces the available surface area for the formation of reactive radicals (such as hydroxyl and superoxide anion radicals), which are essential for the photocatalytic degradation process. As a result, the efficiency of the photocatalytic reaction is compromised.

Table 9. Rate constant K of degradation of MB with different initial concentration of dye.

Concentration (mg/L)	10	20	30
E%	74	48	29
k (10^{-3} min^{-1})	9.57	3.05	1.68

Furthermore, at higher dye concentrations, the increased absorption of light by the dye molecules reduces the amount of light available for the photocatalyst, leading to lower photon absorption. This phenomenon, known as light shielding or light blocking, further reduces the efficiency of the photocatalytic process. As the concentration of MB increases, the dye molecules form a layer on the surface of the photocatalyst, obstructing the effective interaction between the catalyst and the light energy [78–80].

Overall, these results demonstrate that the photocatalytic degradation of MB is less effective at higher initial dye concentrations. The key factors contributing to this decrease in efficiency include the competition for active sites on the catalyst surface, reduced production of reactive radicals, and diminished light absorption due to photon shielding effects. Therefore, to optimize the photocatalytic degradation process, a moderate dye concentration is preferable, as it ensures maximum active site availability and efficient photon absorption by the photocatalyst.

2.2.6. Effect of Scavengers

In the photocatalytic degradation process, reactive species such as hydroxyl radicals (•OH), superoxide anions ($•O_2^-$) and electron–hole pairs (e^-/h^+) are generated and play a crucial role in the breakdown of pollutants like MB. To identify which specific reactive species are responsible for the degradation, selective scavengers or inhibitors are used in a test. These scavengers are chemical compounds that can effectively capture or neutralize a specific reactive species, preventing it from participating in the reaction. By introducing these scavengers into the reaction system, it is possible to determine the role of each reactive species in the photocatalytic process.

Typically, the experiment is set up by maintaining the same experimental conditions, including the photocatalyst concentration, plant extract concentration, and UV light intensity. A series of tests are carried out by adding different types of scavengers to the solution. These scavengers are chosen based on their ability to specifically react with and neutralize a particular reactive species. For example, t-butanol is commonly used as a scavenger for hydroxyl radicals, while potassium bromide (KBr) is used to capture superoxide anions.

After the scavenger is added, the photodegradation efficiency of the substrate is measured, usually by tracking the decrease in the concentration of the pollutant (e.g., MB) over time, often using UV–Vis spectroscopy. The comparison of photodegradation efficiency in the presence of each scavenger helps to pinpoint which reactive species are most responsible for the catalytic breakdown of the substrate.

In this way, the experiment allows for a clearer understanding of the mechanisms driving photocatalysis and can help in optimizing the photocatalytic process by highlighting which species need to be generated in larger quantities for more efficient pollutant removal. Additionally, this approach helps to identify potential limitations in the photocatalytic reaction, guiding further improvements in catalyst design or reaction conditions.

This study aimed to understand the roles of various reactive species involved in the photocatalytic degradation of MB dye by selectively trapping specific species such as •OH, e^-, and h^+. Ethanol was employed to capture hydroxyl radicals; $K_2Cr_2O_7$ was used to trap electrons, and ethylenediaminetetraacetic acid (Na_2-EDTA) was utilized to sequester holes. These scavengers were added to the reaction mixture under optimized experimental conditions, with the irradiation duration maintained at a fixed period of 220 min. The results, as illustrated in Figure 11, demonstrate significant variations in photocatalytic efficiency depending on the type of scavenger used.

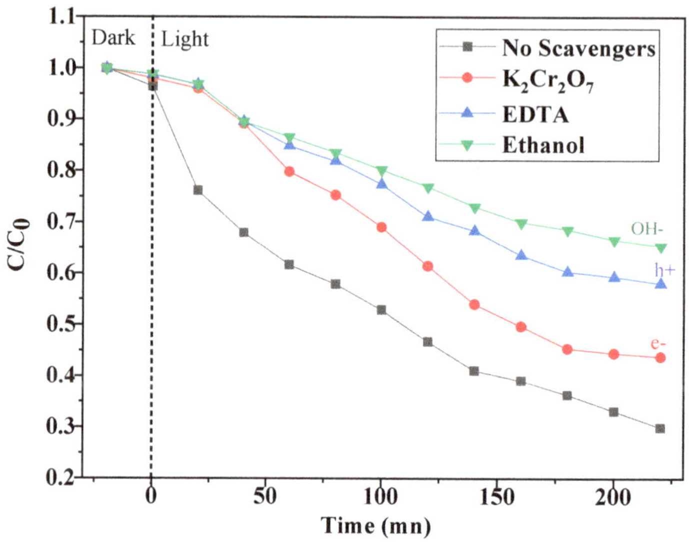

Figure 11. Effect of scavengers on the photocatalytic activity of ZnO NPs.

The purpose of using selective trapping agents was to isolate and quantify the individual contributions of these reactive species to the overall photocatalytic degradation of MB. Figure 12 clearly shows the impact of these scavengers on the degradation process. When ethanol was used to trap •OH, the photocatalytic efficiency decreased significantly, with a degradation rate of only 34%. This is considerably lower than the degradation rate of 67% observed when no scavengers were used, indicating that •OH radicals are primarily responsible for the majority of the degradation process.

In contrast, when h^+ ions were trapped using EDTA, the degradation rate was 41%. This result suggests that the photogenerated holes also play an essential role, although their contribution is less significant than that of the hydroxyl radicals. The presence of $K_2Cr_2O_7$ to trap e^- led to a degradation rate of 54%, showing that electrons also contribute substantially to the photocatalytic activity.

Overall, these findings demonstrate that multiple reactive species, including •OH radicals, h^+ ions, and e^-, are involved in the photocatalytic degradation of MB dye, with each species contributing to varying extents. The primary role of hydroxyl radicals in the degradation process is evident from the significant decrease in degradation efficiency when they were trapped. The results also suggest that optimizing the photocatalytic process could involve enhancing the production or efficiency of these key reactive species to improve degradation rates further.

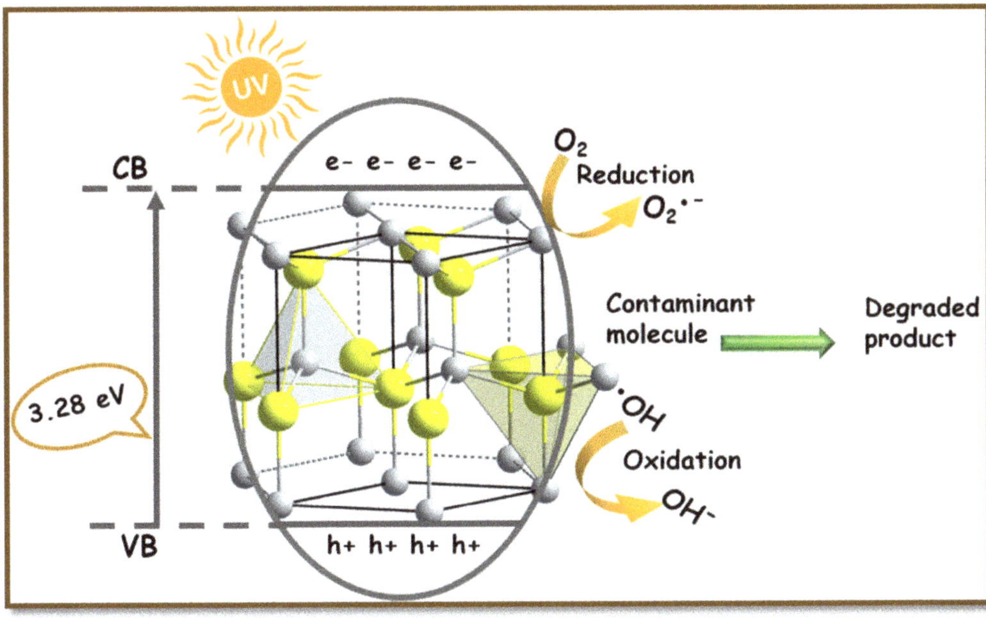

Figure 12. Schematic illustration of the photocatalytic mechanism of ZnO.

2.3. Photocatalytic Mechanism

ZnO functions as an efficient photocatalyst, leveraging light to accelerate various chemical reactions, making it a crucial material in numerous environmental and energy-related applications. Its exceptional photocatalytic properties stem from its wide bandgap and the ability to form electron–hole pairs when exposed to light. These electron–hole pairs are essential for initiating the photocatalytic process, as they participate in redox reactions that degrade pollutants and harness energy from light.

The principle behind ZnO's photocatalytic activity is that when it absorbs ultraviolet (UV) light, the photonic energy matches or exceeds its excitation energy (Eg), electrons in the valence band of ZnO are excited to the conduction band (CB). This excitation creates an electron–hole pair. These electron–hole pairs migrate to the surface of ZnO, where they interact with surrounding molecules, contributing to a series of redox reactions. The h^+ ions are highly reactive and can react with water and hydroxide ions to form •OH, one of the most potent oxidizing agents in photocatalysis. The e^- ions react with molecular O_2 to form $\bullet O_2^-$, which are also reactive and involved in the degradation process.

In addition to these fundamental reactions, H_2O_2 is formed as a byproduct of electron interaction with oxygen, which further contributes to the generation of •OH. These hydroxyl radicals, superoxide anions, and other reactive species collectively degrade pollutants on the surface of the ZnO photocatalyst. This process not only purifies the surrounding environment by breaking down harmful substances but also releases oxygen as a byproduct [81–83]. The consisted response equations are given below as follows:

$$ZnO + h\upsilon \rightarrow h^+ + e^- \quad (9)$$

$$h^+ + H_2O \rightarrow {}^{\bullet}OH + H^+ \quad (10)$$

$$h^+ + OH^- \rightarrow {}^{\bullet}OH \quad (11)$$

$$h^+ + \text{pollutant} \rightarrow (\text{pollutant})^+ \quad (12)$$

$$e^- + O_2 \rightarrow \bullet O_2^- \quad (13)$$

$$\bullet O_2^- + H^+ \rightarrow {}^\bullet OOH \quad (14)$$

$$2{}^\bullet OOH^+ \rightarrow O_2 + H_2O_2 \quad (15)$$

$$H_2O_2 + {}^\bullet O_2^- \rightarrow {}^\bullet OH + OH^- + O_2 \quad (16)$$

$$H_2O_2 + h\upsilon \rightarrow 2{}^\bullet OH \quad (17)$$

$$\text{Pollutant} + ({}^\bullet OH, h^+, {}^\bullet OOH/O_2^-) \rightarrow \text{degraded product} \quad (18)$$

Despite its advantages, ZnO photocatalysts face challenges related to photostability and photocorrosion. Photocorrosion refers to the tendency of ZnO to degrade or dissolve when exposed to UV light, releasing Zn^{2+} ions into the environment. This contamination of the surface not only reduces the catalyst's efficiency but also limits its effectiveness over prolonged use. The geometry and structure of ZnO play a significant role in mitigating photocorrosion. Higher crystallinity, the presence of crystalline planes, and optimized surface structure help prevent the breakdown of ZnO under UV exposure. Additionally, surface oxygen vacancies have been shown to prevent photocorrosion by stabilizing the catalyst surface [62]. To address these limitations and enhance the stability of ZnO photocatalysts, various strategies can be employed. One common method is post-synthesis heat treatment, such as annealing, which improves the crystallinity of ZnO and optimizes the Zn:O ratio. This can significantly enhance the photocatalyst's stability and overall performance. Furthermore, combining ZnO with other materials, particularly carbon-based compounds like graphene or carbon nanotubes, has been shown to reduce photocorrosion. These materials can help improve the electronic properties of ZnO, facilitate electron transfer, and prevent electron–hole recombination, thereby enhancing photocatalytic efficiency [84]. In addition to these conventional strategies, green synthesis approaches are gaining attention. Using plant extracts as capping agents during the synthesis of ZnO nanoparticles can improve their stability and reduce photocorrosion. The phytochemicals present in the plant extracts act as protective agents that stabilize the ZnO structure, reducing the likelihood of degradation during photocatalytic reactions. These green synthesis methods are not only environmentally friendly but also cost-effective, making them an attractive option for the production of stable ZnO photocatalysts.

2.4. Comparison of the Photocatalytic Efficacy of ZnO Nanoparticles Synthetized by Different Methods

Photodegradation of dyes or organic contaminants using ZnO synthesized by "green" methods has attracted increasing interest in recent years due to its ecological and economic benefits. In this study, the photocatalytic performance of our ZnO green synthesized was evaluated by degrading methylene blue under UV light irradiation.

The results show that our green ZnO allowed a photodegradation of 67% of MB after 220 min. This rate is comparable to or higher than some works reported in the literature presented in Table 10.

The results obtained in this study confirm the effectiveness of green synthesis methods to produce ZnO-based photocatalysts. Differences in the literature can be attributed to variations in the synthesis method (nature of biomass used, calcination conditions), crystal structure, and experimental parameters. These variations directly influence the light absorption capacity and the generation of electron–hole pairs, essential for photocatalytic degradation.

Table 10. Comparison of the photocatalytic efficacy of ZnO nanoparticles.

Catalyst	Preparation Method/Plant	Dye	Irradiation	Catalyst Dose/g L^{-1}	Dye Conc/mg L^{-1}	Degradation Efficiency/%	Ref.
ZnO	Green synthesis	MB	UV-lamp	1.5	10	63%	[85]
ZnO	Microwave-assisted urea-nitrate combustion	MB	UV-lamp	0.30	75	>75%	[86]
ZnO	Sol gel	MB	UV (Hg lamp 365 nm)	0.33	10	37%	[87]
ZnO	Green synthesis	MB	UV-lamp	1	10	74%	Present work

2.4.1. Antibacterial Activity

This study investigates the antimicrobial properties of biosynthesized ZnO nanoparticles against a broad spectrum of pathogens, providing crucial insights into their potential as antimicrobial agents. The ZnO NPs were synthesized using a sustainable and environmentally friendly method, with varying concentrations ranging from 0 to 10 g/L. The antimicrobial efficacy of these nanoparticles was evaluated using the MIC method, a standard technique in microbiology. This method assesses the lowest concentration of an antimicrobial agent needed to inhibit the growth of microorganisms, providing quantitative data essential for evaluating the effectiveness of antimicrobial substances.

In this study, the antimicrobial activity of ZnO NPs was tested against both Gram-positive bacteria (*Staphylococcus aureus* and *Bacillus subtilis*), Gram-negative bacteria (*Escherichia coli* and *Salmonella enterica*), and the fungal species Candida albicans. The ability to inhibit the growth of these diverse microorganisms suggests that ZnO nanoparticles could be used as versatile antimicrobial agents. The application of the MIC method offers a reliable means of determining the potency of these nanoparticles, which is crucial for assessing their potential as therapeutic or disinfectant agents.

The microtiter plate technique, a high-throughput method used in this study, enables the simultaneous screening of multiple samples. This technique involves exposing microbial cultures to various concentrations of ZnO nanoparticles and observing growth inhibition, typically marked by the absence of turbidity in the wells. The use of this method allows for an efficient and detailed assessment of the nanoparticles' antimicrobial activity against different pathogens.

The results shown in Table 11 indicate that the biosynthesized ZnO nanoparticles, particularly the ones that were annealed, exhibited significantly better antimicrobial performance compared to those produced by conventional methods. The green synthesis approach, which utilizes plant extracts or other eco-friendly methods, appears to preserve and even enhance the antimicrobial properties of ZnO nanoparticles. This suggests that the green synthesis method not only contributes to environmental sustainability but also plays a role in preserving the antimicrobial properties of the nanoparticles.

However, it is important to note that while the ZnO nanoparticles displayed consistent antibacterial effects, they did not exhibit significant antifungal activity against Candida albicans. This limitation suggests that while these nanoparticles are effective against bacterial pathogens, they may not be as effective against fungal infections. This finding underscores the need for further research to explore potential modifications to the ZnO nanoparticle formulations or to investigate alternative methods for improving their antifungal activity. Such research could help expand the range of applications of these nanoparticles in the treatment of fungal infections.

Table 11. Minimum inhibitory concentrations of green synthesised ZnO NPs.

		Staphylococcus aureus	*Bacillus subtilis*	*Escherichia coli*	*Salmonella enterica*
ZnO NPs Concentrations (g/L)	**Annealing Temperature °C**	**MIC (μg/mL)**	**MIC (μg/mL)**	**MIC (μg/mL)**	**MIC (μg/mL)**
c = 0.0	25	100	25	100	50
	320	25	25	25	12.5
	500	25	6.25	25	50
0.03	25	50	25	100	50
	320	6.25	50	100	12.5
	500	25	12.5	100	12.5
0.1	25	100	50	100	100
	320	25	6.25	100	25
	500	50	6.25	No	25
2	25	No	25	100	100
	320	No	6.25	25	25
	500	No	50	No	No
10	25	25	25	25	25
	320	No	6.25	100	100
	500	No	50	No	100

Despite the absence of significant antifungal effects, this study highlights the considerable promise of biosynthesized ZnO nanoparticles as effective antimicrobial agents for controlling bacterial infections. Their ability to inhibit both Gram-positive and Gram-negative bacteria suggests their potential use in various biomedical and environmental applications. For example, these nanoparticles could be integrated into wound dressings to prevent infections, incorporated into water purification systems to eliminate harmful pathogens, or used as coatings for medical devices to reduce the risk of bacterial contamination.

Moreover, the antibacterial properties of biosynthesized ZnO nanoparticles could be particularly valuable in healthcare settings, where bacterial resistance to conventional antibiotics is an emerging concern. The use of green-synthesized ZnO nanoparticles may offer a safer and more sustainable alternative to synthetic antibiotics, providing a novel approach to combating bacterial infections.

While the biosynthesized ZnO nanoparticles exhibit strong antibacterial activity, further investigations are needed to optimize their formulations for enhanced antifungal performance. This study paves the way for the potential use of ZnO nanoparticles in various applications, including biomedical treatments, environmental remediation, and disinfectant products.

2.4.2. Antibacterial Mechanism

Recent studies have extensively explored the advancements in the field of metal oxide and modified metal oxide nanoparticles, particularly focusing on their antibacterial properties and the eco-friendly synthesis methods used to produce them [88–91]. Among the various metal oxide nanoparticles, ZnO has attracted considerable attention due to its remarkable antibacterial potential. This is primarily because ZnO nanoparticles can release zinc ions (Zn^{2+}) upon interacting with microbial cells, which can penetrate the cell membrane and disrupt crucial cellular functions, leading to the death of the microorganisms [92]. Similarly, ZnO nanoparticles have emerged as promising candidates for antibacterial applications, owing to their capacity to modulate the electrical properties of bacterial membranes. This modulation enhances the reactivity of the nanoparticles' surfaces, facilitating interactions with microbial cells and ultimately resulting in bactericidal effects [93].

In-depth research on the mechanisms behind the antibacterial activity of ZnO nanoparticles has uncovered fascinating details about their mode of action Figure 13. These nanoparticles are believed to form strong interactions with essential biomolecules inside bacterial cells, including RNA and DNA [94]. By interfering with processes like DNA replication, these nanoparticles can induce cell cycle arrest, leading to the death of the bacteria. Besides direct damage to DNA, ZnO nanoparticles can generate oxidative stress within bacterial cells, which disrupts ATP synthesis and leads to the production of reactive oxygen species (ROS). These ROS then initiate a cascade of events that ultimately cause apoptosis in the bacterial cells, further contributing to the antibacterial effects of ZnO nanoparticles [95].

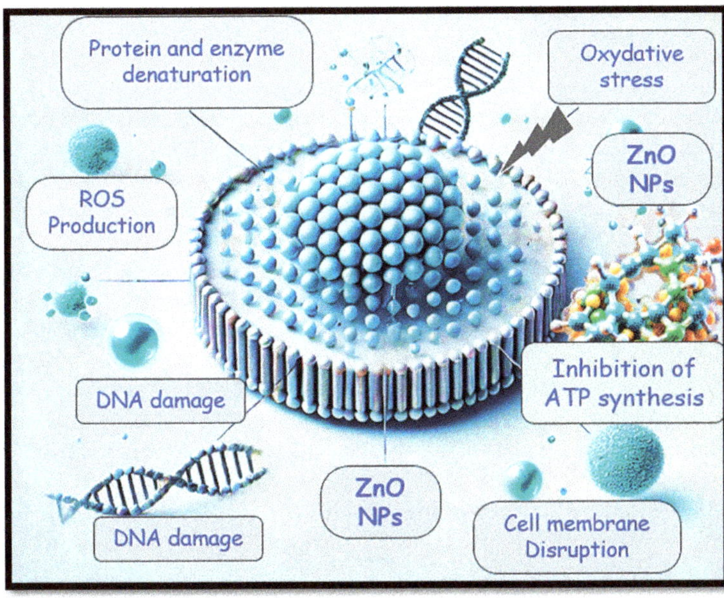

Figure 13. Schematic illustration of the antimicrobial mechanism of ZnO.

With the increasing concerns surrounding the environmental impact of synthetic processes, researchers have shifted their focus toward green synthesis approaches for nanoparticle production [96,97]. These methods utilize renewable natural resources such as plant extracts, bacteria, fungi, algae, and organic products to synthesize nanoparticles. This not only makes the process more environmentally friendly but also enhances the antibacterial properties of the nanoparticles, offering a sustainable alternative to traditional chemical synthesis methods [98]. Plant leaf extracts are particularly valuable for green synthesis, as they contain a wealth of pharmacologically active compounds that help facilitate the interactions between the nanoparticles and bacterial membranes, thus improving their antibacterial efficacy [99,100]. Among these, the green synthesis of ZnO nanoparticles using *Quercus robur* leaf extract has shown promising results, with the nanoparticles demonstrating substantial antimicrobial activity. This aligns with the traditional use of oak bark and leaf extracts in folk medicine to treat microbial infections. The use of *Quercus robur* extract underscores the potential of green synthesis in leveraging the therapeutic properties of natural resources, thus paving the way for environmentally sustainable and effective antimicrobial agents [101,102]. The incorporation of plant-derived compounds in the synthesis process not only enhances the antibacterial properties of the nanoparticles but also reduces the environmental footprint of nanoparticle production. These green synthesis methods offer a practical and sustainable route to producing antibacterial nanoparticles that can be applied in various fields, including medicine, environmental remediation, and material science.

Moreover, the continued research into the green synthesis of ZnO nanoparticles highlights the importance of exploring natural resources to reduce the reliance on synthetic chemicals in the production process. This has significant implications for reducing chemical waste, lowering costs, and improving the efficiency of nanoparticle production. By tapping into the power of natural materials, scientists are working toward developing antimicrobial agents that are both effective and sustainable, ensuring that the benefits of advanced nanotechnology can be enjoyed without compromising the environment.

3. Materials and Methods

3.1. Preparation of Leaf Extract

The selection of *Quercus robur* leaves for this study was based on their abundance and year-round availability, which makes them a practical and sustainable resource for experimental purposes. The leaves used in this research were harvested from the Mashrouha forest region, located in Souk Ahras, Algeria, during the month of January. This timing ensured the collection of healthy, mature leaves that could be effectively used in the synthesis of the extract.

Once collected, the leaves were thoroughly washed with deionized water to eliminate any dust, dirt, or other potential contaminants that might interfere with the quality and purity of the extract. This cleaning step was essential to ensure that only the compounds naturally present in the leaves were involved in the subsequent extraction process.

After the cleaning process, the leaves were dried under controlled conditions—specifically, in the shade at room temperature—for a period of one week. The drying process was performed slowly to prevent the degradation of bioactive compounds within the leaves due to excessive heat or direct sunlight exposure. The leaves were then ground using a blender to increase the surface area and enhance the extraction efficiency when the material was mixed with deionized water.

To prepare the *Quercus robur* extract, varying amounts of ground leaves (0.03, 0.1, 2, and 10 g) were dissolved in 1 L of deionized water. The water was heated to 60 °C and continuously stirred with a magnetic stirrer for 30 min. This heating process helped release the bioactive compounds from the leaf material into the solution, facilitating a more efficient extraction.

After the extraction, the solution was subjected to centrifugation to separate the solid plant matter from the liquid extract. The liquid extract was then filtered to remove any remaining particulates, ensuring a clear solution. Finally, the extract was stored in the refrigerator to preserve its bioactive properties until it was required for further experiments. This methodical preparation of the *Quercus robur* extract allowed for a controlled and reproducible extraction process, enabling the investigation of its potential antimicrobial properties and other relevant bioactivities.

3.2. Green Synthesis of ZnO NPs

To biosynthesize ZnO nanoparticles Figure 14, a precise method involving the combination of zinc acetate dihydrate and *Quercus robur* leaf extract was used. Initially, 6.6 g of zinc acetate dihydrate ($Zn(CH_3COO)_2 \cdot 2H_2O$) was dissolved in 5 mL of distilled water at room temperature while stirring magnetically. This solution was carefully prepared to ensure a uniform distribution of the zinc salt in the solvent, creating the necessary precursor solution for nanoparticle formation. Similarly, 2.85 g of sodium hydroxide (NaOH) was dissolved in water following the same procedure to create the alkaline solution required to adjust the pH during the synthesis process.

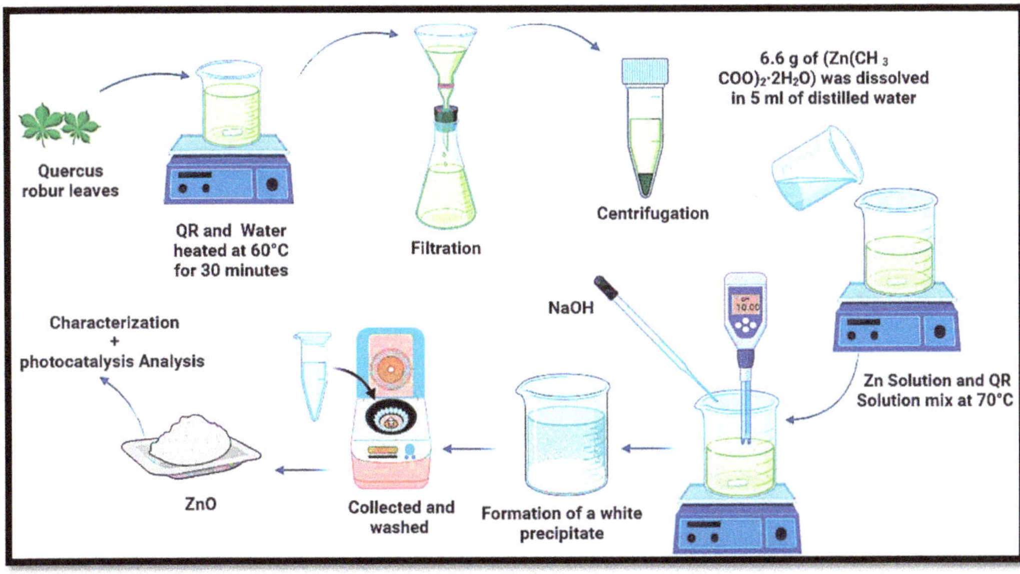

Figure 14. Green synthesis of ZnO NPs using *Quercus robur* leaves extract.

Next, the zinc acetate solution was mixed with 45 mL of *Quercus robur* leaf extract, which serves as a green reducing agent in the biosynthesis of nanoparticles. The leaf extract contains various organic compounds, such as polyphenols, flavonoids, and other phytochemicals, which aid in the transformation of zinc ions (Zn^{2+}) into ZnO nanoparticles through precipitation and subsequent oxidation processes. Once dried, the ZnO nanoparticles were stored in an appropriate container, ready for further characterization and analysis, such as examining their morphology, size distribution, and potential applications in various fields.

The temperatures chosen in the study were based on the previous literature and our experimental optimization process to achieve the desired morphological, structural, and optical properties of ZnO. For instance, research has shown that low-temperature synthesis (~60–100 °C) using hydrothermal or sol–gel methods produces ZnO nanoparticles with controlled morphology and good dispersibility. The mixture was then heated to 70 °C using a magnetic heating stirrer. This temperature was selected to enhance the reaction rate while ensuring that the leaf extract maintained its reducing properties. Once the temperature reached a stable 70 °C, the sodium hydroxide solution was added dropwise over 15 min to adjust the pH of the mixture to the desired level, ensuring that it reached an optimal pH for nanoparticle synthesis.

The solution was kept under continuous stirring and heating for an additional 30 min. During this time, a white precipitate formed, indicating the successful synthesis of ZnO nanoparticles. This precipitate was then separated from the solution by centrifugation at 2500 rpm for 10 min. The solid nanoparticles were carefully washed multiple times with deionized water to remove any residual impurities. After the water wash, the nanoparticles were further cleaned with ethanol to ensure the complete removal of organic substances from the leaf extract. Finally, the obtained ZnO nanoparticles were dried in a laboratory oven at 45 °C for 48 h to remove any remaining solvents and moisture. Once dried, the ZnO nanoparticles were stored in an appropriate container, ready for further characterization and analysis, such as examining their morphology, size distribution, and potential applications in various fields. The percentage yields in this work range from 70% to 85%. The percentage yield can be calculated using the following formula (Equation (19)):

$$\text{Percentage Yield} = \left(\frac{Mass\ of\ ZnO\ obtained}{Theoretical\ mass\ of\ ZnO}\right) \times 100 \tag{19}$$

This green synthesis method not only provides an eco-friendly approach to producing ZnO nanoparticles but also integrates the beneficial properties of plant extracts, which can enhance the stability and functionality of the nanoparticles.

3.3. Photocatalytic Tests

The degradation process of MB using ZnO NPs, synthesized through an eco-friendly green method, was systematically studied by employing UV–Vis spectroscopy. This analytical technique allowed for the tracking of changes in the absorption spectrum, particularly at the wavelength of 664 nm, which is the characteristic λmax for MB. This method enables a precise monitoring of the degradation kinetics over time.

The photocatalytic reactions were carried out under controlled laboratory conditions, where a 100 mL double-walled beaker with a surface area of 32.17 cm^2 was utilized. The beaker was equipped with a cooling water jacket, ensuring that the reaction temperature was consistently maintained at 20 ± 2 °C. This was important to minimize any temperature-related interference in the photocatalytic process.

To provide the necessary UV irradiation, a Haichao T8 15W UV lamp, emitting at 365 nm UVA, was employed. The lamp was positioned 10 cm from the solution, ensuring uniform exposure of the photocatalyst to the UV light. The controlled environment, including the precise regulation of temperature and UV light exposure, helped ensure that the experiment was conducted under optimal and reproducible conditions, thereby allowing for accurate evaluation of the ZnO NPs' photocatalytic degradation efficiency MB. This approach not only ensured effective monitoring of the degradation process but also contributed to a better understanding of the photocatalytic properties of green-synthesized ZnO NPs in the degradation of organic pollutants like MB.

3.4. Antimicrobial Study

The MIC of ZnO was determined through a stepwise two-fold serial dilution technique. Initially, the ZnO NPs were introduced into wells 1 to 10 of row A in a 96-well microplate, where a series of 100 µL dilutions were systematically made across the plate. Each dilution progressively reduced the concentration of the nanoparticles, yielding concentrations of 200, 100, 50, 25, 12.5, 6.25, 3.13, and 1.56 µg/mL. This allowed for the establishment of a concentration gradient, which was essential for identifying the minimum concentration at which the ZnO NPs could effectively inhibit bacterial growth. After setting up the dilutions, the microplate was incubated at 37 °C for a period of 24 h, ensuring that the bacterial strains were exposed to the nanoparticles for sufficient time to allow any potential antibacterial effects to take place.

To evaluate bacterial viability and determine the MIC, 2,3,5-triphenyl tetrazolium chloride (TTC) was used as a viability indicator. TTC is a colorimetric agent that stains living bacterial cells pink due to the reduction in the tetrazolium salt by metabolically active cells. On the other hand, if bacterial growth was inhibited, the TTC remained colorless, indicating a lack of bacterial metabolism. The extent of the color change in each well directly correlates to the degree of bacterial growth inhibition, with the shift from pink to transparent providing clear visual evidence of the antibacterial activity of the ZnO NPs.

The lowest concentration of ZnO NPs at which the bacterial growth was significantly suppressed, as indicated by the absence of pink color in the wells, was identified as the MIC. The results showed that the biosynthesized ZnO NPs exhibited substantial antibacterial activity, demonstrating their potential as effective antimicrobial agents against the tested bacterial strains. The precise correlation between the concentration of ZnO NPs and bacterial inhibition provides critical information for optimizing the use of these nanoparticles in various applications, such as in the development of antimicrobial coatings or medical treatments aimed at combating bacterial infections.

4. Conclusions

This study provides a comprehensive analysis of zinc oxide nanoparticle (ZnO NPs) synthesis and characterization using environmentally friendly methods, highlighting their potential as both antimicrobial agents and photocatalytic catalysts. The research focused on optimizing calcination temperature and examining the effect of plant extract concentration on the structural properties of the nanoparticles. X-ray diffraction (XRD) confirmed the presence of characteristic diffraction peaks corresponding to the hexagonal wurtzite structure of ZnO, with crystallite sizes ranging from 10 to 29 nm, influenced by both extract concentration and annealing temperature. Scanning electron microscopy (SEM) revealed that increasing the leaf extract concentration led to smaller particle sizes; however, higher concentrations also resulted in greater nanoparticle aggregation. Fourier-transform infrared spectroscopy (FTIR) and UV–visible spectroscopy analyses identified functional groups on both the ZnO nanoparticles and the leaf extract, and variations in band gap energy and Urbach energy were observed as a function of extract concentration and calcination temperature. Antimicrobial studies demonstrated that the environmentally synthesized ZnO nanoparticles effectively inhibited the growth of both Gram-positive and G ram-negative bacteria, with minimum inhibitory concentrations ranging from 6.25 to 100 µg/mL depending on the concentration of the leaf extract. However, the antifungal activity against Candida albicans was minimal across all tested concentrations. Photocatalytic degradation tests showed that green-synthesized nanoparticles achieved optimal removal of methylene blue (MB) compared to their chemically synthesized counterparts, with efficiency further enhanced at higher calcination temperatures (320 and 500 °C). In conclusion, the environmentally friendly synthesis of ZnO nanoparticles using leaf extract holds promise for the development of antimicrobial agents and photocatalytic catalysts. While the nanoparticles exhibited notable antibacterial activity, further optimization is required to enhance their antifungal efficacy. Additionally, exploring strategies to improve photocatalytic efficiency—considering factors such as particle size, aggregation, and surface functionalization—should be prioritized. These findings contribute significantly to the advancement of nanoparticle technology for biomedical, environmental, and catalytic applications.

Author Contributions: Conceptualization, S.R., H.G., H.T. and L.A.; data curation, S.R. and H.G.; formal analysis, S.R. and H.G.; funding acquisition, M.S.O. and J.Z.; investigation, S.R., H.G., H.T., L.A., D.C., A.A.A., M.K., J.Z., A.A. and S.L.; methodology, S.R., H.G., H.T., L.A., D.C., M.S.O., J.Z. and A.A.; project administration, H.G., H.T., M.S.O., J.Z. and A.A.; resources, S.R., H.G., H.T., A.A.A., M.K., A.A. and S.L.; supervision, H.G; validation, S.R., H.G. and A.A.; visualization, H.G., H.T., L.A., A.A. and S.L.; writing—original draft, S.R. and H.G.; writing—review and editing, H.G. All authors have read and agreed to the published version of the manuscript.

Funding: Researchers Supporting Project Number (RSPD2025R710), King Saud University, Riyadh, Saudi Arabia.

Data Availability Statement: The data supporting the findings of this manuscript are available from the corresponding authors upon reasonable request.

Acknowledgments: The authors wish to express their thanks to the Algerian Ministry of Higher Education and Scientific Research (MESRS, DGRSDT) for research support. The authors also thank Researchers Supporting Project Number (RSPD2025R710) from King Saud University, Riyadh, Saudi Arabia, for financial support.

Conflicts of Interest: All authors declare that they do not have a conflict of interest.

References

1. Palai, P.; Muduli, S.; Priyadarshini, B.; Sahoo, T.R. A Facile Green Synthesis of ZnO Nanoparticles and Its Adsorptive Removal of Congo Red Dye from Aqueous Solution. *Mater. Today Proc.* **2021**, *38*, 2445–2451. [CrossRef]
2. Kaliraj, L.; Ahn, J.C.; Rupa, E.J.; Abid, S.; Lu, J.; Yang, D.C. Synthesis of Panos Extract Mediated ZnO Nano-Flowers as Photocatalyst for Industrial Dye Degradation by UV Illumination. *J. Photochem. Photobiol. B Biol.* **2019**, *199*, 111588. [CrossRef]
3. Toumi, S.; Lekmine, S.; Touzout, N.; Moussa, H.; Elboughdiri, N.; Boudraa, R.; Benslama, O.; Kebir, M.; Danish, S.; Zhang, J. Harnessing Deep Learning for Real-Time Water Quality Assessment: A Sustainable Solution. *Water* **2024**, *16*, 3380. [CrossRef]
4. Chung, K.T.; Fulk, G.E.; Andrews, A.W. Mutagenicity Testing of Some Commonly Used Dyes. *Appl. Environ. Microbiol.* **1981**, *42*, 641–648. [CrossRef]
5. LaKind, J.S.; Anthony, L.G.; Goodman, M. Review of Reviews on Exposures to Synthetic Organic Chemicals and Children's Neurodevelopment: Methodological and Interpretation Challenges. *J. Toxicol. Environ. Health Part B* **2017**, *20*, 390–422. [CrossRef]
6. Ming, J.; Ni, S.-Q.; Guo, Z.; Wang, Z.-B.; Xie, L. Photocatalytic Material–Microorganism Hybrid Systems in Water Decontamination. *Trends Biotechnol.* 2024; online ahead of print. [CrossRef]
7. Soto-Robles, C.A.; Luque, P.A.; Gómez-Gutiérrez, C.M.; Nava, O.; Vilchis-Nestor, A.R.; Lugo-Medina, E.; Ranjithkumar, R.; Castro-Beltrán, A. Study on the Effect of the Concentration of Hibiscus Sabdariffa Extract on the Green Synthesis of ZnO Nanoparticles. *Results Phys.* **2019**, *15*, 102807. [CrossRef]
8. Tahraoui, H.; Belhadj, A.-E.; Amrane, A.; Toumi, S.; Jaouadi, B.; Zhang, J. Maximizing Diclofenac Bioremoval Efficiency Using *Chlorella vulgaris* Strain H1 and *Chlorella sorokiniana* Strain H2: Unveiling the Impact of Acetic Acid on Microalgae. *J. Taiwan Inst. Chem. Eng.* **2024**, *165*, 105783. [CrossRef]
9. Smara, M.; Khalladi, R.; Moulai-Mostefa, N.; Madi, K.; Mansour, D.; Lekmine, S.; Benslama, O.; Tahraoui, H.; Zhang, J.; Amrane, A. Efficiency of Hydrogen Peroxide and Fenton Reagent for Polycyclic Aromatic Hydrocarbon Degradation in Contaminated Soil: Insights from Experimental and Predictive Modeling. *Processes* **2024**, *12*, 621. [CrossRef]
10. Nedjhioui, M.; Nasrallah, N.; Kebir, M.; Tahraoui, H.; Bouallouche, R.; Assadi, A.A.; Amrane, A.; Jaouadi, B.; Zhang, J.; Mouni, L. Designing an Efficient Surfactant–Polymer–Oil–Electrolyte System: A Multi-Objective Optimization Study. *Processes* **2023**, *11*, 1314. [CrossRef]
11. Kebir, M.; Benramdhan, I.-K.; Nasrallah, N.; Tahraoui, H.; Bait, N.; Benaissa, H.; Ameraoui, R.; Zhang, J.; Assadi, A.A.; Mouni, L. Surface Response Modeling of Homogeneous Photo Fenton Fe (III) and Fe (II) Complex for Sunlight Degradation and Mineralization of Food Dye. *Catal. Commun.* **2023**, *183*, 106780. [CrossRef]
12. Mechati, S.; Zamouche, M.; Tahraoui, H.; Filali, O.; Mazouz, S.; Bouledjemer, I.N.E.; Toumi, S.; Triki, Z.; Amrane, A.; Kebir, M. Modeling and Optimization of Hybrid Fenton and Ultrasound Process for Crystal Violet Degradation Using AI Techniques. *Water* **2023**, *15*, 4274. [CrossRef]
13. Moussa, H.; Hamid, S.; Mameri, A.; Lekmine, S.; Tahraoui, H.; Kebir, M.; Touzout, N.; Dahmoune, F.; Ola, M.S.; Zhang, J. From Green Chemistry to Healthy Environments: Silver Nanoparticles as a Dual Antioxidant and Antibacterial Agents for Advancing Biomedicine and Sustainable Wastewater Treatment. *Bioengineering* **2024**, *11*, 1205. [CrossRef] [PubMed]
14. Tahraoui, H.; Toumi, S.; Boudoukhani, M.; Touzout, N.; Sid, A.N.E.H.; Amrane, A.; Belhadj, A.-E.; Hadjadj, M.; Laichi, Y.; Aboumustapha, M. Evaluating the Effectiveness of Coagulation–Flocculation Treatment Using Aluminum Sulfate on a Polluted Surface Water Source: A Year-Long Study. *Water* **2024**, *16*, 400. [CrossRef]
15. Touahri, S.; Halimi, O.; Zaabat, M.; Mammeri, S.; Boudine, B.; Sebais, M.; Tahraoui, H.; Zhang, J.; Amrane, A. Exploring the Influence of Sr Concentration on the Structural and Catalytic Properties of CuO/$SrSO_4$ Nanocomposites for Organic Dye Degradation. *J. Phys. Chem. Solids* **2024**, *195*, 112299. [CrossRef]
16. Golmohammadi, M.; Honarmand, M.; Ghanbari, S. A Green Approach to Synthesis of ZnO Nanoparticles Using Jujube Fruit Extract and Their Application in Photocatalytic Degradation of Organic Dyes. *Spectrochim. Acta Part A Mol. Biomol. Spectrosc.* **2020**, *229*, 117961. [CrossRef] [PubMed]

17. Farid, M.; Fayyaz, A.; Ahmed, E.; Arooj, M.; Ali, S.; Sarfraz, W.; Abbas, M.; Idrees, A.; Shahzadi, M.; Rasheed, Z. Increase in Food Scarcity, Agricultural Challenges, and Their Management: Pakistan Perspectives. In *Managing Plant Production Under Changing Environment*; Hasanuzzaman, M., Ahammed, G.J., Nahar, K., Eds.; Springer Nature Singapore: Singapore, 2022; pp. 437–458, ISBN 978-981-16-5058-1. [CrossRef]
18. Rajan, C.S.R. Nanotechnology in Groundwater Remediation. *Int. J. Environ. Sci. Dev.* **2011**, *2*, 182–187. [CrossRef]
19. Chelghoum, H.; Nasrallah, N.; Tahraoui, H.; Seleiman, M.F.; Bouhenna, M.M.; Belmeskine, H.; Zamouche, M.; Djema, S.; Zhang, J.; Mendil, A. Eco-Friendly Synthesis of ZnO Nanoparticles for Quinoline Dye Photodegradation and Antibacterial Applications Using Advanced Machine Learning Models. *Catalysts* **2024**, *14*, 831. [CrossRef]
20. Kebir, M.; Bouallouche, R.; Nasrallah, N.; Tahraoui, H.; Elboughdiri, N.; Ait Merzeg, F.; Dergal, F.; Amirouche, S.; Assadi, A.A.; Amrane, A. Sustainable Photodegradation of Amoxicillin in Wastewater with a Nickel Aluminate and ZnO Heterosystem Oxides: Experimental and Gaussian Process Regression Modeling Studies. *Catalysts* **2024**, *14*, 875. [CrossRef]
21. Liu, Y.; Zeng, X.; Hu, X.; Hu, J.; Zhang, X. Two-dimensional Nanomaterials for Photocatalytic Water Disinfection: Recent Progress and Future Challenges. *J. Chem. Technol. Biotechnol.* **2019**, *94*, 22–37. [CrossRef]
22. Teow, Y.H.; Mohammad, A.W. New Generation Nanomaterials for Water Desalination: A Review. *Desalination* **2019**, *451*, 2–17. [CrossRef]
23. Liu, X.; Ma, R.; Wang, X.; Ma, Y.; Yang, Y.; Zhuang, L.; Zhang, S.; Jehan, R.; Chen, J.; Wang, X. Graphene Oxide-Based Materials for Efficient Removal of Heavy Metal Ions from Aqueous Solution: A Review. *Environ. Pollut.* **2019**, *252*, 62–73. [CrossRef] [PubMed]
24. Shan, Z.; Yang, Y.; Shi, H.; Zhu, J.; Tan, X.; Luan, Y.; Jiang, Z.; Wang, P.; Qin, J. Hollow Dodecahedra Graphene Oxide-Cuprous Oxide Nanocomposites with Effective Photocatalytic and Bactericidal Activity. *Front. Chem.* **2021**, *9*, 755836. [CrossRef] [PubMed]
25. Anbarasu, K.; Devarajan, Y. Nanomaterials-Based Wastewater Treatment: Addressing Challenges and Advancing Sustainable Solutions. *BioNanoScience* **2025**, *15*, 149. [CrossRef]
26. Crini, G.; Lichtfouse, E. Advantages and Disadvantages of Techniques Used for Wastewater Treatment. *Environ. Chem. Lett.* **2019**, *17*, 145–155. [CrossRef]
27. Zhou, C.; Lai, C.; Xu, P.; Zeng, G.; Huang, D.; Zhang, C.; Cheng, M.; Hu, L.; Wan, J.; Liu, Y.; et al. In Situ Grown $AgI/Bi_{12}O_{17}Cl_2$ Heterojunction Photocatalysts for Visible Light Degradation of Sulfamethazine: Efficiency, Pathway and Mechanism. *ACS Sustain. Chem. Eng.* **2018**, *6*, 4174–4184. [CrossRef]
28. Lu, K.; Qin, J.; Hu, M.; Hu, L.; Mao, M.; Li, X.; Lin, Z.; Liu, W. High-Efficiency Nickel Recovery from Spent Electroless Nickel Plating Solution: Effective Degradation of High-Concentration Nickel Complexes to Form a Nickel Ferrite Nanomaterial via Fe_3O_4 Catalytic Oxidation. *Environ. Sci. Nano* **2024**, *11*, 900–910. [CrossRef]
29. Ikram, M.; Rashid, M.; Haider, A.; Naz, S.; Haider, J.; Raza, A.; Ansar, M.T.; Uddin, M.K.; Ali, N.M.; Ahmed, S.S.; et al. A Review of Photocatalytic Characterization, and Environmental Cleaning, of Metal Oxide Nanostructured Materials. *Sustain. Mater. Technol.* **2021**, *30*, e00343. [CrossRef]
30. Zhuang, Q.; Li, X.; Lian, X.; Hu, H.; Wang, N.; Wu, J.; Miao, K.; Feng, G.; Luo, X. Catalysis Enhancement of Co_3O_4 through the Epitaxial Growth of Inert ZnO in Peroxymonosulfate Activation: The Catalytic Mechanism of Surface Hydroxyls in Singlet Oxygen Generation. *Cryst. Growth Des.* **2025**, *25*, 319–329. [CrossRef]
31. Ali, T.; Tripathi, P.; Azam, A.; Raza, W.; Ahmed, A.S.; Ahmed, A.; Muneer, M. Photocatalytic Performance of Fe-Doped TiO_2 Nanoparticles under Visible-Light Irradiation. *Mater. Res. Express* **2017**, *4*, 015022. [CrossRef]
32. Hasija, V.; Raizada, P.; Sudhaik, A.; Sharma, K.; Kumar, A.; Singh, P.; Jonnalagadda, S.B.; Thakur, V.K. Recent Advances in Noble Metal Free Doped Graphitic Carbon Nitride Based Nanohybrids for Photocatalysis of Organic Contaminants in Water: A Review. *Appl. Mater. Today* **2019**, *15*, 494–524. [CrossRef]
33. Anpo, M.; Ichihashi, Y.; Takeuchi, M.; Yamashita, H. Design of Unique Titanium Oxide Photocatalysts by an Advanced Metal Ion-Implantation Method and Photocatalytic Reactions under Visible Light Irradiation. *Res. Chem. Intermed.* **1998**, *24*, 143–149. [CrossRef]
34. Guediri, M.K.; Chebli, D.; Bouguettoucha, A.; Bourzami, R.; Amrane, A. Novel Fe_2TiO_5/Reduced Graphene Oxide Heterojunction Photocatalyst with Improved Adsorption Capacity and Visible Light Photoactivity: Experimental and DFT Approach. *Environ. Sci. Pollut. Res.* **2021**, *28*, 8507–8519. [CrossRef]
35. Muthuvel, A.; Jothibas, M.; Mohana, V.; Manoharan, C. Green Synthesis of Cerium Oxide Nanoparticles Using Calotropis Procera Flower Extract and Their Photocatalytic Degradation and Antibacterial Activity. *Inorg. Chem. Commun.* **2020**, *119*, 108086. [CrossRef]
36. Xing, Z.; Zhang, J.; Cui, J.; Yin, J.; Zhao, T.; Kuang, J.; Xiu, Z.; Wan, N.; Zhou, W. Recent Advances in Floating TiO_2-Based Photocatalysts for Environmental Application. *Appl. Catal. B Environ.* **2018**, *225*, 452–467. [CrossRef]
37. Ong, C.B.; Ng, L.Y.; Mohammad, A.W. A Review of ZnO Nanoparticles as Solar Photocatalysts: Synthesis, Mechanisms and Applications. *Renew. Sustain. Energy Rev.* **2018**, *81*, 536–551. [CrossRef]
38. Li, X.; Peng, T.; Zhang, Y.; Wen, Y.; Nan, Z. A New Efficient Visible-Light Photocatalyst Made of SnO_2 and Cyclized Polyacrylonitrile. *Mater. Res. Bull.* **2018**, *97*, 517–522. [CrossRef]

39. Rasheed, H.M.; Aroosh, K.; Meng, D.; Ruan, X.; Akhter, M.; Cui, X. A Review on Modified ZnO to Address Environmental Challenges through Photocatalysis: Photodegradation of Organic Pollutants. *Mater. Today Energy* **2025**, *48*, 101774. [CrossRef]
40. Chikhi, B.; Gouasmi, M.; Mounia, A.; Gasem, L.; Saadi, A.; Mekaoui, N.; Bachari, K.; Boudjemaa, A. Propyl Paraben Removal Using Cu_2O/ZnO-NPs Photocatalyst Elaborated via Green Method. *Environ. Sci. Pollut. Res.* **2025**, *32*, 2107–2122. [CrossRef] [PubMed]
41. Khan, M.M.; Adil, S.F.; Al-Mayouf, A. Metal Oxides as Photocatalysts. *J. Saudi Chem. Soc.* **2015**, *19*, 462–464. [CrossRef]
42. Madi, K.; Chebli, D.; Ait Youcef, H.; Tahraoui, H.; Bouguettoucha, A.; Kebir, M.; Zhang, J.; Amrane, A. Green Fabrication of ZnO Nanoparticles and ZnO/rGO Nanocomposites from Algerian Date Syrup Extract: Synthesis, Characterization, and Augmented Photocatalytic Efficiency in Methylene Blue Degradation. *Catalysts* **2024**, *14*, 62. [CrossRef]
43. Wang, H.; Zhang, P.; Zang, Z. High Performance $CsPbBr_3$ Quantum Dots Photodetectors by Using Zinc Oxide Nanorods Arrays as an Electron-Transport Layer. *Appl. Phys. Lett.* **2020**, *116*, 162103. [CrossRef]
44. Li, C.; Han, C.; Zhang, Y.; Zang, Z.; Wang, M.; Tang, X.; Du, J. Enhanced Photoresponse of Self-Powered Perovskite Photodetector Based on ZnO Nanoparticles Decorated $CsPbBr_3$ Films. *Sol. Energy Mater. Sol. Cells* **2017**, *172*, 341–346. [CrossRef]
45. Dejen, K.D.; Sabir, F.K.; Ananda Murthy, H.C.; Ayanie, G.T.; Shume, M.S.; Bekele, E.T. Green Synthesis of Nanomaterials for Environmental Remediation. In *Green Nanoremediation*; Policarpo Tonelli, F.M., Roy, A., Ananda Murthy, H.C., Eds.; Springer International Publishing: Cham, Switzerland, 2023; pp. 27–65. ISBN 978-3-031-30557-3.
46. Duan, H.; Wang, D.; Li, Y. Green Chemistry for Nanoparticle Synthesis. *Chem. Soc. Rev.* **2015**, *44*, 5778–5792. [CrossRef]
47. Iravani, S. Green Synthesis of Metal Nanoparticles Using Plants. *Green Chem.* **2011**, *13*, 2638. [CrossRef]
48. Bandeira, M.; Giovanela, M.; Roesch-Ely, M.; Devine, D.M.; Da Silva Crespo, J. Green Synthesis of Zinc Oxide Nanoparticles: A Review of the Synthesis Methodology and Mechanism of Formation. *Sustain. Chem. Pharm.* **2020**, *15*, 100223. [CrossRef]
49. Jena, S.K.; Sadasivam, R.; Packirisamy, G.; Saravanan, P. Nanoremediation: Sunlight Mediated Dye Degradation Using Electrospun PAN/CuO–ZnO Nanofibrous Composites. *Environ. Pollut.* **2021**, *280*, 116964. [CrossRef]
50. Alavi-Tabari, S.A.R.; Khalilzadeh, M.A.; Karimi-Maleh, H. Simultaneous Determination of Doxorubicin and Dasatinib as Two Breast Anticancer Drugs Uses an Amplified Sensor with Ionic Liquid and ZnO Nanoparticle. *J. Electroanal. Chem.* **2018**, *811*, 84–88. [CrossRef]
51. Chandran, S.P.; Chaudhary, M.; Pasricha, R.; Ahmad, A.; Sastry, M. Synthesis of Gold Nanotriangles and Silver Nanoparticles Using Aloe Vera Plant Extract. *Biotechnol. Prog.* **2006**, *22*, 577–583. [CrossRef]
52. Song, J.Y.; Kim, B.S. Rapid Biological Synthesis of Silver Nanoparticles Using Plant Leaf Extracts. *Bioprocess Biosyst. Eng.* **2009**, *32*, 79–84. [CrossRef]
53. Rathnasamy, R.; Thangasamy, P.; Thangamuthu, R.; Sampath, S.; Alagan, V. Green Synthesis of ZnO Nanoparticles Using Carica Papaya Leaf Extracts for Photocatalytic and Photovoltaic Applications. *J. Mater. Sci. Mater. Electron.* **2017**, *28*, 10374–10381. [CrossRef]
54. Sangeetha, G.; Rajeshwari, S.; Venckatesh, R. Green Synthesis of Zinc Oxide Nanoparticles by Aloe Barbadensis Miller Leaf Extract: Structure and Optical Properties. *Mater. Res. Bull.* **2011**, *46*, 2560–2566. [CrossRef]
55. Ramesh, M.; Anbuvannan, M.; Viruthagiri, G. Green Synthesis of ZnO Nanoparticles Using Solanum Nigrum Leaf Extract and Their Antibacterial Activity. *Spectrochim. Acta Part A Mol. Biomol. Spectrosc.* **2015**, *136*, 864–870. [CrossRef]
56. Thema, F.T.; Manikandan, E.; Dhlamini, M.S.; Maaza, M. Green Synthesis of ZnO Nanoparticles via Agathosma Betulina Natural Extract. *Mater. Lett.* **2015**, *161*, 124–127. [CrossRef]
57. Pugazhendhi, A.; Mathimani, T.; Alharbi, S.A.; Chinnathambi, A.; Karuppusamy, I.; Brindhadevi, K.; Kheawfu, K.; Pikulkaew, S. In Vitro Efficacy of Green Synthesized ZnO Nanoparticles against Biofilm and Virulence of Serratia Marcescens. *Prog. Org. Coat.* **2022**, *166*, 106781. [CrossRef]
58. Khoso, S.; Mehar, S.; Anam; Naheed, N.; Saeed, F.; Khan, N.; Abbasi, B.H. Green synthesis of ZnO nanoparticles from foeniculum vulgare, its characterization and biological potential against bacteria. *J. Anim. Plant Sci.* **2021**, *32*, 229–237. [CrossRef]
59. Etape, E.P.; Foba-Tendo, J.; Ngolui, L.J.; Namondo, B.V.; Yollande, F.C.; Nguimezong, M.B.N. Structural Characterization and Magnetic Properties of Undoped and Ti-Doped ZnO Nanoparticles Prepared by Modified Oxalate Route. *J. Nanomater.* **2018**, *2018*, 9072325. [CrossRef]
60. Bourzami, R.; Guediri, M.K.; Chebli, D.; Bouguettoucha, A.; Amrane, A. Bottom-up Construction of Reduced-Graphene-Oxide-Anchored Spinel Magnet $Fe_{2.02}Ni_{1.01}O_{3.22}$, Anatase TiO_2 and Metallic Ag Nanoparticles and Their Synergy in Photocatalytic Water Reduction. *J. Environ. Chem. Eng.* **2021**, *9*, 105307. [CrossRef]
61. Demissie, M.G.; Sabir, F.K.; Edossa, G.D.; Gonfa, B.A. Synthesis of Zinc Oxide Nanoparticles Using Leaf Extract of *Lippia adoensis* (Koseret) and Evaluation of Its Antibacterial Activity. *J. Chem.* **2020**, *2020*, 7459042. [CrossRef]
62. Mallika, A.N.; Reddy, A.R.; Reddy, K.V. Annealing Effects on the Structural and Optical Properties of ZnO Nanoparticles with PVA and CA as Chelating Agents. *J. Adv. Ceram.* **2015**, *4*, 123–129. [CrossRef]
63. Chen, Y.-S.; Liao, C.-H.; Chueh, Y.-L.; Lai, C.-C.; Chen, L.-Y.; Chu, A.-K.; Kuo, C.-T.; Wang, H.-C. High Performance Cu_2O/ZnO Core-Shell Nanorod Arrays Synthesized Using a Nanoimprint GaN Template by the Hydrothermal Growth Technique. *Opt. Mater. Express* **2014**, *4*, 1473. [CrossRef]

64. Caliman, F.A.; Robu, B.M.; Smaranda, C.; Pavel, V.L.; Gavrilescu, M. Soil and Groundwater Cleanup: Benefits and Limits of Emerging Technologies. *Clean Technol. Environ. Policy* **2011**, *13*, 241–268. [CrossRef]
65. Oniya, E.O.; Olubi, O.E.; Ibitoye, A.; Agbi, J.I.; Agbeni, S.K.; Faweya, E.B. Effect of Milling Equipment on the Level of Heavy Metal Content of Foodstuff. *Phys. Sci. Int. J.* **2018**, *20*, 1–8. [CrossRef]
66. Naiel, B.; Fawzy, M.; Halmy, M.W.A.; Mahmoud, A.E.D. Green synthesis of zinc oxide nanoparticles using Sea Lavender (*Limonium pruinosum* L. Chaz.) extract: Characterization, evaluation of anti-skin cancer, antimicrobial and antioxidant potentials. *Sci. Rep.* **2022**, *12*, 20370. [CrossRef] [PubMed]
67. Sharma, V.; Sharma, J.K.; Kansay, V.; Sharma, V.D.; Sharma, A.; Kumar, S.; Sharma, A.K.; Bera, M.K. The Effect of Calcination Temperatures on the Structural and Optical Properties of Zinc Oxide Nanoparticles and Their Influence on the Photocatalytic Degradation of Leather Dye. *Chem. Phys. Impact* **2023**, *6*, 100196. [CrossRef]
68. Agnieszka, K.-R.; Ewa, M.; Teofil, J. Structural Characterisation of ZnO Particles Obtained by the Emulsion Precipitation Method. *J. Nanomater.* **2012**, 656353. [CrossRef]
69. Khan, S.H.; R, S.; Pathak, B.; Fulekar, M. Photocatalytic Degradation of Organophosphate Pesticides (Chlorpyrifos) Using Synthesized Zinc Oxide Nanoparticle by Membrane Filtration Reactor under UV Irradiation. *Front. Nanosci. Nanotechnol.* **2015**, *1*, 23–27. [CrossRef]
70. Christobel, G.J. Vibrational Spectroscopy of ZnO-ZnS Nanoparticles. *Int. J. Sci. Res.* **2013**, *5*, 2228–2230.
71. Veiga, A.; Toledo, M.D.G.T.; Rossa, L.S.; Mengarda, M.; Stofella, N.C.F.; Oliveira, L.J.; Gonçalves, A.G.; Murakami, F.S. Colorimetric Microdilution Assay: Validation of a Standard Method for Determination of MIC, IC50%, and IC90% of Antimicrobial Compounds. *J. Microbiol. Methods* **2019**, *162*, 50–61. [CrossRef]
72. Bourzami, R.; AitYoucef, H.C.; Hamdouni, N.; Sebais, M. Synthesis, Crystal Structure, Vibrational Spectra and Thermal Properties of Novel Ionic Organic-Inorganic Hybrid Material. *Chem. Phys. Lett.* **2018**, *711*, 220–226. [CrossRef]
73. Ouksel, L.; Chafaa, S.; Bourzami, R.; Hamdouni, N.; Sebais, M.; Chafai, N. Crystal Structure, Vibrational, Spectral Investigation, Quantum Chemical DFT Calculations and Thermal Behavior of Diethyl [Hydroxy (Phenyl) Methyl] Phosphonate. *J. Mol. Struct.* **2017**, *1144*, 389–395. [CrossRef]
74. Redjili, S.; Ghodbane, H.; Bourzami, R.; Mayouf, F.; Chebli, D.; Boudechicha, A.; Djelloul, C. Green Synthesis of Silver Oxide Nanoparticles: Eco-Friendly Approach for Sustainable Solutions. *Ind. Crops Prod.* **2025**, *223*, 120168. [CrossRef]
75. Boudiaf, M.; Messai, Y.; Bentouhami, E.; Schmutz, M.; Blanck, C.; Ruhlmann, L.; Bezzi, H.; Tairi, L.; Eddine Mekki, D. Green Synthesis of NiO Nanoparticles Using Nigella Sativa Extract and Their Enhanced Electro-Catalytic Activity for the 4-Nitrophenol Degradation. *J. Phys. Chem. Solids* **2021**, *153*, 110020. [CrossRef]
76. Liu, H.R.; Shao, G.X.; Zhao, J.F.; Zhang, Z.X.; Zhang, Y.; Liang, J.; Liu, X.G.; Jia, H.S.; Xu, B.S. Worm-Like Ag/ZnO Core–Shell Heterostructural Composites: Fabrication, Characterization, and Photocatalysis. *J. Phys. Chem. C* **2012**, *116*, 16182–16190. [CrossRef]
77. Garg, R. Characterization and Performance Evaluation of Synthesized ZnO Nanoflowers, Nanorods, and Their Hybrid Nanocomposites with Graphene Oxide for Degradation of Orange G. *Environ. Sci. Pollut. Res.* **2021**, *28*, 57009–57029. [CrossRef] [PubMed]
78. Chen, X.; Wu, Z.; Liu, D.; Gao, Z. Preparation of ZnO Photocatalyst for the Efficient and Rapid Photocatalytic Degradation of Azo Dyes. *Nanoscale Res. Lett.* **2017**, *12*, 143. [CrossRef] [PubMed]
79. Bagheri, M.; Mahjoub, A.R.; Mehri, B. Enhanced Photocatalytic Degradation of Congo Red by Solvothermally Synthesized $CuInSe_2$–ZnO Nanocomposites. *RSC Adv.* **2014**, *4*, 21757. [CrossRef]
80. Kumar, R.; Kumar, G.; Umar, A. ZnO Nano-Mushrooms for Photocatalytic Degradation of Methyl Orange. *Mater. Lett.* **2013**, *97*; 100–103. [CrossRef]
81. Mutalib, A.A.A. ZnO Photocatalysts Applications in Abating the Organic Pollutant Contamination: A Mini Review. *Total. Environ. Res. Themes* **2022**, *3*, 100013. [CrossRef]
82. Naha, U. Rapid Precipitation-Free Synthesis of Zinc Oxide Nanowire Arrays. *Mater. Today Proc.* **2023**, *79*, 132–138. [CrossRef]
83. Yashni, G.; Al-Gheethi, A.; Al-Kahtani, A.A.; Al-Sahari, M.; Hazhar, N.J.N.; Noman, E.; Alkhadher, S. Bio-Inspired ZnO NPs Synthesized from Citrus Sinensis Peels Extract for Congo Red Removal from Textile Wastewater via Photocatalysis: Optimization, Mechanisms, Techno-Economic Analysis. *Chemosphere* **2021**, *281*, 130661. [CrossRef]
84. Gawade, V.V.; Gavade, N.L.; Shinde, H.M.; Babar, S.B.; Kadam, A.N.; Garadkar, K.M. Green Synthesis of ZnO Nanoparticles by Using Calotropis Procera Leaves for the Photodegradation of Methyl Orange. *J. Mater. Sci. Mater. Electron.* **2017**, *28*, 14033–14039. [CrossRef]
85. Madhumitha, G.; Fowsiya, J.; Gupta, N.; Kumar, A.; Singh, M. Green Synthesis, Characterization and Antifungal and Photocatalytic Activity of Pithecellobium Dulce Peel–Mediated ZnO Nanoparticles. *J. Phys. Chem. Solids* **2019**, *127*, 43–51. [CrossRef]
86. Sherly, E.D.; Vijaya, J.J.; Selvam, N.C.S.; Kennedy, L.J. Microwave Assisted Combustion Synthesis of Coupled ZnO–ZrO_2 Nanoparticles and Their Role in the Photocatalytic Degradation of 2,4-Dichlorophenol. *Ceram. Int.* **2014**, *40*, 5681–5691. [CrossRef]
87. Shore, A. Expression of Concern: One-Pot Sol–Gel Synthesis of Reduced Graphene Oxide Uniformly Decorated Zinc Oxide Nanoparticles in Starch Environment for Highly Efficient Photodegradation of Methylene Blue. *RSC Adv.* **2018**, *8*, 10332. [CrossRef]

88. Kumar Sur, U.; Ankamwar, B.; Karmakar, S.; Halder, A.; Das, P. Green Synthesis of Silver Nanoparticles Using the Plant Extract of Shikakai and Reetha. *Mater. Today Proc.* **2018**, *5*, 2321–2329. [CrossRef]
89. Suppiah, D.D.; Julkapli, N.M.; Sagadevan, S.; Johan, M.R. Eco-Friendly Green Synthesis Approach and Evaluation of Environmental and Biological Applications of Iron Oxide Nanoparticles. *Inorg. Chem. Commun.* **2023**, *152*, 110700. [CrossRef]
90. Kavitha, A.; Doss, A.; Praveen Pole, R.P.; Pushpa Rani, T.P.K.; Prasad, R.; Satheesh, S. A Mini Review on Plant-Mediated Zinc Oxide Nanoparticles and Their Antibacterial Potency. *Biocatal. Agric. Biotechnol.* **2023**, *48*, 102654. [CrossRef]
91. Gudkov, S.V.; Burmistrov, D.E.; Serov, D.A.; Rebezov, M.B.; Semenova, A.A.; Lisitsyn, A.B. Do Iron Oxide Nanoparticles Have Significant Antibacterial Properties? *Antibiotics* **2021**, *10*, 884. [CrossRef]
92. Manjamadha, V.P.; Muthukumar, K. Ultrasound Assisted Green Synthesis of Silver Nanoparticles Using Weed Plant. *Bioprocess Biosyst. Eng.* **2016**, *39*, 401–411. [CrossRef]
93. Raffi, M.; Hussain, F.; Bhatti, T.M.; Akhter, J.I.; Hameed, A.; Hasan, M.M. Antibacterial Characterization of Silver Nanoparticles against *E. coli* ATCC-15224. *J. Mater. Sci. Technol.* **2008**, *24*, 192–196.
94. Panáček, A.; Kvítek, L.; Prucek, R.; Kolář, M.; Večeřová, R.; Pizúrová, N.; Sharma, V.K.; Nevěčná, T.; Zbořil, R. Silver Colloid Nanoparticles: Synthesis, Characterization, and Their Antibacterial Activity. *J. Phys. Chem. B* **2006**, *110*, 16248–16253. [CrossRef] [PubMed]
95. Xia, T.; Kovochich, M.; Brant, J.; Hotze, M.; Sempf, J.; Oberley, T.; Sioutas, C.; Yeh, J.I.; Wiesner, M.R.; Nel, A.E. Comparison of the Abilities of Ambient and Manufactured Nanoparticles to Induce Cellular Toxicity According to an Oxidative Stress Paradigm. *Nano Lett.* **2006**, *6*, 1794–1807. [CrossRef] [PubMed]
96. Song, H.Y.; Ko, K.K.; Oh, I.H.; Lee, B.T. Fabrication of Silver Nanoparticles and Their Antimicrobial Mechanisms. *Eur. Cells Mater.* **2006**, *11* (Suppl. S1), 58.
97. Sambhy, V.; MacBride, M.M.; Peterson, B.R.; Sen, A. Silver Bromide Nanoparticle/Polymer Composites: Dual Action Tunable Antimicrobial Materials. *J. Am. Chem. Soc.* **2006**, *128*, 9798–9808. [CrossRef] [PubMed]
98. Menon, S.; Agarwal, H.; Kumar, S.R.; Kumar, S.V. Green synthesis of silver nanoparticles using medicinal plant acalypha indica leaf extracts and its application as an antioxidant and antimicrobial agent against foodborne pathogens. *Int. J. Appl. Pharm.* **2017**, *9*, 42–50. [CrossRef]
99. Yadav, J.P.; Kumar, S.; Budhwar, L.; Yadav, A.; Yadav, M. Characterization and Antibacterial Activity of Synthesized Silver and Iron Nanoparticles Using Aloe Vera. *J. Nanomed. Nanotechnol.* **2016**, *7*, 3. [CrossRef]
100. Agarwal, H.; Venkat Kumar, S.; Rajeshkumar, S. A Review on Green Synthesis of Zinc Oxide Nanoparticles—An Eco-Friendly Approach. *Resour.-Effic. Technol.* **2017**, *3*, 406–413. [CrossRef]
101. Benyagoub, E.; Nabbou, N.; Boukhalkhel, S.; Dehini, I. The In Vitro Evaluation of the Antimicrobial Activity of *Quercus robur* L. Methanolic and Aqueous Leaves' Extracts, from the Algerian High Plateaus Against Some Uropathogenic Microbial Strains. *Phytothérapie* **2020**, *18*, 262–274. [CrossRef]
102. Šukele, R.; Skadiņš, I.; Koka, R.; Bandere, D. Antibacterial Effects of Oak Bark (*Quercus robur*) and Heather Herb (*Calluna vulgaris* L.) Extracts against the Causative Bacteria of Bovine Mastitis. *Vet. World* **2022**, *15*, 2315–2322. [CrossRef]

catalysts

MDPI

Article

Fabrication of a Novel Silica–Alumina-Based Photocatalyst Incorporating Carbon Nanotubes and Nanofiber Nanostructures Using an Unconventional Technique for Light-Driven Water Purification

Osama Saber [1,*], Abdullah Alshehab [1], Nagih M. Shaalan [1], Asmaa M. Hegazy [2], Fatimah K. Aljasem [3] and Aya Osama [1]

1 Department of Physics, College of Science, King Faisal University, P.O. Box 400, Al-Ahsa 31982, Saudi Arabia; alshehab@kfu.edu.sa (A.A.); nmohammed@kfu.edu.sa (N.M.S.); ayamohamed19984@yahoo.com (A.O.)
2 Clinical Laboratory, Sciences Department, Applied Medical Sciences College, University of Ha'il, Ha'il 55424, Saudi Arabia; a.hegazy@uoh.edu.sa
3 Fab Lab Abdulmonem Al Rashed Humanitarian Foundation, Al-Ahsa 31982, Saudi Arabia; fofo25kj@gmail.com
* Correspondence: osmohamed@kfu.edu.sa; Tel.: +966-13-589-9440

Abstract: The advancement of optical materials has garnered significant interest from the global scientific community in the pursuit of efficient photocatalysts for the purification of water using light. This challenge, which cannot be addressed using traditional methods, is tackled in the present study utilizing unconventional approaches. This study presents the fabrication of an effective photocatalyst using an unconventional approach that employs explosive reactions. This method successfully produces 3D nanostructures composed of carbon nanotubes (CNTs), carbon nanofibers (CNFs), and silica–alumina nanoparticles at temperatures below 270 °C. Gold-supported silica–alumina–CNT–CNF nanostructures were synthesized and characterized using XRD, TEM, SEM, and EDX, in addition to mapping images. To study and determine the photoactivity of these produced nanostructures, two well-known photocatalysts—titanium dioxide and zinc oxide—were synthesized at the nanoscale for comparison. The results showed that the presence of CNTs and CNFs significantly reduced the band gap energy from 5.5 eV to 1.65 eV and 3.65 eV, respectively, after modifying the silica–alumina structure. In addition, complete degradation of green dye was achieved after 35 min of light exposure using the modified silica–alumina structure. Additionally, the surface properties of the modified silica–alumina had a positive role in accelerating the photocatalytic decomposition of the green dye NGB. A kinetic study confirmed that the modified silica–alumina functions as a promising additive for optical applications, accelerating the photocatalytic degradation of NGB to a rate three times faster than that of the prepared titanium dioxide and six times that of the prepared zinc oxide.

Keywords: silica–alumina photocatalyst; carbon nanofibers and nanotubes; low-temperature synthesis for CNTs and CNFs; explosive method; surface characterization; low band gap energy; purification of water; pollutant removal using light

1. Introduction

Dyes represent a significant source of colored organic compounds that are released as waste during the textile dyeing process. The high levels of organic materials in wastewater, combined with the stability of contemporary synthetic dyes, render traditional biological methods for wastewater treatment inadequate for complete color removal and the degradation of organic pollutants [1]. Conventional techniques for removing color from water, such as coagulation, filtration, adsorption [2–4], and ozone treatment [5], have their own unique advantages and disadvantages. However, these standard methods exhibit several limitations and are not universally effective against all types of pollutants. As a result, photocatalysis has emerged as a promising solution for addressing water contamination, utilizing light as the driving force to eliminate pollutants. Numerous studies [6–11] have explored the use of light in photocatalytic degradation processes to purify water containing industrial contaminants. In these investigations, researchers have primarily aimed to develop efficient photocatalysts to enhance the effectiveness of photocatalytic processes [12–17]. However, the market potential of photocatalytic degradation techniques is limited due to the restricted applications and shortcomings of conventional photocatalysts such as titanium dioxide and zinc oxide. Therefore, there is a pressing need for alternative photocatalysts to overcome the challenges associated with traditional photocatalytic methods [18].

Silica–alumina structures are among the most well-known porous materials used in catalytic reactions. However, these structures are not sensitive to light and have not traditionally been applied in optical applications due to their large band gap energies, which typically range from 6.5 to 7.5 eV [19]. Therefore, the conversion of silica–alumina from non-optical materials to optical structures is a challenge for optical applications such as photocatalytic processes and solar cells. Previous studies have only considered silica–alumina structures and alumina–silicate-based systems, such as zeolite, as supporting materials for semiconductor photocatalysts. With reference to the cited literature and to the best of the authors' knowledge, previous studies have rarely considered using silica–alumina nanoparticles for the photocatalytic degradation of dyes.

In 2024, Baha et al. [20] investigated the photocatalytic decomposition of MB dye using bayerite/zeolite with TiO_2. In this study, 100% degradation efficiency was achieved after 4 h of light radiation. Additionally, in 2023, zeolite structures with different silica/alumina ratios were used to support titanium dioxide for the photocatalytic degradation of the pharmaceutical compound atenolol [21]. In this study, 50% photocatalytic degradation of atenolol was achieved after 70 min of light irradiation. Hutsul et al. [22] prepared ZnO-supported zeolite for the photocatalytic degradation of methylene blue dye. The dyes exhibited incomplete decomposition even after 1 h of UV irradiation. Recently, a copper oxide–carbon nitride–zeolite catalyst prepared by Ohn et al. [23] exhibited a high efficiency for degrading MB and CV dyes. In 2022, silica was used as a photocatalyst for the degradation of methylene blue dye, where 94% degradation efficiency was achieved after 5 h of light radiation [24].

The texture of silica–alumina structures plays an important role in catalytic processes, given their high thermal and chemical stability, high porosity, and large specific surface area. The mesoporous structure of silica–alumina plays a vital role for increasing the efficiency of photocatalysts by trapping pollutants inside. In addition, the high surface area of silica–alumina is suitable for creating new optical active sites through integration with carbon nanotubes and nanofibers in addition to supporting gold nanoparticles.

To enable optical applications of porous silica–alumina structures, gold nanoparticles can be introduced, as gold has a broad spectrum of light absorption and can function as an electron trap in photocatalysts. In addition, gold nanoparticles offer several benefits to photocatalysts, including functioning as light-trapping receptors due to their strong surface plasmon resonance (SPR) effect. A locally enhanced electric field is then produced close to the gold nanoparticles as a result of the photoexcitation of the SPR. Furthermore, SPR extends light absorption to longer wavelengths and increases light dispersion [10].

Using carbon nanofibers and nanotubes, three-dimensional (3D) porous silica–alumina nanocomposites were engineered to enhance photocatalytic efficiency by creating new optically active sites and lowering the optical band gap energy. In this study, explosive reactions of solid fuel were initiated within the porous silica–alumina matrix, leading to nanoscale structural development and the growth of carbon nanotubes (CNTs) and nanofibers (CNFs). Additionally, the high temperature and pressure generated from the explosive reactions create strong atomic-scale interactions between CNTs and CNFs in the silica–alumina framework. Crack bridging was achieved through the growth of CNTs and CNFs with Al_2O_3 during the explosive processes, according to the following equation: ($2Al_2O_3 + 6C \rightarrow Al_4C_3 + 3CO_2$). Crack bridging is supported by a strong CNT–Al_2O_3 interface, as suggested by many researchers [25,26]. Balani and Agarwal [25] confirmed the possibility of enhanced interfacial bonding between Al and C, indicating that crystals with high surface energy try to adhere to new surfaces to minimize their overall energy.

The growth of 1D nanostructures such as CNTs and CNFs with 0D silica–alumina nanoparticles to form 3D porous nanostructures presents a promising new additive for photocatalytic and catalytic applications. Therefore, X-ray diffraction, SEM, TEM, imaging mapping, and EDX were used to identify this novel nanocomposite. The surface characteristics and porous structures of the prepared nanocomposite were evaluated using N_2 adsorption–desorption isotherms at −196 °C. Additionally, the specific surface area, pore volume, and average pore radius were determined using the Brunauer–Emmett–Teller (BET) equation and the t-method. The pore size distribution was analyzed using desorption data following the method of Barrett, Joyner, and Halenda. To determine the optical efficiency of this novel nanocomposite, well-known photocatalysts such as zinc oxide and titanium dioxide were prepared for comparative analysis. In addition, conventional silica–alumina structures were used to highlight the benefits of the explosive method. UV–Vis spectroscopy was employed to thoroughly investigate the optical properties of the synthesized nanomaterials. In addition, the photocatalytic activity of these materials was evaluated based on their ability to degrade green dye. Furthermore, the photocatalytic activity of the prepared materials was investigated using kinetic models of photocatalytic processes.

2. Results

2.1. X-Ray Diffraction

The X-ray diffraction pattern of ZO-1 particles is displayed in Figure 1a. The X-ray diffraction pattern displayed four peaks at angles of 32.11°, 34.23°, 36.44°, and 47.67°. These diffraction lines were identified as wurtzite (JCPDS 36-1451) ZnO crystals by comparing them with the reflection patterns (100), (002), (101), and (102) of the zincite phase. The observed wurtzite crystal structure along with the broad diffraction peaks indicate that the ZnO particles are nanoscale in size.

Figure 1b depicts the X-ray diffraction pattern of TO-2. It exhibits three wide peaks at 2θ = 25.39°, 37.92°, and 48.11°. These peaks are due to the reflection planes (101), (004), and (200) of titanium dioxide. Furthermore, the observed diffraction patterns are consistent with the anatase phase, as confirmed by comparison with the full standard diffraction diagram of JCPDS 21-1272, as shown in Figure 1b. The broadness of the peaks suggests that the synthesized titanium dioxide particles are crystallized at the nanoscale.

The XRD pattern of the prepared silica–alumina sample SA-3 is shown in Figure 1c. A weak peak was observed at 2θ = 22.45°, indicating that SA-3 exhibits an amorphous structure, as shown in Figure 1c. Following explosive reactions, the crystalline structure of silica–alumina was modified as shown in Figure 1d. A prominent and broad peak for SAGC-3 is noted at 2θ = 25.55°. This peak corresponds to a d-spacing of 0.348 nm attributed to the plane reflection of graphite (002), which is characteristic of carbon nanotubes and nanofibers (CNTs and CNFs) [27,28]. Additionally, the faint peak at 2θ = 22.33° indicates the presence of an amorphous silica–alumina structure [29].

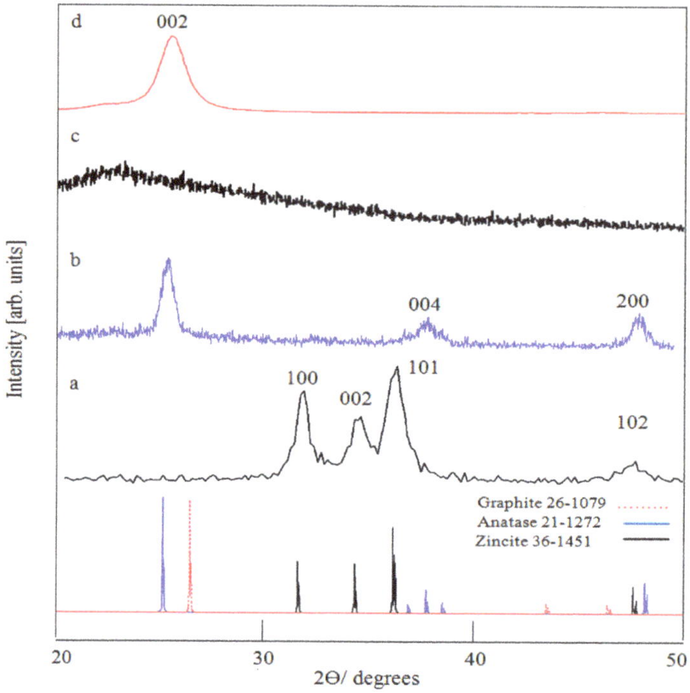

Figure 1. X-ray diffraction patterns of (**a**) ZO-1, (**b**) TO-2, (**c**) SA-3, and (**d**) SAGC-3.

2.2. Scanning and Transmission Electron Microscopy

The morphology of the ZO-1 sample was observed using TEM. Figure 2a shows SEM images of ZO-1. TEM images of ZO-1 indicated that the zinc oxide particles are nanoscale, consistent with the XRD results. As shown in Figure 2b, the individual zinc oxide particles exhibited irregular shapes, with an average diameter of 20 nm. The morphology of the TO-2 sample was observed using SEM. Figure 3 shows SEM images of TO-2. A large collection of nanoparticles was observed for TO-2, as shown in Figure 3a. In addition, SEM images indicated that the titanium dioxide nanoparticles aggregate to create porous structures, as shown in Figure 3a. Figure 3b shows that the titanium dioxide nanoparticles are predominantly spherical in shape, with diameters less than 50 nm.

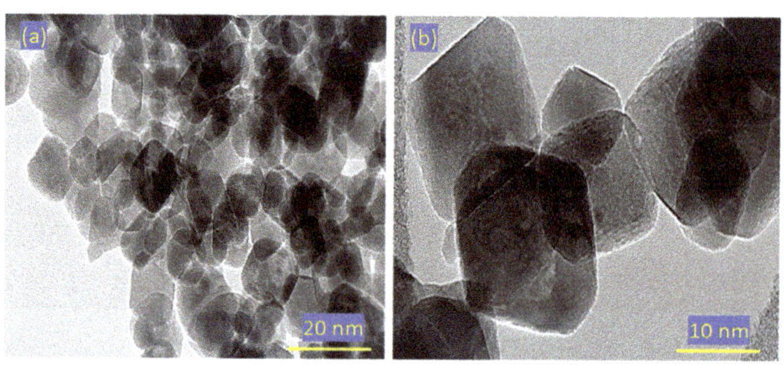

Figure 2. Transmission electron microscopy (TEM) images of ZO-1: (**a**) 20 nm and (**b**) 10 nm.

Figure 3. Scanning electron microscopy (SEM) images of TO-2: (**a**) 1µm and (**b**) 100 nm.

Figure 4 shows TEM images of the SAGC-3 sample. Carbon nanotubes and nanofibers were observed inside the silica–alumina matrix, as shown in Figure 4a. The average CNT diameter is 5 nm, whereas that of CNF is less than 5 nm. In addition, Figure 4b shows the dispersion and distribution of CNTs and CNFs in the silica–alumina matrix.

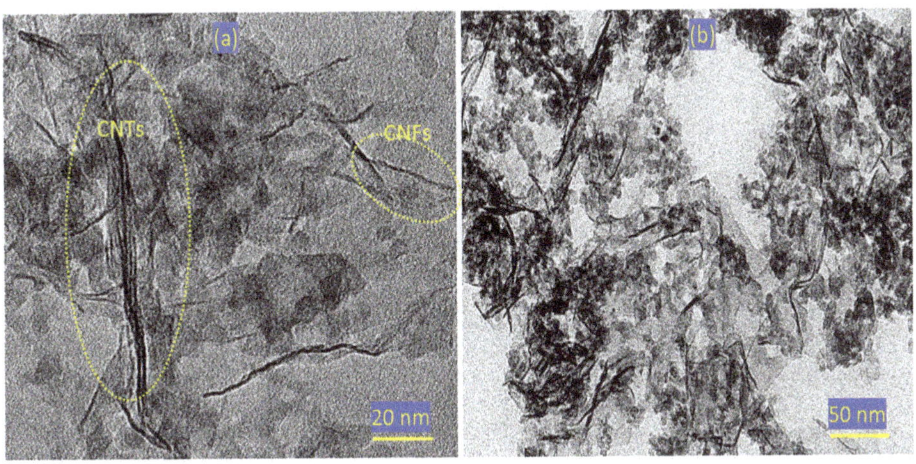

Figure 4. Transmission electron microscopy (TEM) images of SAGC-3: (**a**) 20 nm and (**b**) 50 nm.

2.3. Mapping Images and EDX Analysis

The dispersed silicon, aluminum, oxygen, and gold molecules in the SAGC-3 sample are visible in the element mapping images shown in Figure 5. These elements are represented as colored dots dispersed across specific regions. Additionally, as seen in Figure 5a,b, the violet dots in the silicon chart match the yellow dots in the aluminum chart, indicating the formation of a united silica–alumina structure. Similarly, the red dots in the oxygen chart (Figure 5c) align with the silicon and aluminum distributions, indicating the formation of the silica–alumina structure. Furthermore, as shown in Figure 5d, the gold mapping image reveals the dispersion of gold nanoparticles, which are symbolized by green dots on the gold chart.

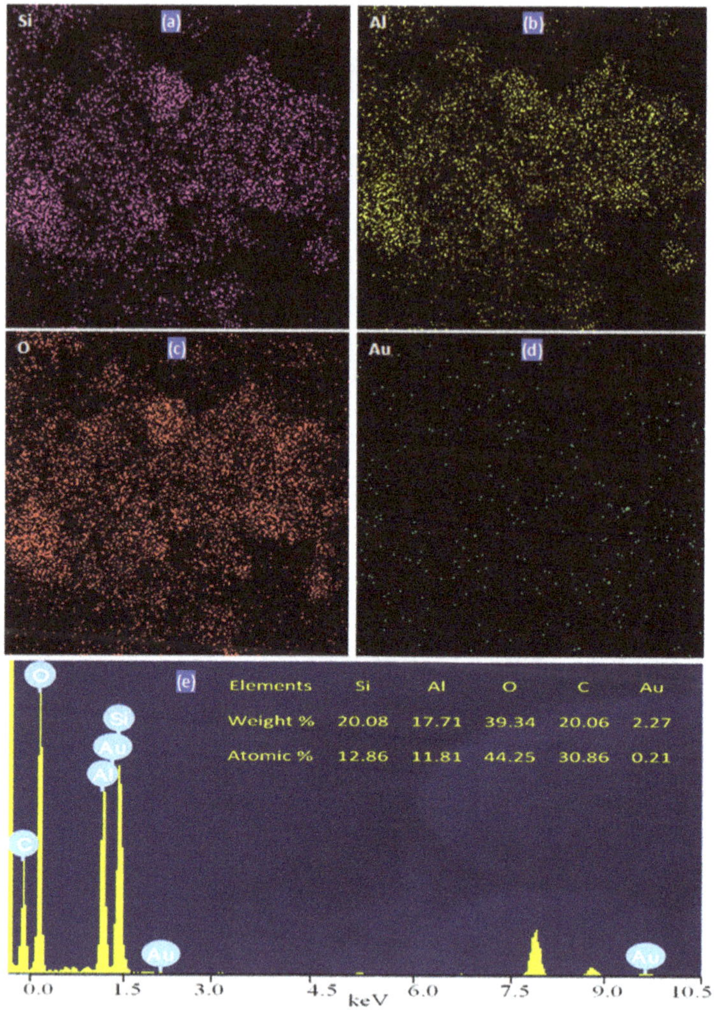

Figure 5. Mapping images of SAGC-3: (**a**) silicon, (**b**) aluminum, (**c**) oxygen, (**d**) gold, and (**e**) EDX spectrum with corresponding elemental composition.

The elemental composition of SAGC-3 and its EDX spectrum are displayed in Figure 5e. The EDX spectrum showed sharp signals for silicon, aluminum, and oxygen nanoparticles, consistent with the TEM images. These findings indicate that the main matrix of SAGC-3 is a silica–alumina structure. In addition, small signals were observed for the gold nanoparticles in Figure 5e, indicating low gold content. Additionally, the presence of carbon nanotubes and nanofibers in the SAGC-3 sample was confirmed by the observation of a distinct signal at low energy.

The basic composition of SAGC-3 was determined using EDX analysis, as seen in Figure 5e (inset). The atomic percentages of both silicon and aluminum are approximately equal, consistent with the proportions used during synthesis. The atomic percentage of oxygen is nearly double that of Si and Al, confirming the formation of the silica–alumina structure. The atomic percentage of carbon is 30.86%, indicating a high content of CNTs and CNFs inside the silica–alumina matrix. Additionally, the measured gold content is approximately 2 wt.%, aligning with the intended composition during preparation.

2.4. Surface Characteristics of Au/Silica–Alumina–CNT–CNF Nanoparticles

The texture and porous structures of silica–alumina are crucial factors in enhancing its catalytic activity. Therefore, nitrogen adsorption–desorption isotherms at 77 K were obtained for SAGC-3, as shown in Figure 6.

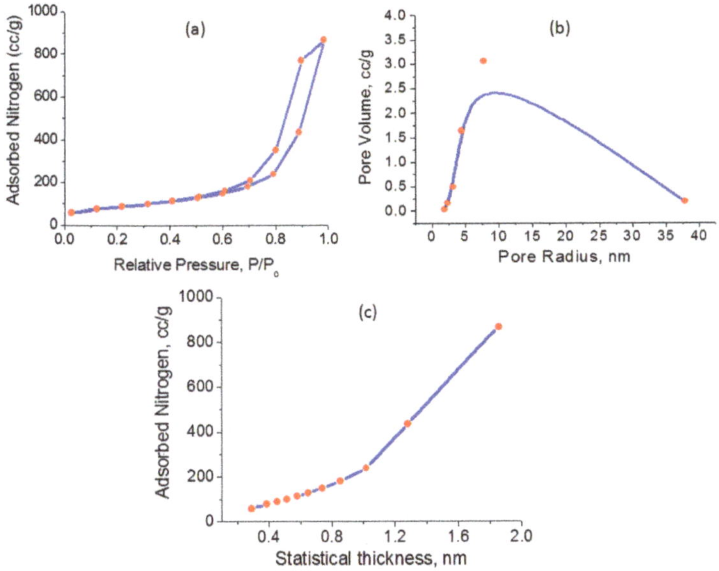

Figure 6. Surface properties of SAGC-3 nanocomposite: (**a**) nitrogen adsorption–desorption isotherms, (**b**) pore size distribution, and (**c**) V_{l-t} plot.

The adsorption data for the SAGC-3 isotherm were analyzed using the BET, De Boer, and BJH methods to determine the total pore volume, the specific surface area, and the average pore size. According to Brunauer and Emmett's classification, SAGC-3 exhibits a type IV isotherm, as illustrated in Figure 6a. However, due to the absence of a clear plateau at high p/p_o, this isotherm is classified as pseudo-type II or an intermediate between types II and IV. In addition, a hysteresis loop is observed and closes at a relative pressure of approximately 0.5.

The lower branch of the hysteresis loop represents nitrogen gas addition, while the upper branch corresponds to progressive withdrawal. This hysteresis loop is caused by the filling and emptying of mesopores, which occur as a result of capillary condensation. According to the IUPAC classification, the loop is classified as H3, indicating that SAGC-3 is made up of porous aggregates, with a lattice of cross-linked pores forming the interior free space. The computed surface characteristics from this isotherm revealed that SAGC-3 has a large specific surface area (S_{BET} = 307.0 m^2/g).

SAGC-3 has a total pore volume (V_p) of 1.39 cc/g and an average pore size (r_p) of 7.78 nm. SAGC-3's large total pore volume and pore size can be achieved by producing void surface area (pores) or clusters with a high surface-to-volume ratio. The V_{l-t} graph in Figure 6c verified SAGC-3's mesoporous microstructure, with upward departure at $t > 0.9$. The V_{l-t} figure shows a linear segment with an upward deviation, indicating mesoporous texture. The presence of capillary condensation in the mesopores produces the upward deviation. Once the mesopores are filled with nitrogen adsorbate, further adsorption proceeds via a multilayer mechanism, resulting in a linear segment on the t-plot. The BJH method was applied to the desorption branch of the isotherm within the hysteresis region to determine the pore size distribution, surface area, and pore volume. The results indicate that the majority of SAGC-3 pores have diameters limited to the restricted range of 5 to 8 nm (radius), as seen in Figure 6b.

2.5. Optical Properties

A nondestructive technique was used to measure the optical characteristics of the produced samples with a UV–Vis–NIR spectrophotometer. ZnO is an oxide semiconductor with significant exciton binding energy (60 meV) and an optical band gap of approximately 3.37 eV [1]. The absorbance spectrum of the produced ZO-1 nanoparticles in the wavelength range of 220–360 nm is displayed in Figure 7a.

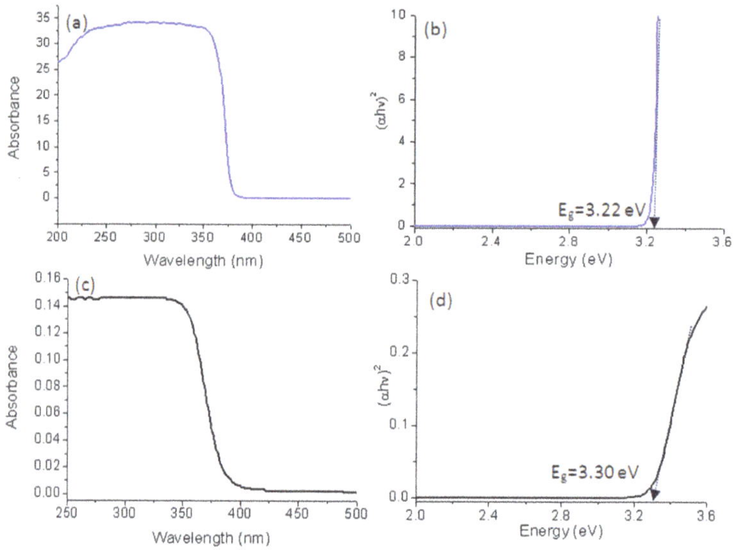

Figure 7. Optical properties: (**a**) UV–Vis absorbance of ZO-1, (**b**) band gap energy of ZO-1, (**c**) UV–Vis absorbance of TO-2, and (**d**) band gap energy of TO-2.

The excitonic transition is linked to the rapid rise in absorbance. Direct electronic transitions from the valence band to the conduction band determine the cut-off behavior at the blue end of the spectrum. As a result, the direct optical band gap, E_g, can be calculated using absorption data collected at short wavelengths. The theoretical relationship between E_g and the absorption coefficient for direct band gap energy is expressed as $(\alpha h\nu)^2$ = constant $(h\nu - E_g)^{1/2}$. The absorbance factor is α, and the value (ν) represents the frequency of radiation. Figure 7b depicts a plot of $(\alpha h\nu)^2$ versus photon energy ($h\nu$). The optical band gap, E_g, is calculated by extrapolating the plot's linear section. The obtained value for E_g is 3.22 eV. A minimal shift from 3.37 eV to 3.22 eV was observed. This is feasible given the nanoscale dimensions of the synthesized zinc oxide and the quantum confinement area.

Figure 7c,d show the UV–visible absorbance spectrum and band gap energy of TO-2. Figure 7c depicts the absorbance spectrum of TO-2 between 250 nm and 340 nm. These findings indicate that the TO-2 nanoparticles are sensitive in the UV region, which is consistent with previously published titanium dioxide results. This conclusion was supported by computing the band gap energy for TO-2, as seen in Figure 7d. TO-2's band gap energy was 3.30 eV. The band gap energies are comparable to those of bulk titanium oxide. This is acceptable given that the titanium dioxide nanoparticles generated in this study are much larger than the quantum confinement zone.

The conventional value E_g of silica–alumina is 7.5 eV [19]. This large band gap energy observed for silica–alumina indicates that these structures are not light-sensitive and, therefore, are not inherently suitable as optical materials. These data were confirmed by measuring the UV–Vis absorbance of the prepared silica–alumina nanoparticles (SA-3) as shown in Figure 8a. The UV–Vis absorbance spectrum of SA-3 exhibited an absorption cut-off below 275 nm. Also, the band gap energy of SA-3, shown in Figure 8b, was calculated to be 5.5 eV, which is characteristic of pure silica–alumina. To enhance light sensitivity, this study has used a new approach for preparing silica–alumina in the presence of explosive reactions. SAGC-3 was prepared and investigated using a UV–Vis absorption technique.

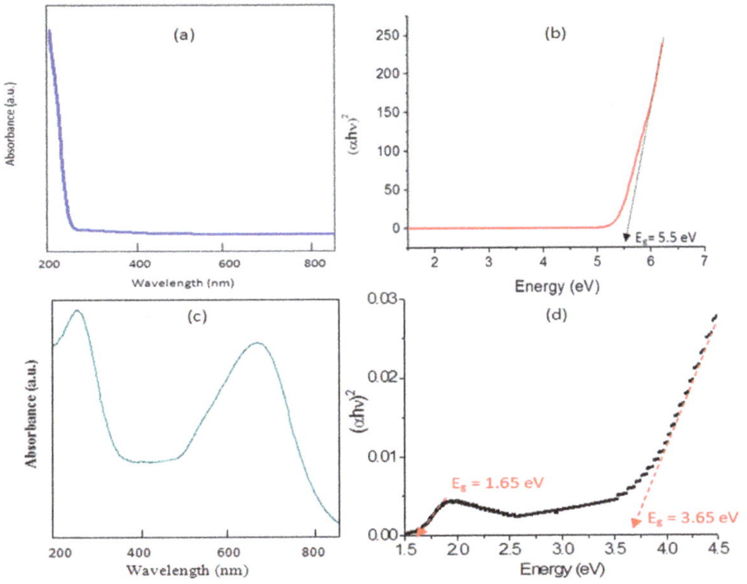

Figure 8. Optical properties of (**a**) UV–Vis absorbance of SA-3, (**b**) band gap energy of SA-3, (**c**) UV–Vis absorbance of SAGC-3, and (**d**) band gap energy of SAGC-3.

Figure 8c,d show the UV–Visible absorbance spectrum and the band gap energy of SAGC-3. Figure 8c shows significant absorption in both the visible and UV regions, with two pronounced maxima at 250 nm and 650 nm. Additionally, an absorption cut-off is observed near 800 nm. This finding indicates that, following modification via explosive reactions, the silica–alumina structure becomes sensitive to UV radiation and visible light. The enhancement in optical properties can be attributed to the incorporation of CNTs, CNFs, and gold nanoparticles.

The band gap energy was calculated to verify this improvement. SAGC-3's band gap energy is displayed in Figure 8d. Two band gap energies were detected at 1.65 eV and 3.65 eV. From 5.5 eV to 1.65 eV and 3.65 eV, the band gap energy was reduced compared to conventional silica–alumina. The significant red shift and considerable narrowing of the band gap energy can be attributed to the introduction of new electronic states within the band gap, resulting from defects in CNFs and CNTs defects.

The absorption edge of the prepared materials is defined by the cut-off wavelength ($\lambda_{cut\text{-}off}$), which corresponds to the wavelength at which the material begins to absorb light. The energy of the band gap (E_g) can be used to compute the $\lambda_{cut\text{-}off}$, as indicated by the following expression: $\lambda_{cut\text{-}off} = 1243/E_g$ [30,31]. The $\lambda_{cut\text{-}off}$ of ZO-1 occurred at 386.0 nm, which is consistent with the pure zinc oxide value presented in Table 1.

Table 1. Band gap energy and cut-off wavelength of the prepared photocatalysts.

Sample	Bang Gap Energy (eV)	Cut-off Wavelength (nm)
ZO-1	3.22	386.0
TO-2	3.30	376.7
SA-3	5.50	226.0
SAGC-3	1.65, 3.65	753.3, 340.6

For the pure titanium oxide, the $\lambda_{cut\text{-}off}$ of TO-2 was observed at 376.7 nm. The cut-off wavelengths of both ZO-1 and TO-2 indicated that they are sensitive to light at the UV region. For SA-3, the cut-off wavelength was observed at a low value of 226 nm, indicating that the conventional silica–alumina is only responsive to a very narrow range of UV light. When modified via explosive reactions, the silica–alumina structure exhibited a strong red shift for $\lambda_{cut\text{-}off}$ from 226.0 nm to 753.3 nm. According to Table 1, the $\lambda_{cut\text{-}off}$ for SAGC-3 occurred at 753.3 nm and 340.6 nm, confirming that SAGC-3 is active in a wide range of light from the UV to the visible region, indicating that the optical characteristics of silica–alumina are significantly improved by the explosive reactions. The shift in the cut-off wavelength and the reduction in band gap energy enhanced the light sensitivity of the prepared nanocomposite SAGC-3 across the visible and ultraviolet spectra.

2.6. Water Purification

The optical activity of the modified silica–alumina was compared with that of conventional silica–alumina and the well-known photocatalysts ZnO and TiO_2 through photocatalytic degradation of naphthol green B (NGB) dye in water. The photocatalytic degradation of NGB was examined while exposing the aqueous solution of NGB to UV light in the presence of the nanomaterials. The absorbance of the colored solution was measured at various intervals following light exposure. NGB degradation is shown by a decrease in absorption at λ_{max} = 714 nm. Other absorption peaks at 320, 283, and 232 nm reveal degradation of naphthyl rings in NGB, as seen in Figures 9 and 10. Blank controls were performed without a photocatalyst. NGB was stable under light irradiation. The photocatalytic degradation of NGB was studied as a function of light irradiation time in the presence of one of the nanomaterials (ZO-1, TO-2, SA-3, and SAGC-3). As shown in Figure 9a, minimal degradation was noted when the samples were combined with NGB solution and placed in a dark environment for 30 min. This time point is noted as 0 min irradiation and used for comparison. The photocatalytic degradation of NGB was detected upon light exposure in aqueous solutions of NGB that contained the synthesized nanomaterials. The results are presented in Figures 9 and 10.

In the absence of light, Figure 9a illustrates how the adsorption ability of ZO-1 affects the NGB concentration. The dye concentration was not altered, suggesting that the adsorption process had no impact on the photocatalytic degradation of NGB. In the presence of light, the photocatalytic decomposition of NGB increased as the radiation time increased. After 150 min of exposure to light, 100% of NGB was degraded, as shown in Figure 9a.

As shown in Figure 9b, the TO-2 sample yielded minimal changes in the NGB concentration in the absence of light, indicating that only adsorption occurred under these conditions. With light, the photocatalytic degradation of NGB increased as shown in Figure 9b. Complete disappearance of the green color of the dye was observed after 110 min of light radiation.

In the presence of SAGC-3, fast photocatalytic degradation was observed as shown in Figure 10. Light radiation caused a clear reduction in the dye concentration after 5 min. This reduction continued to increase with ongoing light exposure, achieving 100% dye removal after 35 min. This indicates that the SAGC-3 sample accelerated the photocatalytic degradation of NGB. This acceleration was notable based on comparisons with the conventional silica–alumina SA-3. Figure 11 shows the photocatalytic degradation of NGB after 35 min of light radiation in the presence of the conventional silica–alumina SA-3 and the modified silica–alumina SAGC-3.

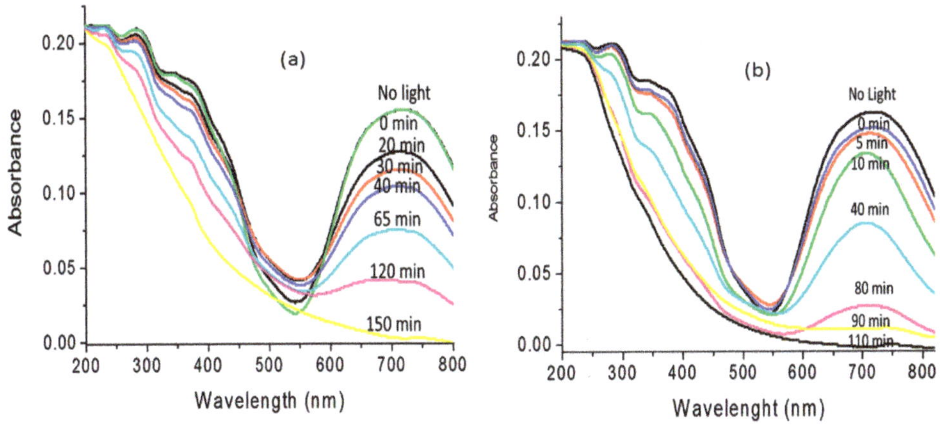

Figure 9. Optical UV–Vis absorbance of NGB dye after exposure to UV-light radiation for different periods of time in the presence of (**a**) ZnO-1 and (**b**) TO-2.

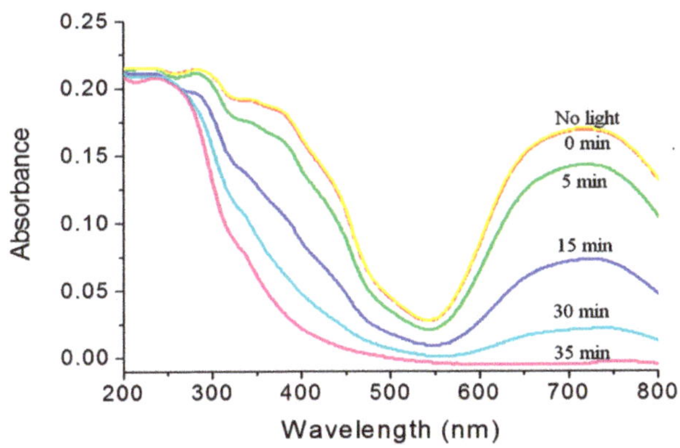

Figure 10. Optical UV–Vis absorbance of NGB dye after exposure to UV-light radiation for different periods of time in the presence of SAGC-3.

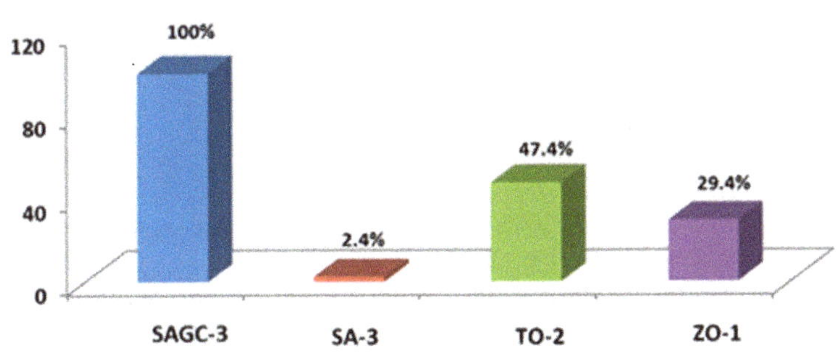

Figure 11. Photocatalytic degradation percentages of NGB after 35 min of light radiation in the presence of the prepared materials.

In the presence of SA-3, minimal variation in the dye concentration was noted after 35 min of light radiation. Figure 11 shows a low removal rate of green dye (2.4%). This indicates that the conventional silica–alumina is ineffective in the presence of light. A comparison between the results before and after modification reveals that the growth of CNTs and CNFs through explosive reactions plays a positive role in producing an efficient photocatalyst.

Additionally, Figure 11 presents a comparison between the photocatalytic decomposition of NGB after 35 min of light irradiation in the presence of the modified silica–alumina and the well-known photocatalysts titanium dioxide (TO-2) and zinc oxide (ZO-1). This comparison indicates that the modified silica–alumina is more effective under light irradiation than titanium dioxide and zinc oxide.

2.7. Kinetic Study

Kinetic models have been used to measure the rate of photocatalytic degradation. The Langmuir–Hinshelwood mechanism has been identified as a suitable model for heterogeneous photocatalysis. The photocatalytic degradation of NGB solutions in water is characterized as a bimolecular reaction. Throughout this process, the concentration of water remains constant, while the dye concentration varies significantly over time. Consequently, this reaction resembles a first-order (monomolecular) reaction. The kinetic reactions involved in the degradation of the NGB dye were analyzed using the following equation:

$$\ln C_o/C_t = kt \tag{1}$$

The initial concentration of the dye (C_o) was recorded at time zero, while the concentrations at various subsequent times (C_t) were also measured. The variable (k) represents the reaction's rate constant. Kinetic curves illustrating the photocatalytic breakdown of NGB with the synthesized photocatalysts ZO-1, TO-2, and SAGC-3 are presented in Figure 12.

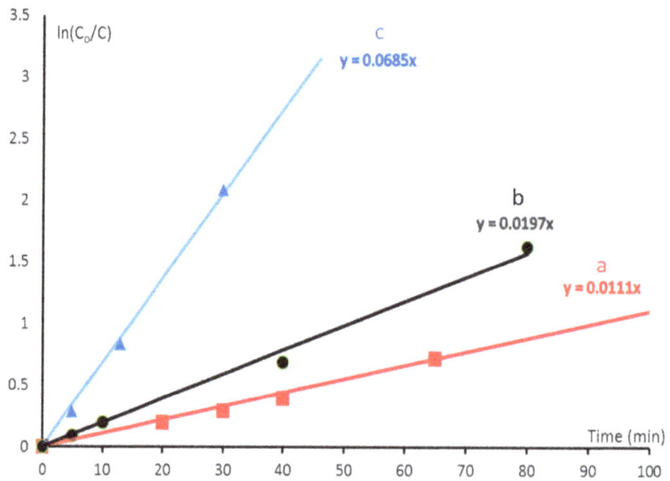

Figure 12. Kinetic models of photocatalytic reactions of NGB dye in the presence of light with the following photocatalysts: (a) ZO-1, (b) TO-2, and (c) SAGC-3.

Figure 12a illustrates the linear relationship observed in the decomposition of dyes when utilizing ZO-1, indicating that this process follows a pseudo-first-order kinetic reaction. Furthermore, the rate constant for this reaction is determined to be 0.0111 min^{-1}. In Figure 12b, the linear relationship for the dye decomposition reaction using TO-2 is presented, with a rate constant of 0.0197 min^{-1}. The decomposition reaction's rate constant increased to 0.0685 min^{-1} for SAGC-3, as shown in Figure 12c. In addition to the support of gold, the kinetic parameters of SAGC-3 demonstrate the beneficial effects of CNTs and CNFs in accelerating the rate of photocatalytic degradation of NGB.

The analysis of the modified silica–alumina (SAGC-3) in relation to traditional photocatalysts, such as zinc oxide (ZO-1) and titanium dioxide (TO-2), demonstrated the successful development of a novel and efficient photocatalyst based on silica–alumina for the removal of colored contaminants from water. Furthermore, as illustrated in Figure 12, the modified silica–alumina significantly enhanced the photocatalytic degradation rate of NGB, achieving a speed three times greater than that of the synthesized titanium dioxide and six times faster than the synthesized zinc oxide.

3. Discussion

The modified silica–alumina SAGC-3 demonstrated outstanding photoactivity, as evidenced by its rapid photocatalytic degradation of NGB when exposed to light. The impressive performance of SAGC-3 can be attributed to three key factors that enhance the photocatalytic degradation reaction. The mechanism of these photocatalytic degradation reactions is driven by three essential processes as follows:

$$\text{Separation: SAGC-3} + \text{Light} \rightarrow h^+ + e^- \quad (2)$$

$$\text{Accumulation of electrons: } e^- \rightarrow \text{Conduction band} \quad (3)$$

$$\text{Recombination: } h^+ + e^- \rightarrow \text{Catalyst} \quad (4)$$

The first factor depends on the growth of CNTs and CNFs with silica–alumina nanoparticles based on explosive reactions. The effective incorporation of silicon and aluminum oxides with the crystalline structure of carbon species created new optically active centers within the photocatalysts, leading to a reduction in its band gap energy. This reduction enhances its effectiveness under light because of the low band gap energy of T-carbon (2.273 eV) [32]. Therefore, the band gap energy of aluminum oxide nanoparticles decreased from 5.5 eV (SA-3 pure silica–alumina nanoparticles) to 1.65 eV and 3.65 eV for SAGC-3. This factor accelerated the separation process (Equation (2)) between the electrons and holes. By accelerating the reaction described in Equation (2), the excited electrons are collected in the conduction band, as shown in Equation (3). In the presence of a large number of electrons in the conduction band, the electrons tend to return to the valance band due to their mutual repulsion. In the recombination process (Equation (4)), the electrons recombine with holes in the valance band, which slows down the photocatalytic process. Therefore, these electron accumulation and recombination processes must be prevented to accelerate photocatalytic reactions. The presence of CNTs and CNFs offers another advantage in this regard. CNTs and CNFs prevented the accumulation and recombination processes through a fast transfer of electrons from the conduction band to other spots over the surface of the photocatalyst because the carbon nanotubes and nanofibers are good conductors that function like electrical wires.

The second factor focuses on the mesoporous structure of silica–alumina and its high surface area. The nanopores of SAGC-3 (pore radius = 7.78 nm) serve as effective traps for capturing pollutants, facilitating rapid degradation. In addition, the high surface area of SAGC-3 (S_{BET} = 307.0 m^2/g) increases the available surface area for light exposure and enhances the efficiency of the photocatalytic reactions.

Gold is the third factor that accelerates photocatalytic degradation through several beneficial effects on the photocatalyst. Gold has a wide range of light absorption and can act as electron traps. Additionally, it possesses strong surface plasmon resonance (SPR), which functions as a light-trapping receptor. Image mapping of SAGC-3 revealed a high dispersion of gold nanoparticles over the photocatalyst surface.

Based on these factors, the fast photocatalytic degradation of NGB was achieved through the production of numerous oxidizing agents, as shown in Figure 13.

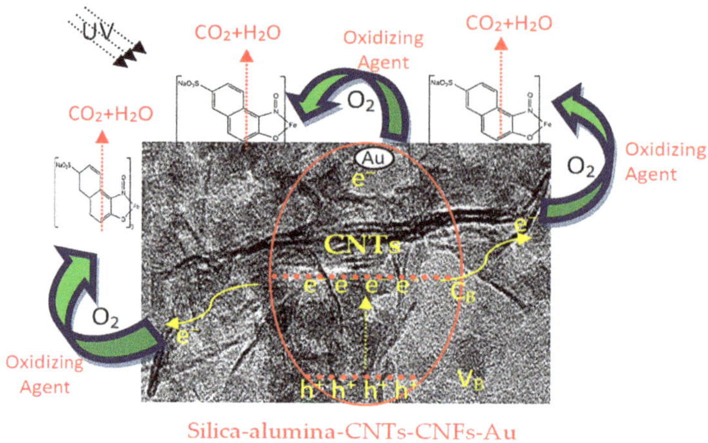

Figure 13. Schematic representation of the production of oxidizing agents and photocatalytic degradation of the dye in the presence of the optimal photocatalyst SAGC-3.

In Figure 13, the separation process was accelerated by lowering the band gap energy and creating new active sites in addition to the effect of gold nanoparticles. At the same time, the carbon nanotubes and nanofibers facilitate the transfer of excited electrons across the surface of the photocatalyst. These excited electrons then react with oxygen molecules, generating a significant number of oxidizing agents. Because of the high surface area and the mesoporous structure of SAGC-3, a significant number of NGB molecules were trapped on the catalyst's surface, where they react with oxidizing agents. Through advanced oxidation reactions, the dye was degraded and converted to carbon dioxide and water after 35 min of exposure to light radiation.

This mechanism was confirmed by photoluminescence (PL) measurements, which are a useful technique for studying the efficiency of photogenerated electrons and hole migration and separation. In the PL spectrum of SAGC-3 (Figure 14), three emission peaks were observed at 412 nm, 503 nm, and 750 nm with low intensity. The emission peaks were related to the separation and recombination rate of the photogenerated carriers.

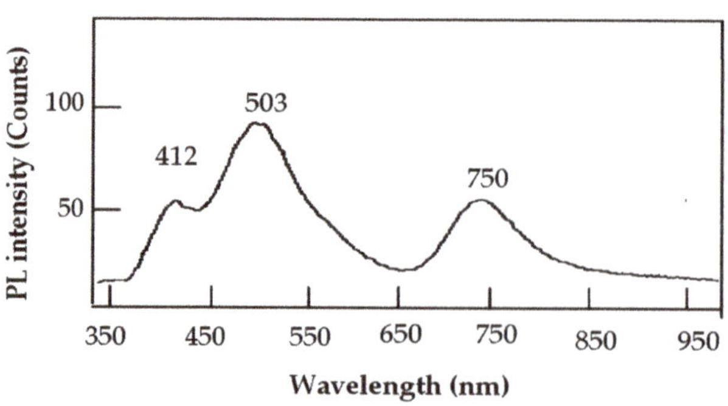

Figure 14. Photoluminescence measurements of SAGC-3.

Consistent with the results of Zhu et al. [33] and Liu et al. [34], the low intensity of the peaks suggests a high separation efficiency of the photogenerated electrons and holes, with a low recombination rate. Additionally, in agreement with the results of Xiang et al. [35], these peaks indicated that the photogenerated carriers have a prolonged lifetime, which is favorable for the photocatalytic reaction. The low intensity of the emission peaks could be due to trapping states or oxygen vacancies within the aluminum and silicon oxide structure, which are saturated by the carbon nanotubes and nanofibers.

In addition, the kinetic study confirmed the mechanism behind the rapid degradation of pollutants, showing that the modified silica–alumina (SAGC-3) accelerated the photocatalytic degradation rate of NGB three times faster than the prepared titanium dioxide and six times faster than the prepared zinc oxide.

4. Materials and Methods

4.1. Synthesis of the Modified Silica–Alumina

The modified silica–alumina nanocomposite was prepared in three steps. The first step depended on the reaction of $Al(NO_3)_3$ with NH_4HCO_3 to produce aluminum hydroxide($Al(OH)_3$) saturated with ammonium nitrate (NH_4NO_3), which can be used as explosive material. Then, the explosive material was mixed with an alcoholic solution of tetraethyl orthosilicate. In the second step, explosive reactions were performed using the alcoholic mixture inside an autoclave, as shown in Figure 15. The last step focuses on supporting gold nanoparticles onto the prepared nanocomposite.

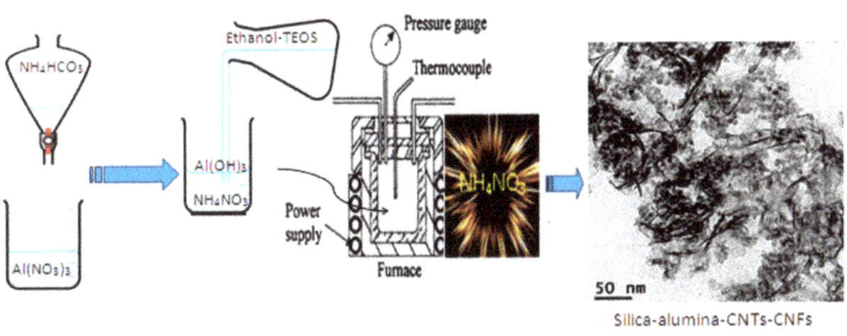

Figure 15. Steps for the preparation of silica–alumina–carbon nanotubes and nanofibers through explosive reactions.

4.1.1. Synthesis of Explosive Compounds

Aluminum nitrate $Al(NO_3)_3{\cdot}9H_2O$ and tetraethyl orthosilicate (TEOS) were provided from Sigma Aldrich and used as sources for producing silica and alumina, respectively. Cetyl tri-methyl ammonium bromide (CTAB) surfactant was used to control the nano-size of the products. Ammonium bicarbonate (NH_4HCO_3) was used as a source for fabricating the explosive material, namely ammonium nitrate. Using the sol–gel technique, aluminum hydroxide was produced and combined with ammonium nitrate to work as a solid fuel. Initially, aluminum nitrate (0.03 M) and CTAB (0.0003 M) were mixed to produce one liter of aqueous solution. This solution was reacted with ammonium bicarbonate (10 wt.% solution) to produce a gel at pH = 6.7 as shown in the following equation:

$$Al(NO_3)_3 + 3NH_4HCO_3 \rightarrow 3NH_4NO_3 + Al(OH)_3 + 3CO_2$$

This white gel was mixed with an alcoholic solution of 0.03 mole of TEOS. This mixture was stirred for 6 h to form a homogenous mixture. The gel was separated by filtration and dried at room temperature.

4.1.2. Explosive Reactions for Silica–Alumina CNT and CNF Growth

An appropriate amount of the prepared gel, synthesized in the previous step, was stirred with 250 mL of ethyl alcohol. The resulting alcoholic mixture was thermally treated by gradually heating it inside a pressurized vessel (autoclave). Ammonium nitrate is sensitive to temperature and explodes at temperatures above 250 °C. At 260 °C, the pressure increased to 100 bar due to the explosion of ammonium nitrate, as shown in the following equations:

At explosion (260 °C):

$$8NH_4NO_3 \text{ (8 mole; P = 1 bar)} \rightarrow 5N_2 + 4NO + 2NO_2 + 16H_2O \text{ (27 mole; P = 100 bar)}$$

$$Al(OH)_3 + Si(OC_2H_5)_4 + C_2H_5OH \rightarrow SiO_2\text{-}Al_2O_3\text{-CNT-CNF} + H_2O$$

By gradually releasing the gases from the autoclave, the pressure was reduced while introducing an inert gas (argon) to remove any remaining gases. The resulting product was calcined at 500 °C in the presence of air to produce silica–alumina–CNT–CNF nanocomposites.

4.1.3. Gold Support

To prepare gold nanoparticles, gold chloride was mixed with 50 mL of ethylene glycol. This solution was stirred with heating at 65 °C for 60 min. Then, sodium hydroxide was added until the solution became alkaline. After stirring 2 h, a few drops of hydrazine hydrate were added, and the temperature was maintained at 75 °C for 2 h.

The gold was supported on the prepared silica–alumina–CNT–CNF nanocomposites using the dry impregnation process. Typically, 1 g of the prepared nanocomposite was mixed with deionized water. The nanocomposite was weighted after water impregnation to determine the amount of water absorbed, calculated by the difference in the weight of titanium dioxide before and after water impregnation. Then, 20 mL of the aqueous gold solution, which was prepared in the previous step, was mixed with 2 g of the prepared SiO_2–Al_2O_3–CNT–CNF nanocomposite to achieve a gold content of 2 wt.% in the final product. After stirring for 20 min, the product was kept under vacuum for 24 h. The resulting material was designated as SAGC-3.

4.2. Synthesis of Silica–Alumina Nanoparticles

Silica–alumina nanocomposites with comparable molar ratios were synthesized using the solvent thermal technique. Initially, specific quantities of aluminum tri-sec-butoxide and tetra-ethyl orthosilicate were dissolved in a surplus of ethyl alcohol. This mixture was stirred for two hours and subsequently transferred to a pressurized vessel (autoclave). The reaction occurred at a temperature of 260 °C and a pressure of 100 bars. After the reaction was complete, the solvent vapors were evacuated and replaced with an inert gas. The resulting product was then calcined at 500 °C in an air atmosphere, yielding the silica–alumina nanocomposite designated as SA-3.

4.3. Synthesis of Zinc Oxide Nanoparticles

Zinc oxide nanoparticles were synthesized using a solvent thermal method. Zinc acetate served as the zinc precursor, while methanol acted as the solvent. An adequate quantity of zinc acetate (16 g) was dissolved in a surplus of methanol (120 mL). The mixture was stirred for 12 h. Subsequently, the solution was transferred into a pressurized vessel. The processes of dehydroxylation and drying were conducted under supercritical conditions, specifically at a temperature of 260 °C and a pressure of 100 bar, utilizing 350 mL of ethanol. Supercritical drying was performed directly with the gel containing the solvent. The temperature of the autoclave was gradually increased until it exceeded the critical threshold.

As a result, the pressure exceeds the critical threshold. At this stage, the autoclave is vented by carefully opening its valve. Once the pressure inside the autoclave nears atmospheric levels, it is purged with an inert gas, specifically argon, to remove any remaining gases. Subsequently, the autoclave is allowed to cool to room temperature. This purging process can occur during the cooling phase, as long as the temperature remains sufficiently high to prevent liquid condensation in the smallest pores of the gel. Since the solvent vapor is replaced with an inert gas, liquid condensation does not occur. The product underwent thermal treatment at 450 °C for 4 h and was designated as ZO-1.

4.4. Synthesis of Titanium Dioxide Nanoparticles

The sol–gel method was used to create titanium dioxide nanoparticles based on alcohol-based procedures. A total of 20 milliliters of titanium isopropoxide and 125 milliliters of ethyl alcohol were reacted to create the alcoholic mixes. Titanium isopropoxide molecules are extremely reactive and polar due to the strong electronegativity between titanium and oxygen. Through simultaneous hydrolysis and condensation procedures, 125 mL of the deionized water was added to the alcoholic titanium isopropoxide mixtures to create the gel form. The white product was separated using filtration and washing procedures with ethyl alcohol and deionized water. The product was vacuum-dried for a full day at room temperature. For three hours, the product underwent heat treatment at 450 °C. The product was classified as TO-2.

4.5. Characterization of the Prepared Samples

To analyze the crystalline structures, X-ray powder diffraction (XRD) was conducted using a Rigaku RINT 2200 (Tokyo, Japan) with CuK (filtered) as the radiation source. The device was operated at a wavelength of 0.154 nm across a 2θ range of 20–50°. A JEOL JEM-2100F (Tokyo, Japan) was used to image the prepared samples using transmission electron microscopy (TEM) and to determine the elemental composition of the prepared samples using energy-dispersive X-ray spectroscopy (EDX). Scanning electron microscopy (SEM) was performed using a JEOL JSM-6330F at 15 kV and 12 mA. The porous structures and surface characteristics were evaluated based on the complete adsorption–desorption isotherm of nitrogen gas at 77 K. These processes were conducted using a Quanta-chrome Nova sorption system. The optical properties of the synthesized nanomaterials were assessed using the Shimadzu 3600 UV diffuse reflection method. Reflectance data were converted to absorption values using spectrophotometer UV 3600 software (Shimadzu, Columbia, MD, USA). For solid samples, an ISR-603 spectrophotometer equipped with an integrated ball attachment (Shimadzu, Columbia, MD, USA) was utilized. Liquid samples were analyzed for absorption coefficients using conventional UV–Vis techniques.

4.6. Photocatalytic Processes

The photocatalytic decomposition of naphthol green B (NGB) was employed to purify water contaminated by industrial pollutants using light. To assess the efficiency and photoactivity of the synthesized materials, photocatalytic reactions were conducted in a quartz immersion well reactor (RQ400) equipped with a 400 W mercury lamp, which emits a broad spectrum of wavelengths in both the visible and ultraviolet ranges. The 3040/PX0686 mercury lamp (Camberley, Surrey, UK), which has a 400 W medium pressure, was combined with a 400 mL standard reaction flask (model 3308) to perform the photocatalytic reactions. The major portion of the radiation for a medium pressure lamp focuses on the range from 365 to 366 nm. Moreover, it produces small amounts in the ultraviolet region at 254, 265, 270, 289, 297, 302, 313, and 334 nm. In addition, significant amounts of radiation are produced in the visible region at 405–408, 436, 546, and 577–579 nm.

The NGB concentration was 0.4×10^{-3} M. To study the adsorption effect of the photocatalyst, the experiment was performed in the absence of light before performing the standard test. In addition, the same experiment was performed without a photocatalyst in the presence of light to measure the stability of the dye in light. In the main experiment, the concentration of dye was followed by measuring and analyzing the absorption of NGB in accordance with the Beer–Lambert law. The change in concentration was tracked by monitoring the green band at 714 nm associated with NGB. Small samples of the contaminated water were taken after light radiation at various time intervals, and the UV spectrophotometer was used to measure and follow the concentration of the remaining dye.

5. Conclusions

The current study successfully met several objectives. The primary goal was to transform non-optical silica–alumina into a functional photocatalyst using explosive methods involving solid fuels, specifically, ammonium nitrate and aluminum hydroxide. This transformation was accomplished by reducing the band gap energy of silica–alumina from 5.5 eV to 1.65 eV and 3.65 eV through the growth of carbon nanofibers and nanotubes, respectively. The secondary objective centered on the synthesis of carbon nanotubes and nanofibers at a comparatively low temperature of 260 °C. The third goal was to enhance the degradation of industrial pollutants after brief exposure to light radiation. In this context, all green pollutants were eliminated after only 35 min of illumination. Consequently, the findings suggest that silica–alumina modified through explosive reactions serves as a novel additive for nanocomposites designed for optical applications, including photocatalysis. Furthermore, the zero-dimensional nanoparticles and one-dimensional carbon nanotubes and nanofibers are effective materials for constructing three-dimensional mesoporous nanostructures with a large surface area.

Author Contributions: Conceptualization, O.S. and A.O.; methodology, O.S., A.A., A.M.H. and N.M.S.; software, A.O. and F.K.A.; validation, O.S., A.A., A.M.H. and N.M.S.; formal analysis, O.S., A.O. and F.K.A.; investigation, O.S., A.O. and F.K.A.; resources, O.S., A.A., A.M.H. and N.M.S.; data curation, A.O. and F.K.A.; writing—original draft preparation, O.S., A.O. and F.K.A.; writing—review and editing, O.S., A.A., A.M.H. and N.M.S.; visualization, O.S. and A.A.; supervision, O.S., F.K.A. and A.O.; project administration, O.S.; funding acquisition, O.S. All authors have read and agreed to the published version of the manuscript.

Funding: This research was funded by King Faisal University (grant number KFU251262). The APC was funded by the grant number KFU251262.

Data Availability Statement: Data will be provided by the authors upon request.

Acknowledgments: The work was supported by the Deanship of Scientific Research, Vice Presidency for Graduate Studies and Scientific Research at King Faisal University (Grant No. KFU251262). The authors extend their appreciation to the Abdulmonem Alrashad Humanitarian Foundation (FSP-2 No. 0003).

Conflicts of Interest: The authors declare no conflicts of interest.

References

1. Saber, O.; El-Brolossy, T.; Al Jaafari, A.A. Improvement of Photocatalytic Degradation of Naphthol Green B Under Solar Light Using Aluminum Doping of Zinc Oxide Nanoparticles. *Water Air Soil Pollut.* **2012**, *223*, 4615. [CrossRef]
2. Gupta, V.K.; Gupta, B.; Rastogi, A.; Agarwal, S.; Nayak, A. A comparative investigation on adsorption performances of mesoporous activated carbon prepared from waste rubber tire and activated carbon for a hazardous azo dye Acid Blue. *J. Hazard. Mater.* **2011**, *186*, 891–901. [CrossRef] [PubMed]
3. Gupta, V.K.; Jain, R.; Nayak, A.; Agarwal, S.; Shrivastava, M. Removal of the hazardous dye—Tartrazine by photodegradation on titanium dioxide surface. *Mater. Sci. Eng. C* **2011**, *31*, 1062–1067. [CrossRef]

4. Gupta, V.K.; Jain, R.; Agarwal, S.; Shrivastava, M. Kinetics of photo-catalytic degradation of hazardous dye Tropaeoline 000 using UV/TiO_2 in a UV reactor. *Colloids Surf. A Physicochem. Eng. Asp.* **2011**, *378*, 22–26. [CrossRef]
5. Lorimer, J.P.; Mason, T.J.; Plattes, M.; Phull, S.S.; Walton, D.J. Degradation of dye effluent. *Pure Appl. Chem.* **2001**, *12*, 1957–1968. [CrossRef]
6. Chen, N.; Xia, J.; Li, L.; Lv, Q.; Zhao, K.; Ahmad, M.; Xiao, Z.; Wang, S.; Ye, F.; Zhang, Q. Comprehensive enhancement of photocatalytic H_2O_2 generation and antibacterial efficacy on carbon nitride through a straightforward polydopamine coating strategy. *Surf. Interfaces* **2025**, *56*, 105566. [CrossRef]
7. Saber, O.; Osama, A.; Shaalan, N.M.; Osama, M. Engineering the Integration of Titanium and Nickel into Zinc Oxide Nanocomposites through Nanolayered Structures and Nanohybrids to Design Effective Photocatalysts for Purifying Water from Industrial Pollutants. *Catalysts* **2024**, *14*, 340. [CrossRef]
8. Saber, O.; Osama, A.; Shaalan, N.M.; Osama, M. Nanolayered Structures and Nanohybrids Based on a Ternary System Co/Ti/Zn for Production of Photo-Active Nanocomposites and Purification of Water Using Light. *Nanomaterials* **2024**, *14*, 93. [CrossRef]
9. Saber, O.; Osama, M.; Alshoaibi, A.; Shaalan, N.M. Doaa Osama Designing inorganic–magnetic–organic nanohybrids for producing effective photocatalysts for the purification of water. *RSC Adv.* **2022**, *12*, 18282–18295. [CrossRef]
10. Saber, O.; Awada, C.; Osama, A.; Shaalan, N.M.; Alshoaibi, A.; Mansour, S.A.; Osama, M.; Farha, A.H. Boosting the Efficiency of Titanium Oxides and Accelerating Industrial Pollutant Removal Through Triple-Action Effects to Purify Water by Light. *Catalysts* **2024**, *14*, 772. [CrossRef]
11. Albeladi, A.; Khan, Z.; Al-Thabaiti, S.A.; Patel, R.; Malik, M.A.; Meht, S. A Fe_3O_4-CdO Nanocomposite for Organic Dye Photocatalytic Degradation: Synthesis and Characterization. *Catalysts* **2024**, *14*, 71. [CrossRef]
12. Lau, G.E.; Abdullah, C.A.C.; Ahmad, W.A.N.; Assaw, W.S.; Zheng, A.L.T. Eco-Friendly Photocatalysts for Degradation of Dyes. *Catalysts* **2020**, *10*, 1129. [CrossRef]
13. Hauser, K.; Hassou, S.O.; Riaz, U. Ultrasound-Assisted Photocatalytic Degradation of Azo Dyes under Visible Light Irradiation Using Polythiophene-Decorated $CoFe_2O_4$ Nanohybrids. *Catalysts* **2024**, *14*, 495. [CrossRef]
14. Palma Soto, E.; Rodriguez Gonzalez, C.A.; Luque Morales, P.A.; Reyes Blas, H.; Carrillo Castillo, A. Degradation of Organic Dye Congo Red by Heterogeneous Solar Photocatalysis with Bi_2S_3, Bi_2S_3/TiO_2, and Bi_2S_3/ZnO Thin Films. *Catalysts* **2024**, *14*, 589. [CrossRef]
15. Mbuyazi, T.B.; Ajibade, P.A. Photocatalytic Degradation of Organic Dyes by Magnetite Nanoparticles Prepared by Co-Precipitation. *Int. J. Mol. Sci.* **2024**, *25*, 7876. [CrossRef]
16. Kalaycıoğlu, Z.; Ozuğur Uysal, B.; Pekcan, O.; Erim, F.B. Efficient Photocatalytic Degradation of Methylene Blue Dye from Aqueous Solution with Cerium Oxide Nanoparticles and Graphene Oxide-Doped Polyacrylamide. *ACS Omega* **2023**, *8*, 13004–13015. [CrossRef]
17. Khan, S.; Noor, T.; Iqbal, N.; Yaqoob, L. Photocatalytic Dye Degradation from Textile Wastewater: A Review. *ACS Omega* **2024**, *9*, 21751–21767. [CrossRef]
18. Long, Z.; Li, Q.; Wei, T.; Zhang, G.; Ren, Z. Historical development and prospects of photocatalysts for pollutant removal in water. *J. Hazard. Mater.* **2020**, *395*, 122599. [CrossRef]
19. Tahir, D.; Kwon, H.; Shin, H.; Oh, S.; Kang, H.; Heo, S.; Chung, J.; Lee, J.; Tougaard, S. Electronic and optical properties of Al_2O_3/SiO_2 thin films grown on Si substrate. *J. Phys. D Appl. Phys.* **2010**, *43*, 255301. [CrossRef]
20. Baha, A.A.; Ait-Karra, A.; Idouhli, R.; Tabit, K.; Zakir, O.; Dikici, B.; Khadiri, M.E.; Abouelfida, A. Synergistic Photocatalysis of Bayerite/Zeolite Loaded TiO_2 Nanocomposites for Highly Efficient Degradation of Organic Pollutants in Aqueous Environments. *Silicon* **2024**, *16*, 4843–4856. [CrossRef]
21. Stojanović, S.; Vranješ, M.; Šaponjić, Z.; Rac, V.; Rakić, V.; Ignjatović, L.; Damjanović-Vasilić, L. Photocatalytic performance of TiO_2/zeolites under simulated solar light for removal of atenolol from aqueous solution. *Int. J. Environ. Sci. Technol.* **2023**, *20*, 1–16. [CrossRef]
22. Hutsul, K.; Ivanenko, I.; Patrylak, L.; Pertko, O.; Kamenskyh, D. ZnO/Zeolite composite photocatalyst for dyes degradation. *Appl. Nanosci.* **2023**, *13*, 7601–7609. [CrossRef]
23. John, A.; Rajan, M.S.; Thomas, J. Sunlight Driven Degradation of Drug Residues Using CuO Incorporated- Zeolite Supported-Graphitic Carbon Nitride. *Catal. Lett.* **2025**, *155*, 23. [CrossRef]
24. Zarazúa-Morín, M.E.; Galindo-Luna, A.S.; Gallegos-Sánchez, V.J.; Juárez-Ramírez, I. Chemical exfoliation of silica filters used on methylene blue degradation by photocatalysis. *Chem. Pap.* **2022**, *76*, 5627–5642. [CrossRef]
25. Balani, K.; Agarwal, A. Wetting of carbon nanotubes by aluminum oxide. *Nanotechnology* **2008**, *19*, 165701. [CrossRef]
26. Keshri, A.K.; Huang, J.; Singh, V.; Choi, W.; Seal, S.; Agarwal, A. Synthesis of aluminum oxide coating with carbon nanotube reinforcement produced by chemical vapor deposition for improved fracture and wear resistance. *Carbon* **2010**, *48*, 431–442. [CrossRef]
27. Fan, Y.Y.; Li, F.; Cheng, H.M.; Su, G.; Yu, Y.D.; Shen, Z.H. Preparation, morphology, and microstructure of diameter-controllable vapor-grown carbon nanofibers. *J. Mater. Res.* **1998**, *8*, 13. [CrossRef]

28. Luhrs, C.C.; Moberg, M.; Maxson, A.; Brewer, L.; Menon, S. IF-WS2/Nanostructured Carbon Hybrids Generation and Their Characterization. *Inorganics* **2014**, *2*, 211–232. [CrossRef]
29. Toyama, N.; Inoue, H.; Ohki, S.; Tansho, M.; Shimizu, T.; Umegaki, T.; Kojima, Y. Influence of hollow silica-alumina composite spheres prepared using various amount of L(+)-arginine on their activity for hydrolytic dehydrogenation of ammonia borane. *Adv. Mater. Lett.* **2016**, *7*, 339–343. [CrossRef]
30. Sethi, R.; Ahmad, S.; Aziz, A.; Siddiqui, A.M. Structural, optical and electrical properties of tin oxide thin films for application as a wide band gap semiconductor. *AIP Conf. Proc.* **2015**, *1675*, 030039.
31. Soumya, S.S. Optical, Dielectric and Structural Properties of SnO2 Nanoparticles via Sol-Gel Method. *J. Phys. Conf. Ser.* **2023**, *2426*, 012039. [CrossRef]
32. Sun, P.; Bai, L.; Kripalani, D.R.; Zhou, K. A new carbon phase with direct bandgap and high carrier mobility as electron transport material for perovskite solar cells. *Comput. Mater.* **2019**, *5*, 9. [CrossRef]
33. Zhu, X.; Yang, F.; Liu, J.; Zhou, G.; Chen, D.; Liu, Z.; Fang, J. Design and Architecture of P-O Co-Doped Porous g-C_3N_4 by Supramolecular Self-Assembly for Enhanced Hydrogen Evolution. *Catalysts* **2022**, *12*, 1583. [CrossRef]
34. Liu, J.; Li, X.; Hou, H.; Zhou, M. Facile Synthesis of Anatase–Rutile Diphase N-doped TiO2 Nanoparticles with Excellent Visible Light Photocatalytic Activity. *Catalysts* **2020**, *10*, 1126. [CrossRef]
35. Xiang, D.; Wang, Z.; Xu, J.; Shen, H.; Zhang, X.; Liu, N. Defective TiO_2/MIL-88B(Fe) Photocatalyst for Tetracycline Degradation: Characterization and Augmented Photocatalytic Efficiency. *Catalysts* **2024**, *14*, 528. [CrossRef]

catalysts

MDPI

Article

β-Cyclodextrin Functionalization of Nitrogen-Doped Graphene to Enhance Dispersibility and Activate Persulfate for Trace Antibiotic Degradation in Water

Min Yao and Nan Wu *

School of Environment and Municipal Engineering, Lanzhou Jiaotong University, Lanzhou 730070, China; yaomin0952@126.com

* Correspondence: wunan0516@163.com

Abstract: The functionalization of nitrogen-doped graphene with β-Cyclodextrin (designated β/N-rGO) was employed to enhance the dispersibility of graphene materials and to establish an adsorption-catalytic oxidation system using peroxymonosulfate (PMS) for the removal of trace antibiotics from water. The experimental results indicated that β-Cyclodextrin was effectively dispersed on the support structure of nitrogen-doped graphene, which enhanced the specific surface area and dispersibility of the material. The adsorption-catalytic oxidation system comprising β/N-rGO and PMS degraded 92.35% of sulfamethoxazole within 12 min and exhibited significant removal efficiency for sulfonamides, quinolones, macrolides, tetracyclines, β-lactams, and chloramphenicol antibiotics across a pH range of 3–12. The reaction time was reduced by over 10% compared to the unmodified material, with a more pronounced improvement in treatment efficiency, particularly under low-pH conditions. The activation energy of β/N-rGO was estimated to be approximately 4.5 kJ/mol, and elevated temperatures accelerated the reaction, with the removal rate remaining above 85% after five regeneration cycles. Quenching experiments and EPR spectra confirmed that the primary pathway for PMS activation by β/N-rGO was a non-radical process dominated by singlet oxygen. These findings broaden the research and application scope of cyclodextrin–graphene interactions and provide a feasible approach for the removal of trace antibiotics from water.

Keywords: graphene; β-Cyclodextrin; persulfate; antibiotics; disperse

1. Introduction

The advantages and characteristics of graphene, including its large specific surface area, high electrical conductivity, excellent catalytic properties, high carrier density, good optical transmittance, and chemical stability, are well recognized. These properties enable its widespread application in fields such as sensors, electronics, composites, optoelectronics, bioengineering, and energy production and storage, demonstrating significant promise [1–4]. Graphene, with its unique electronic structure and tunability, holds significant potential for catalytic applications and is utilized to activate oxidants such as H_2O_2, KIO_4, O_3, and persulfate (PS) for the degradation of organic pollutants in water. In the presence of graphene-based materials, the degradation efficiency of oxidants is significantly enhanced, with peroxymonosulfate (PMS) being the most commonly used oxidant that forms a catalytic oxidation system with graphene [5–9]. The catalytic mechanisms of the graphene-based carbon catalyst/PMS system involve both radical and non-radical pathways, with key differences reflected in oxidative selectivity, oxidative potential, the nature of active oxidizing species, and the pH effect. However, the hydrophobicity of graphene and the strong Van der Waals forces between its layers promote agglomeration, which significantly impairs the catalytic performance of the material and limits its application in composites [10–13].

The methods to overcome the agglomeration of graphene-based materials and improve their dispersion performance primarily involve two categories: physical and chemical methods. Physical methods include mechanical stirring, ball milling, ultrasonic assistance, and other techniques that improve the dispersion of graphene in aqueous solutions but do not alter the internal molecular structure of graphene nor do they fundamentally address the issue of graphene agglomeration [14–17]. Chemical methods can fundamentally address graphene agglomeration and enable the functional modification of graphene materials, primarily through covalent bonding to graft free radicals or chemical reagents, oxygen-containing groups, and organic functional groups onto the edges or bases of C-C bonds. Additionally, Van der Waals forces, π–π interactions, hydrogen bonding, and other effects between surfactants and graphene can increase the repulsive forces between graphene layers, mitigating agglomeration [18–22]. β-Cyclodextrin (β-CD) and graphene are among the commonly used functionalized materials that can form supramolecular hybrids, either covalently or non-covalently, and are widely applied in water pollution treatment and catalysis. β-CD is a cyclic oligosaccharide composed of seven glucose units linked by α-1,4-glycosidic bonds, characterized by a specific cavity structure, external hydrophilic, and internal hydrophobic properties. These properties enable β-CD to selectively bind various organic, inorganic, and biological guest molecules into its cavities, forming stable host–guest inclusion complexes or nanostructured supramolecular assemblies in hydrophobic environments, exhibiting high molecular selectivity and enantioselectivity [23–25]. β-CD is water soluble, and its combination with other materials can enhance the solubility and stability of target functional materials in aqueous solutions. Therefore, modifying various graphene materials with β-CD can combine the common characteristics of both, resulting in new materials with unique properties, including a large specific surface area, high catalytic and enrichment abilities, and enhanced solubility and stability. It was reported that β-CD can effectively reduce the agglomeration of graphene and enhance its adsorption performance [26–31]. Cyclodextrin-modified graphene oxide with a three-dimensional porous structure (CD-GAM), as prepared by Nie et al., exhibits an excellent broad-spectrum adsorption of organic micropollutants. Its adsorption capacities for 2,4-dichlorophenol, propranolol hydrochloride, ethyl estradiol, and bisphenol a reached equilibrium within 10 min, and the regeneration adsorption capacity of CD-GAM remained above 80% over five cycles [32]. At present, the influence of the interaction between cyclodextrin and graphene was primarily investigated in terms of collaborative adsorption and electrochemical sensing applications. Although studies showed that cyclodextrin mitigates graphene layer agglomeration, the underlying mechanism remains unclear [33–35].

Environmental problems such as drug resistance, persistent pollution, and the challenging degradation of these compounds due to antibiotic overuse have garnered significant attention. Currently, over 250 types of antibiotic drugs are registered, among which sulfonamides (e.g., sulfamethoxazole, SMZ), quinolones (e.g., levofloxacin, LVX), macrolides (e.g., clindamycin, CLI), tetracyclines (e.g., tetracycline, TC), β-lactams (e.g., penicillin, PEN), and chloramphenicol (e.g., chloramphenicol, CHL) are frequently detected in aquatic environments across a broad spectrum of species [36,37]. Existing studies showed that conventional water treatment processes are not capable of completely degrading antibiotics, particularly given the diverse range of antibiotic compounds found in natural water systems. This is primarily because antibiotics in aqueous environments often exist at trace concentrations, which limits the efficiency of conventional treatment methods. Therefore, addressing antibiotic contamination at micro or trace levels necessitates either enhancing current treatment methods or developing novel, practical technologies to mitigate the adverse effects of antibiotic resistance on human health and ecosystems [38–40]. In a previous study, an adsorption-catalyzed oxidation (ACO) system was developed, employing nitrogen-doped reduced graphene (N-rGO) with various nitrogen-doping forms and PMS, which demonstrated high efficiency in degrading several of the aforementioned trace-level antibiotics in aqueous environments [41,42]. In this study, building on the use of β-CD to mitigate the agglomeration of N-rGO, we first investigated whether β-CD doping influences the catalytic performance of N-rGO and the efficiency of the ACO system. Secondly, we examined whether β-CD's ability to enhance the dispersion of graphene materials is affected in a strong oxidation environment. Given the widespread occurrence and high concentrations of sulfonamide drugs in the environment, they often coexist with various impurities at trace levels in both solid and liquid matrices, and their metabolites exhibit potential toxicity. Therefore, this study employed SMZ as the primary model compound to investigate the effects of enhanced oxidation systems on antibiotic degradation and the role of matrix components in aqueous environments [43]. This study expands the research and application scope of the interaction between cyclodextrins and graphene, addressing the detrimental effects of agglomeration and dispersion on the performance of graphene-based materials.

2. Results and Discussion

2.1. Properties of Catalytic Materials

From the elemental composition in Table 1, it can be observed that the primary elemental compositions and their respective contents in β/N-rGO and N-rGO are similar, with the nitrogen percentage being particularly close, thereby providing comparable conditions for an analysis of the two materials. As shown in the SEM image of Figure 1a, β/N-rGO retained similar structural characteristics to those of N-rGO (Figure 1b), exhibiting folds on both surfaces, a smooth appearance, and a semi-transparent state. However, the stacking phenomenon in β/N-rGO became more pronounced following the addition of β-CD, with the folds between the lamellae becoming more prominent. Additionally, the defects at the edges were more apparent, accompanied by a layer of rough coverage and significant stacking, indicating that β-CD bonded with the graphene [32]. The elemental composition and SEM images of N-rGO presented in Table 1 and Figure 1 were derived from the previous work conducted by our research group [41].

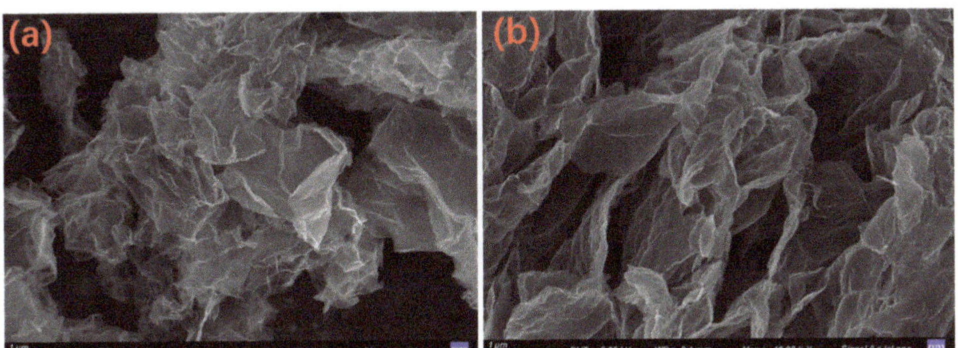

Figure 1. SEM of β/N-rGO (**a**) and N-rGO (**b**).

Table 1. Elemental composition of various nitrogen-doped graphene materials.

	C, At, %	O, At, %	N, At, %
β/N-rGO	63.73	2.93	32.78
N-rGO	61.85	2.99	34.75

As shown in Figure 2a, the absorption peaks at 3390 cm^{-1}, 1627 cm^{-1}, 1397 cm^{-1}, and 1032 cm^{-1} observed in N-rGO were also present in β/N-rGO. The peak at 3390 cm^{-1} is attributed to the stretching vibrations of -OH and -COOH groups, while the peaks at 1627 cm^{-1} and 1032 cm^{-1} became sharper, corresponding to the anti-symmetric glycosidic (C-O-C) stretching vibration and another characteristic feature of the β-CD monomer, respectively [44]. Additionally, β/N-rGO exhibited a new absorption peak at 2924 cm^{-1}, attributed to the aliphatic -C-H stretching vibration, a characteristic feature of β-CD. In conclusion, the appearance of the above characteristic peaks in the FTIR spectra confirmed the successful grafting of β-CD onto N-rGO [45]. Figure 2b shows characteristic shifts in the D and G bands for both graphene-based materials, indicating that the modification of N-rGO by β-CD did not alter the integrity of the graphene structure. However, the I_D/I_G ratio decreased, suggesting that β/N-rGO contained a higher proportion of sp^2-hybridized carbon atoms. This may be attributed to the introduction of sp^2-hybridized carbon atoms from β-CD during grafting. A higher proportion of sp^2-hybridized carbon atoms correlates with fewer defects, which was shown in the literature to enhance the π–π interactions between the material and contaminants [46]. The infrared and Raman images of N-rGO presented in Figure 2 were derived from the previous work conducted by our research group [41].

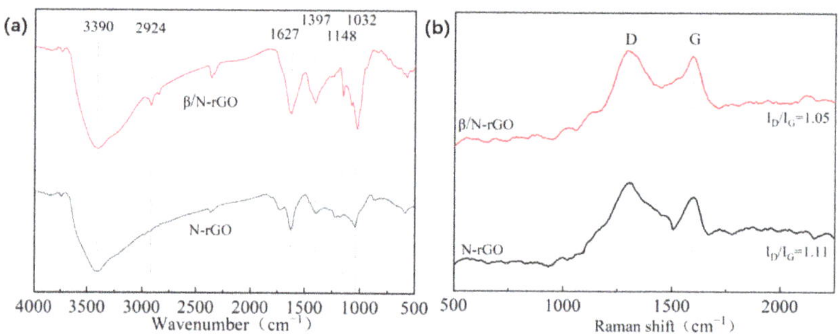

Figure 2. The FTIR spectra (**a**), Raman spectra (**b**) of different materials.

Figure 3a presents the full-scan XPS spectra of β/N-rGO and N-rGO. The C1s, N1s, and O1s peaks at 298 eV, 400 eV, and 533 eV, respectively, confirmed the successful doping of nitrogen atoms in both materials, consistent with previous literature reports. However, the O1s peak at 533 eV in β/N-rGO was significantly higher than in N-rGO, which may be attributed to the higher O:C ratio of β-CD, leading to an increased oxygen content in β/N-rGO [47]. Figure 3b,c reveal that β-CDβ-CD significantly altered the carbon and nitrogen structures of β-CD-functionalized N-rGO compared to N-rGO, despite similar nitrogen content. The percentage of graphite-type N increased by 5.26% and pyrrole-type N rose by 27.28% while pyridine-type nitrogen decreased by more than 30%.

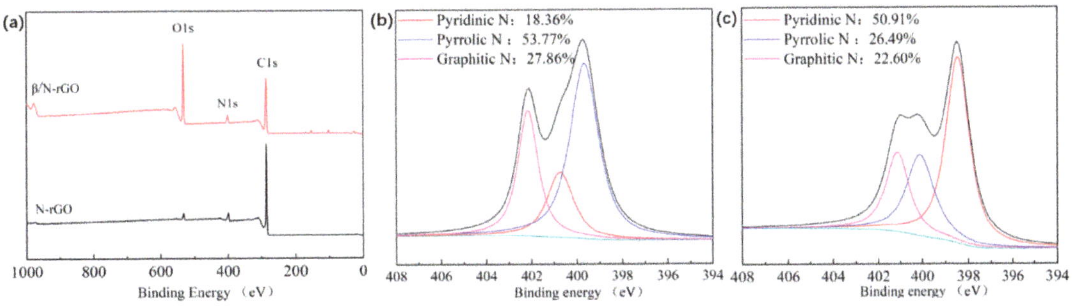

Figure 3. The XPS survey (**a**), N1s scan of β/N-rGO (**b**), and N-rGO (**c**).

2.2. Degradation Kinetics Analysis

Numerous studies demonstrated that β-CD can increase the specific surface area and active sites of graphene, thereby enhancing its adsorption capacity for pollutants [30,31]. Additionally, research showed that metal catalysts modified by β-CD can improve the catalytic effectiveness of PMS [28,48]. To investigate the effect of β-CD on PMS activation by N-rGO in the "adsorption-catalytic oxidation" system, β/N-rGO was introduced into the system to analyze its kinetic process. According to the elemental analysis in the previous section, the nitrogen content of β/N-rGO was 2.93%, closely matching that of N-rGO (2.99%). The effect of β-CD on the degradation of SMZ in the "adsorption-catalytic oxidation" system was then compared under both materials with similar nitrogen content (Figure 4a). β/N-rGO accelerated the catalytic reaction of PMS, enhancing the production of ROS and improving the degradation rate of antibiotics in the "adsorption-catalytic oxidation" system. SMZ degradation was completed within 12 min, which was more than 10% faster than the N-rGO + PMS system. Additionally, as shown in the inset of Figure 4a, the specific surface area of the β-CD-modified N-rGO was 267 m^2/g, 15.09% higher than that of N-rGO, and its adsorption capacity increased by approximately 5%, thereby enhancing the material's adsorption performance.

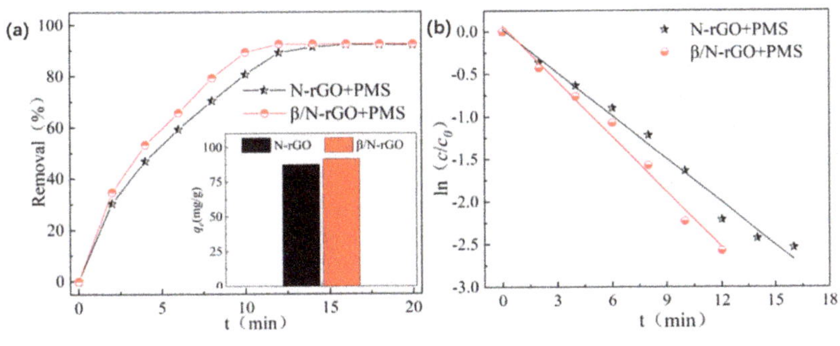

Figure 4. Degradation efficiency (**a**) and its first-order kinetic fit (**b**) of the two material systems for sulfamethoxazole (C_0 = 10 μg/L, $C_{β/N\text{-}rGO}$ = 0.1 mg/L, $C_{N\text{-}rGO}$ = 0.1 mg/L, C_{PMS} = 1 mg/L, T = 288.15 K), and the inset shows the adsorption capacity of the two materials for sulfamethoxazole.

To further investigate the characteristics of the "adsorption-catalytic oxidation" system, the degradation rates of the different systems were fitted to the first-order kinetic equation. The fitting equation is provided below, with the results shown in Figure 4b and the kinetic model parameters listed in Table 2. The infrared and Raman images of N-rGO presented in Figure 2 were derived from the previous work conducted by our research group [41].

$$lnc/c_0 = -k_1 t$$

where c and c_0 are the antibiotic concentration at moment t and the initial moment of the reaction, respectively, mg/L; k^1 is the first-order kinetic rate constant, min^{-1}.

Table 2. First-order kinetic model parameters of different degradation modes.

	β/N-rGO + PMS	N-rGO + PMS
k_1 (min^{-1})	0.2165	0.1686
$t_{1/2}$	3.20	4.11
R^2	0.9859	0.9842

The fitting results indicated that the R^2 values of the equations were all greater than 0.98, confirming that the degradation process closely followed the first-order kinetic model. Comparison of the reaction kinetic parameters revealed that the half-life for PMS activation by β/N-rGO was 22.14% shorter than that of N-rGO, indicating a faster PMS degradation rate for SMZ catalyzed by β/N-rGO at all concentrations.

2.3. *Analysis of Influencing Factors*

2.3.1. Influence of pH Value

As shown in Figure 5a, β/N-rGO exhibited superior PMS catalytic ability after β-CD modification compared to N-rGO across the pH range of 3–12, overcoming the pH dependence inherent in traditional advanced oxidation processes to some extent. Additionally, since the structures of various antibiotics are closely related to pH conditions, the degradation efficiency of the β/N-rGO + PMS "adsorption-catalytic oxidation" system was further examined for a range of antibiotics, including quinolones (LVX), macrolides (CLI), tetracyclines (TC), β-lactams (PEN), and chloramphenicol (CHL), across different pH values, with the results presented in Figure 5b–c. The removal rates of SMZ, LVX, TC, and PEN were significantly reduced under low-pH conditions, dropping to approximately 85%. This reduction is likely due to the competitive effect of the excess H^+ ions at low pH, which interacted with the catalyst's active sites and the negatively charged chain-catalyzed reaction intermediates (e.g., HSO_5^-), prematurely terminating the chain reaction. This led to a decrease in ROS and a subsequent reduction in degradation efficiency [12]. As shown in Figure 5c, compared to the N-rGO + PMS system, the degradation efficiency of the modified "adsorption-catalytic oxidation" system for the above antibiotics was improved to varying degrees, with a significant enhancement in the degradation of SMZ, LVX, TC, PEN, and CHL under low-pH conditions. Therefore, the functionalization of N-rGO with β-CD enhanced the catalyst's adaptability to varying pH conditions. Notably, antibiotic degradation efficiency increased with rising pH levels, as alkaline activation is a primary mechanism for PMS activation. Alkaline conditions promote the decomposition of PMS into reactive species such as superoxide radicals and singlet oxygen. Singlet oxygen is a highly selective oxidant that readily reacts with electron-rich compounds, including antibiotics. In practical water treatment scenarios, wastewater typically exhibits a neutral pH. Enhanced removal efficiency under neutral or alkaline conditions facilitates practical application, particularly in treating alkaline wastewaters such as those from textile and medical industries. Furthermore, the inherent characteristics of such wastewater can further enhance antibiotic degradation within the treatment system. [49,50].

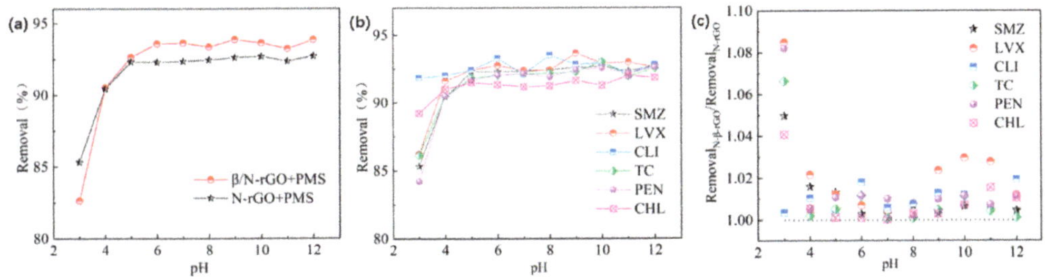

Figure 5. (**a**) Degradation of SMZ by two material systems at different pH values (C_0 = 10 μg/L, $C_{β/N\text{-}rGO}$ = 0.1 mg/L, C_{PMS} = 1 mg/L, T = 288.15 K), (**b**) degradation of different antibiotics by β/N-rGO-catalyzed PMS at different pH values (C_0 = 10 μg/L, $C_{β/N\text{-}rGO}$ = 0.1 mg/L, C_{PMS} = 1 mg/L, T = 288.15 K), (**c**) the ratio of the removal rate between cyclodextrin-functionalized and unfunctionalized "adsorption-catalytic oxidation" systems.

2.3.2. Effect of Temperature

As shown in Figure 6, the time to reach reaction equilibrium was significantly reduced with increasing temperature, indicating that higher temperatures positively promote the β/N-rGO-catalyzed PMS process. At 308.15 K, the degradation of antibiotics can be completed within 8 min, which is more than 30% faster than the degradation reaction time at ambient temperature (298.15 K). The change in reaction rate is further illustrated by fitting the apparent kinetic constants at different temperatures (see Table 3). The reaction rate constant at 308.15 K was over 50% higher than that at 288.15 K. These results indicated that increasing temperature rapidly enhanced the reaction rate, likely due to the temperature dependence of the carrier mobility in β/N-rGO. This aligns with the temperature effect observed in nitrogen-doped carbon-based catalysts (e.g., carbon nanotubes), as reported in previous studies [51]. Additionally, the activation energy of β/N-rGO, calculated using the Arrhenius equation, was approximately 4.15 kJ/mol, much lower than that of cobalt- or manganese-loaded nanocarbon catalysts reported in the literature [41]. This suggests that more molecules can cross the activation energy barrier at a given temperature for β/N-rGO catalysts, enhancing catalytic efficiency.

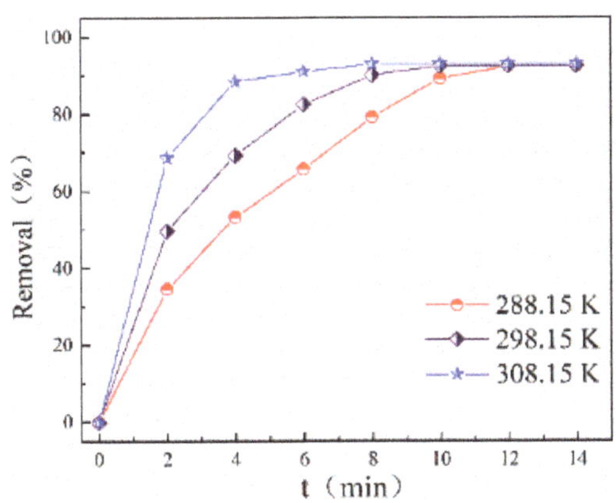

Figure 6. Effect of temperature on antibiotic degradation efficiency (C_0 = 10 μg/L, $C_{β/N\text{-}rGO}$ = 0.1 mg/L, C_{PMS} = 1 mg/L)

Table 3. Calculation results of activation energy and related parameters.

	Temperature (K)			E kJ/mol	R^2
	288.15	298.15	308.15		
k_1 (min^{-1})	0.2165	0.2625	0.3293	4.15	0.9844
R^2	0.9882	0.9893	0.8964		

2.3.3. Effect of Initial Antibiotic Concentration on Degradation Effect

In natural waters and municipal wastewater, the concentrations of various antibiotics typically range from 1 to 100 μg/L. To evaluate the system's efficacy in treating low-antibiotic concentrations, a maximum concentration of 20 μg/L was selected [52]. As shown in Figure 7, the "adsorption-catalytic oxidation" system comprising β/N-rGO and PMS effectively degraded trace antibiotics at concentrations ranging from 1 to 20 μg/L within 12 min. This indicated that the catalyst's degradation efficiency was unaffected by the initial antibiotic concentration. In simple adsorption, high initial pollutant concentrations reduce the number of available adsorption sites on the adsorbent, limiting its capacity to adsorb pollutants at higher concentrations. In the "adsorption-catalytic oxidation" system, the active sites on the surface of β/N-rGO serve dual roles of adsorption and catalysis. Excess antibiotics are adsorbed onto the material's surface, and the PMS present rapidly removes the excess pollutants through ROS produced by PMS catalysis on β/N-rGO. This process mitigates the issues caused by excessive adsorption. Therefore, in the "adsorption-catalytic oxidation" system, the coexistence of adsorption and catalysis allows β/N-rGO to sustain a continuous process of adsorption, catalysis, and oxidative degradation of trace antibiotics in water. As shown in Table 4, the "adsorption-catalytic oxidation" system demonstrated the advantages of a rapid removal rate and high removal efficiency for trace-level antibiotics in water compared to other treatment methods.

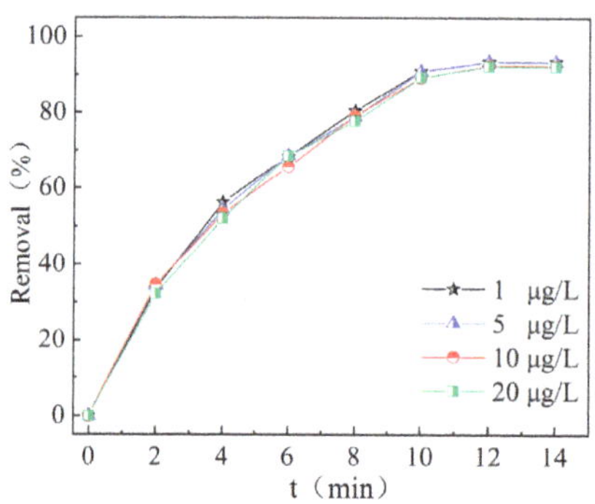

Figure 7. Effect of initial antibiotic concentration on degradation effect (C_{PMS} = 1 mg/L, $C_{\beta/N\text{-}rGO}$ = 0.1 mg/L, T = 288.15 K).

Table 4. Comparison with other treatments.

Method	Treatment Time (min)	Removal Rate (%)	Ref.
β/N-rGO + PMS	<15	>90	This study.
N-rGO Adsorption	<60	>80	[53]
Sch/BB Fenton	<25	>95	[54]
Co-based MOF + PMS	<60	>85	[55]

2.3.4. Effect of PMS Concentration

As shown in Figure 8, the concentration of PMS played a crucial role in influencing the degradation efficiency. At lower PMS concentrations, the limited amount of oxidant significantly reduced the generation of reactive oxygen species through catalytic activation, leading to insufficient oxidation capacity and diminished degradation performance. The removal rate of SMZ by the system decreased by 4.55%, and the reaction equilibrium time was prolonged by 33.33%. Figure 8 also clearly illustrates that, during the initial phase of the reaction, the degradation curve rose sharply, whereas it leveled off in the later stages. This observation supports the conclusion that, although the overall oxidant dosage was low, the relatively higher PMS concentration in the early phase promoted the generation of reactive species, leading to an enhanced degradation efficiency at the beginning of the reaction. When the concentration of PMS increased to 5 mg/L, the reaction rate and removal rate did not decrease with the increase in the oxidant content; instead, they decreased. This may be attributed to two primary factors: First, the catalyst dosage was fixed, and the limited number of active sites created a bottleneck in the catalytic oxidation process. Secondly, an excess of PMS can initiate a chain reaction of free radicals, resulting in self-quenching. In this scenario, the free radicals were consumed through side reactions, triggering a scavenging effect that reduced the concentration of reactive radicals or generated fewer reactive species, thereby diminishing the overall degradation efficiency [56].

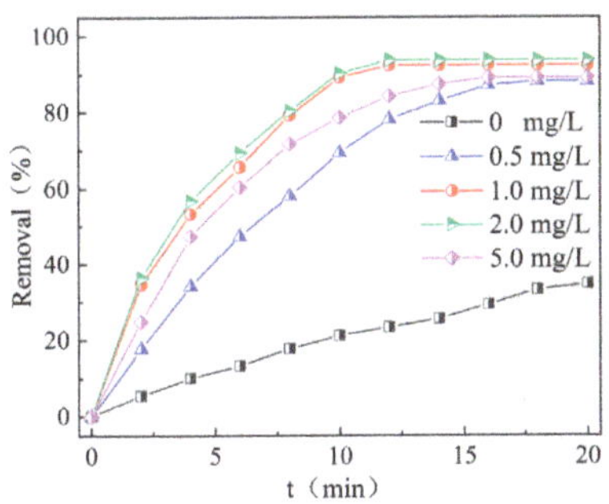

Figure 8. Effect of PMS concentration degradation (C_0 = 10 μg/L, $C_{\beta/N\text{-}rGO}$ = 0.1 mg/L, T = 288.15 K).

2.3.5. Effect of Anions on Degradation Effect

As shown in Figure 9, the removal of SMZ by the "adsorption-catalytic oxidation" system consisting of β/N-rGO and PMS was unaffected by the addition of excessive anions, suggesting that PMS activation by β/N-rGO may proceed via a non-radical pathway. In comparison to the free radicals generated by conventional PMS activation (e.g., $\cdot SO_4^-$,$\cdot OH$), the inhibition or enhancement of these free radicals did not affect the degradation rate of antibiotics when sufficient amounts of carbonate (HCO_3^-, CO_3^{2-}), nitrate (NO_3^-), or phosphate (HPO_4^{2-}, $H_2PO_4^-$) were added. Furthermore, it was reported that the addition of chloride ions (Cl^-) can activate PMS to produce free radicals, such as $\cdot Cl_2$, $\cdot Cl$, and $\cdot ClOH$, which possess an oxidizing capacity similar to that of $\cdot SO_4^-$ and $\cdot OH$, thereby promoting pollutant degradation. Additionally, when $\cdot Cl$ is not involved in organic matter degradation, it may act as a scavenger for $\cdot SO_4^-$, reducing the degradation rate of organic matter. However, in this experiment, the addition of Cl^- did not affect the degradation rate of organic matter. In conclusion, the degradation of antibiotics in the "adsorption-catalytic oxidation" system comprising β/N-rGO and PMS primarily followed a non-radical pathway [57–60].

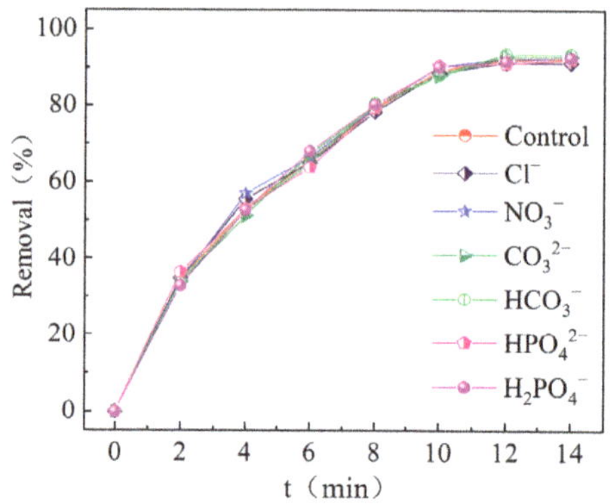

Figure 9. Effect of different anions on degradation effect (C_0 = 10 μg/L, $C_{\beta/N\text{-}rGO}$ = 0.1 mg/L, C_{PMS} = 1 mg/L, C_{anion} = 100 mmol/L, T = 288.15 K).

2.3.6. Effect of Humic Acid on Degradation Effects

Humic acid contains a significant number of phenolic hydroxyl and carboxyl groups, which may adsorb to the surface of β/N-rGO, thereby obstructing the catalytically active sites and diminishing degradation efficiency. Thus, we further investigated the impact of humic acid on degradation efficiency. As shown in Figure 10, the presence of humic acid at concentrations ranging from 1 to 10 mg/L had less than a 1% impact on the degradation of SMZ in the "adsorption-catalytic oxidation" system, suggesting that its influence on antibiotic degradation by this catalyst was negligible. One possible explanation is that a large number of ROS were generated during SMZ degradation in the "adsorption-catalytic oxidation" system, leading to the rapid removal of phenolic hydroxyl and carboxyl groups in humic acid, thereby maintaining the continuity of the "adsorption-catalytic oxidation" process. This active oxygen encompasses not only free radicals such as $\cdot SO_4^-$ and $\cdot OH$ but also a significant number of non-free radical species. Therefore, compared with pure free radical reactions, the β/N-rGO-catalyzed PMS system exhibited stronger resistance to interference from humic acid, thereby overcoming the limitations of individual adsorption or free radical reactions, which are significantly influenced by natural organic matter.

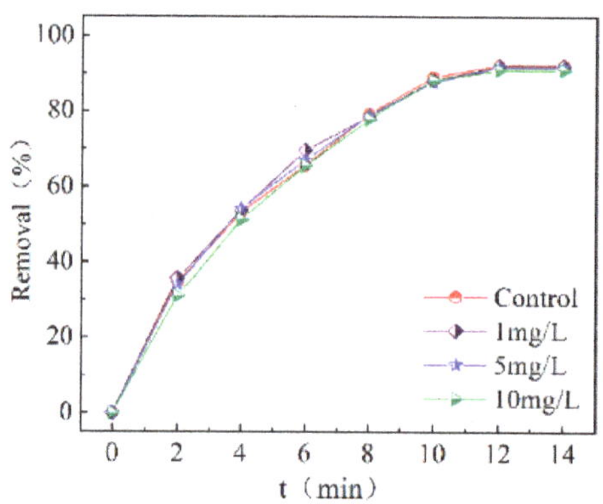

Figure 10. Effect of humic acid on degradation effect (C_0 = 10 μg/L, $C_{\beta/N\text{-}rGO}$ = 0.1 mg/L, C_{PMS} = 1 mg/L, T = 288.15 K).

2.3.7. Impacts of Recycling and Regeneration

Figure 11 illustrates the effects of the two recycling methods on the catalytic performance of the β/N-rGO. It was evident that the performance of the regenerated catalysts surpassed that of direct recycling, with this difference becoming more pronounced as the operation cycle was extended. When β/N-rGO was directly recycled without regeneration, although antibiotics and their degradation intermediates adsorbed on the catalyst surface can be rapidly removed by catalytically generated ROS, the delayed removal of pollutants over extended operation cycles led to a reduction in the active sites on the β/N-rGO surface due to coverage, which in turn weakened its interaction with PMS. After regeneration, the surface of β/N-rGO was cleaned and the active sites were reactivated, resulting in superior performance compared to the unregenerated material. Although the regeneration process may alter the material's structure, leading to the inactivation of some active sites, it is noteworthy that the antibiotic removal rate by regenerated β/N-rGO decreased by only about 5%, indicating that the "adsorption-catalytic oxidation" system has significant potential for regeneration and recycling. The recovery rate of the catalyst after regeneration was calculated to be 91.37%.

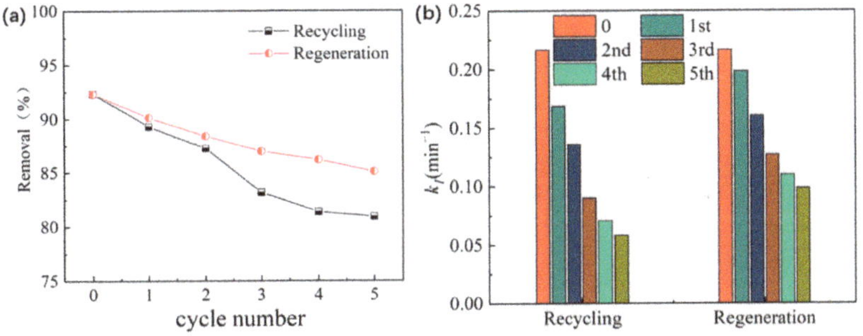

Figure 11. Effect of cycling mode and number of cycles on degradation of SMZ (**a**) removal rate, (**b**) reaction constant.

2.4. Analysis of the Effect of β-CD in Improving Graphene Agglomeration

As shown in Figure 12a–f, compared with N-rGO, the dispersibility of β/N-rGO remained stable over time, with no signs of agglomeration or settling, indicating that β-CD modification significantly enhanced the dispersibility of graphene in water. Characterization analysis indicated that β-CD modified graphene oxide primarily through covalent or non-covalent linkages. On one hand, covalent bonding results in the formation of a β-CD–graphene oxide hybrid, enhancing its water solubility. On the other hand, non-covalent interactions, such as Van der Waals forces, π–π stacking, and hydrogen bonding, cause β-CD to adsorb onto the surface of N-rGO, increasing the repulsive forces between graphene sheets. When this repulsive force exceeds the intrinsic intermolecular forces of graphene, the β/N-rGO sheets separate, leading to the formation of large edge defects and exposing more functional groups [27,32]. Furthermore, the potential direct effect of β-CD on PMS activation was investigated. As shown in Figure 12g, the EPR spectra of the reaction between β-CD and PMS did not show characteristic peaks for free radicals, such as $\cdot SO_4^-$, ·OH, or singlet oxygen, suggesting that the primary catalytic effect in PMS activation by β/N-rGO arose from the active sites within the structure of nitrogen-doped graphene itself. β-CD itself did not activate PMS; its primary effect on catalytic performance was enhancing the dispersion of nitrogen-doped graphene, thereby exposing more catalytic active sites and improving overall catalytic performance.

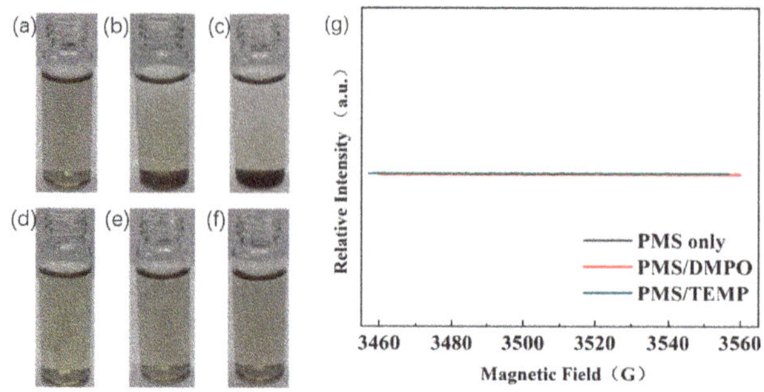

Figure 12. Dispersion effects of N-rGO (**a–c**) and β/N-rGO (**d–f**) at 0, 5, and 10 min moments, (**g**) EPR spectra of the reaction of β-CD with PMS.

2.5. Quenching Experiments and EPR Spectra Analysis

As shown in Figure 13a, the addition of free radical scavengers, such as tert-butanol (TBA), methanol (MeOH), and p-benzoquinone (PBQ), had minimal impact on the degradation efficiency of SMZ, whereas β-carotene significantly reduced the degradation efficiency, highlighting the dominance of the non-radical oxidative pathway in the β/N-rGO + PMS system, where singlet oxygen plays a crucial role. Furthermore, the electron paramagnetic resonance (EPR) spectra in Figure 13b support the inference regarding the type of ROS activated by β/N-rGO in the PMS system. Triple peaks with greater intensity were observed in the presence of TEMP, a typical singlet oxygen signal, whereas the characteristic peaks captured by DMPO were weak, indicating that the concentrations of $\cdot SO4^{-}$ and $\cdot\cdot OH$ in the system were minimal [61].

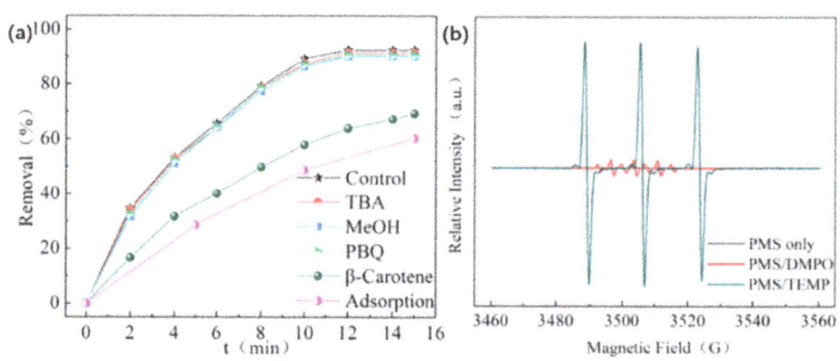

Figure 13. (**a**) Quenching experiment (C_0 = 10 μg/L, $C_{\beta/N\text{-}rGO}$ = 0.1 mg/L, C_{PMS} = 1 mg/L, pH = 7), (**b**) EPR spectra.

3. Conclusions

In this study, β-CD was shown to functionalize graphene materials, effectively addressing the issue of agglomeration while enhancing their catalytic activity and application potential. Structural characterization confirmed that β-CD disperses effectively within the support structure of nitrogen-doped graphene, increasing the material's specific surface area and enhancing its dispersibility. The "adsorption-catalytic oxidation" system consisting of β/N-rGO and PMS can degrade 92.35% of SMZ in 12 min, with the reaction time reduced by over 10% compared to the unmodified system. The degradation of various antibiotics, including sulfonamides, quinolones, macrolides, tetracyclines, β-lactams, and chloramphenicol, was observed to be effective across a pH range of 3 to 12, with a more pronounced improvement in degradation efficiency under acidic conditions when compared to the unmodified material. The system's activation energy was calculated to be 4.5

kJ/mol, and common anions and humic acids in water did not impact the degradation of antibiotics by the β/N-rGO + PMS "adsorption-catalytic oxidation" system, which exhibited a long service life, with a removal rate of 85% after five regeneration cycles. Quenching experiments and EPR spectra confirmed that the primary reaction pathway for PMS activation by β/N-rGO is a non-radical pathway dominated by singlet oxygen. Therefore, β-CD functionalization is a key method for improving the agglomeration behavior of graphene-based materials, enhancing not only the dispersion of graphene but also the interfacial interactions, thereby promoting the advancement of material applications.

4. Experiments and Methods

4.1. Catalyst Preparation

Graphene oxide (GO) and nitrogen-doped reduced graphene oxide (N-rGO) were synthesized following the same procedure as described in the previous study [42]. After preparing a 0.5 g/L N-rGO and 4 g/L β-CD aqueous solution, 100 mL of this solution was transferred into a 250 mL conical flask. Then, 1.5 mL of ammonia ($NH_4{\cdot}H_2O$) was added, and the mixture was thoroughly mixed. The solution was then sonicated for 5 min and placed in a 60 °C water bath for 3 h of reaction. After the reaction, the mixture was centrifuged, and the resulting graphene was dried at 60 °C for 12 h. Upon completion of the reaction, the powder was ground and passed through a 100-mesh sieve to obtain the β-CD-nitrogen-doped reduced graphene oxide powder, labeled β/N-rGO.

4.2. Experimental Methods

A total of 80 mL of 12.5 μg/L antibiotic solution and 10 mL of β/N-rGO dispersion (1 mg/L) were added to a 120 mL amber vial. Without allowing adsorption equilibrium to be reached, 10 mL of 10 mg/L PMS solution was immediately added, and the mixture was briefly shaken before initiating the catalytic reaction on a shaker. The pH range of the reaction was 3–12, with temperatures of 288.15, 298.15, and 308.15 K. The anion concentration was 100 mmol/L, and humic acid concentrations were 1, 5, and 10 mg/L. The catalysts were not regenerated during the cycling experiments. After each degradation process, the reactants were directly added for the next cycle, which was repeated five times. In the regeneration cycle experiment, 10 mL of a 100 μg/L antibiotic solution and 120 mg of potassium PMS were added as reactants. Eluents included 0.1 mol/L hydrochloric acid, 0.1 mol/L sodium hydroxide, and deionized water.

Tert-butanol (TBA), methanol (MeOH), p-benzoquinone (PBQ), and β-carotene were employed as blockers of reactive oxygen species (ROS) reactions, and overdosing was applied to evaluate whether the ROS reaction was inhibited. Then, 5,5-dimethyl-1-pyrroline-N-oxide (DMPO) was employed as a trapping agent for sulfate and hydroxyl radicals, while 2,2,6,6-tetramethylpiperidine (TEMP) was used to trap singlet oxygen. Electron paramagnetic resonance (EPR) was utilized to capture and characterize ROS.

4.3. Antibiotic Concentration Testing

The concentration of antibiotics was determined using the HPLC method with a C18 column, and the experimental conditions were as follows:

(1) Sulfamethoxazole (SMZ): The mobile phase consisted of water, acetonitrile, and triethylamine in a volume ratio of 799:200:1. Sodium hydroxide or glacial acetic acid was used to adjust the pH to 5.9. The flow rate was 1 mL/min, and the wavelength was 240 nm.

(2) Levofloxacin (LVX): The mobile phase consisted of 0.05 mol/L potassium dihydrogen phosphate (pH adjusted to 2.8 with phosphoric acid) and acetonitrile (*v*/*v*: 82:18). The flow rate was 1 mL/min, and the wavelength was 293 nm.

(3) Clindamycin (CLI): The mobile phase consisted of 0.68 g/L potassium dihydrogen phosphate (pH adjusted to 7.5 with 250 g/L sodium hydroxide) and acetonitrile (v/v: 5:45). The flow rate was 1 mL/min, and the wavelength was 210 nm.
(4) Tetracycline (TC): The mobile phase consisted of methanol and 0.02 mol/L potassium dihydrogen phosphate (pH 2.0 adjusted with nitric acid) in a volume ratio of 12:88. The flow rate was 1 mL/min, and the wavelength was 280 nm.
(5) Penicillin G sodium (PEN): The mobile phase consisted of 0.1 mol/L potassium dihydrogen phosphate (pH 2.5 adjusted with phosphoric acid) and acetonitrile in a volume ratio of 70:30. The flow rate was 1 mL/min, and the wavelength was 225 nm.
(6) Chloramphenicol (CHL): The mobile phase consisted of methanol and water in a volume ratio of 6:4. The flow rate was 0.5 mL/min, and the wavelength was 272 nm.

4.4. Characterization Methods

The material morphology was characterized using scanning electron microscopy (SEM) (ZEISS, Suzhou, China). Fourier-transform infrared (FTIR) (BRUKER, Ettlingen, Germany) spectroscopy was employed to analyze the distribution and changes in functional groups in graphene materials. X-ray photoelectron spectroscopy (XPS) (PANalytical, Almelo, Netherlands) was used to characterize the C-N bonding structure and the bonding characteristics of oxygen-containing functional groups. Raman spectroscopy (Renishaw, Gloucestershire, Britain) was employed to accurately distinguish defects in the graphene structure.

Author Contributions: Data curation, M.Y.; formal analysis, M.Y. and N.W.; writing—original draft, M.Y.; conceptualization, N.W.; resources, N.W.; writing—review and editing, N.W. All authors have read and agreed to the published version of the manuscript.

Funding: This research was funded by the Natural Science Foundation of Gansu Province (No. 24JRRA240), the Innovation Project of Gansu Province University Faculty Fund (No. 2024A-034), the Project Supported by the Young Scholars Science Foundation of Lanzhou Jiaotong University (2022045), and Tianyou Youth Talent Lift Program of Lanzhou Jiaotong University.

Data Availability Statement: Data will be made available on request.

Conflicts of Interest: The authors declare no conflicts of interest.

References

1. Guo, M.; Yuan, H.; Ni, K.; Ye, C.; Pan, F.; Xiong, J.; Zhu, Y. Structural repair of reduced graphene oxide promoted by single-layer graphene. *Adv. Sci.* **2025**, *12*, 2410088. [CrossRef] [PubMed]
2. Zhang, Y.; Zhang, H.; Wang, X.; Shi, X.; Zhao, Z.; Wang, Y.; Liu, J.; Tang, C.; Wang, G.; Li, L. Graphene nanoribbons enhancing the electronic conductivity and ionic diffusion of graphene electrodes for high-performance microsupercapacitors. *Energy Environ. Mater.* **2024**, *7*, e12681. [CrossRef]
3. Fan, X.; He, C.; Ding, J.; Gao, Q.; Ma, H.; Lemme, M.C.; Zhang, W. Graphene MEMS and NEMS. *Microsyst. Nanoeng.* **2024**, *10*, 154. [CrossRef] [PubMed]
4. Gronkiewicz, K.; Richter, L.; Knechtel, F.; Pyrcz, P.; Leidinger, P.; Günther, S.; Ploetz, E.; Tinnefeld, P.; Kamińska, I. Expanding the range of graphene energy transfer with multilayer graphene. *Nanoscale* **2024**, *16*, 13464–13470. [CrossRef]
5. Gu, W.; Yang, M.; Chen, Z.; Cao, T.; Zhang, Y.; Li, Y.; Zhang, R. New insights into enhanced electrochemical advanced oxidation mechanism of B-doped graphene aerogel: Experiments, molecular dynamics simulations and DFT. *J. Hazard. Mater.* **2023**, *443*, 130331. [CrossRef]
6. Zhang, Y.; Zhang, Z. A sustainable and metal-free advanced oxidation process for efficient water purification by electrified reduced graphene oxide membrane. *Chem. Catal.* **2024**, *4*, 101056. [CrossRef]
7. Wang, H.; Xu, Z.; Yang, X. Probing the formation mechanisms of reactive oxygen species in graphene oxide-catalyzed ozone advanced oxidation processes. *Carbon* **2025**, *233*, 119831. [CrossRef]
8. Su, P.; Zhou, M.; Lu, X.; Yang, W.; Ren, G.; Cai, J. Electrochemical catalytic mechanism of N-doped graphene for enhanced H_2O_2 yield and in-situ degradation of organic pollutant. *Appl. Catal. B Environ.* **2019**, *245*, 583–595. [CrossRef]

9. Long, Y.; Dai, J.; Zhao, S.; Su, Y.; Wang, Z.; Zhang, Z. Atomically dispersed cobalt sites on graphene as efficient periodate activators for selective organic pollutant degradation. *Environ. Sci. Technol.* **2021**, *55*, 5357–5370. [CrossRef]
10. Li, X.; Chen, X.; Yan, Y.; Wang, F.; Feng, L.; Chen, Y. Nitrogen-doped graphene for tetracycline removal via enhancing adsorption and non-radical persulfate activation. *Environ. Res.* **2023**, *235*, 116642. [CrossRef]
11. Liu, X.; Zhang, Y. Nitrogen-doped reduced graphene oxide (N-rGO) three-dimensional electrode electrochemically activates persulfate for the degradation of tetracycline. *Carbon Lett.* **2024**, *34*, 865–34879. [CrossRef]
12. Hirani, R.A.K.; Asif, A.H.; Rafique, N.; Shi, L.; Zhang, S.; Saunders, M.; Tian, W.; Wang, S.; Sun, H. Heterogeneous activation of persulfate by macroscopic nitrogen-doped graphene oxide cubes for the degradation of antibiotic contaminants in water. *Sep. Purif. Technol.* **2023**, *319*, 124110. [CrossRef]
13. Jiang, X.; Wang, R.; Yang, Q.; Guo, W.; Ren, X. Surface-regulated Fe_xO_y/NG catalysts enhancing persulfate activation to remove tetracycline via a singlet oxygen-dominated pathway. *Sep. Purif. Technol.* **2025**, *361*, 131570. [CrossRef]
14. Luo, J.; Yang, L.; Sun, D.; Gao, Z.; Jiao, K.; Zhang, J. Graphene oxide "surfactant"-directed tunable concentration of graphene dispersion. *Small* **2020**, *16*, 2003426. [CrossRef]
15. Sonin, A.S.; Churochkina, N.A.; Kaznacheev, A.V.; Golovanov, A.V. Mesomorphism of graphene oxide dispersions. *Colloid J.* **2021**, *83*, 161–182. [CrossRef]
16. Zhang, N.; Ma, P.; Yang, Z.; Fang, Y.; Zhang, Z. Physical dispersion method and mechanism of graphene. *J. Superhard Mater.* **2023**, *45*, 186–191. [CrossRef]
17. Li, L.; Zhou, M.; Jin, L.; Mo, Y.; Xu, E.; Chen, H.; Liu, L.; Wang, M.; Chen, X.; Zhu, H. Green preparation of aqueous graphene dispersion and study on its dispersion stability. *Materials* **2020**, *13*, 4069. [CrossRef]
18. Wang, Q.; Zhan, D.; Qi, G.; Wang, Y.; Zheng, H. Impact of the microstructure of polycarboxylate superplasticizers on the dispersion of graphene. *New Carbon Mater.* **2020**, *35*, 547–558. [CrossRef]
19. Qamar, S.; Ramzan, N.; Aleem, W. Graphene dispersion, functionalization techniques and applications: A review. *Synth. Met.* **2024**, *307*, 117697. [CrossRef]
20. Feng, B.; Wang, Z.; Suo, W.; Wang, Y.; Wen, J.; Li, Y.; Suo, H.; Liu, M.; Ma, L. Performance of graphene dispersion by using mixed surfactants. *Mater. Res. Express* **2020**, *7*, 095009. [CrossRef]
21. Wilczewski, S.; Skórczewska, K.; Tomaszewska, J.; Lewandowski, K.; Szulc, J.; Runka, T. Manufacturing homogenous PVC/graphene nanocomposites using a novel dispersion agent. *Polym. Test.* **2020**, *91*, 106868. [CrossRef]
22. Zhao, W.; Sugunan, A.; Gillgren, T.; Larsson, J.A.; Zhang, Z.; Zhang, S.; Nordgren, N.; Sommertune, J.; Ahniyaz, A. Surfactant-free stabilization of aqueous graphene dispersions using starch as a dispersing agent. *ACS Omega* **2021**, *6*, 12050–12062. [CrossRef]
23. Sun, L.; Xu, G.; Tu, Y.; Zhang, W.; Hu, X.; Yang, P.; Wu, D.; Liang, Y.; Wei, D.; Li, A.; et al. Multifunctional porous β-cyclodextrin polymer for water purification. *Water Res.* **2022**, *222*, 118917. [CrossRef]
24. Bie, W.; Zhang, S.; Zhang, L.; Li, H.; Sun, X.; Cai, T.; Wang, Z.; Kong, F.; Wang, W. Thioether-functionalized porous β-cyclodextrin polymer for efficient removal of heavy metal ions and organic micropollutants from water. *Carbohydr. Polym.* **2024**, *324*, 121509. [CrossRef]
25. Zhang, W.; Sun, P.; Liu, D.; Zhao, Q.; Zou, B.; Zhou, L.; Ye, Z. Method to fabricate porous multifunction β-cyclodextrin modified resin for ultrafast and efficient removal of Cu(II) and bisphenol A. *J. Taiwan Inst. Chem. E* **2021**, *119*, 286–297. [CrossRef]
26. Xiong, T.; Zou, C.; Lin, S. β-Cyclodextrin magnetic graphene oxide (β-CD@MGO) for efficient removal of Ni(II) and phenol: Adsorption performance, reusability and adsorption mechanism studies. *Diam. Relat. Mater.* **2023**, *139*, 110269. [CrossRef]
27. Verma, M.; Hong, Y.; Kumar, V.; Bisht, B.M.; Vlaskin, S.; Chae, S.; Shah, K.J.; Kim, H. Enhanced removal and stepwise recovery of inorganic and organic micropollutants from water using novel graphene oxide doped multifunctional β−cyclodextrin chitosan polymer. *J. Phys. Chem. Solids* **2023**, *182*, 111615. [CrossRef]
28. Almajidi, Y.Q.; Majeed, A.A.; Ali, E.; Abdullaev, S.; Koka, N.A.; Bisht, Y.S.; Fenjan, M.N.; Alawadi, A.; Alsalamy, A.; Saleh, L.H. A versatile magnetic nanocomposite based on cellulose-cyclodextrin hydrogel embedded with graphene oxide and Cu_2O nanoparticles for catalytic application. *Int. J. Biol. Macromol.* **2024**, *260*, 129367. [CrossRef] [PubMed]
29. Zhu, M.; Liang, H.; Gong, X. A novel magnetic β-cyclodextrin-modified graphene oxide and chitosan composite as an adsorbent for trace extraction of four bisphenol pollutants from environmental water samples and food samples. *Environ. Res.* **2024**, *29*, 867. [CrossRef]
30. Ouyang, J.; Gao, J.; Shen, J.; Wei, Y.; Wang, C. Preparation of magnetic graphene/β-cyclodextrin polymer composites and their application for removal of organic pollutants from wastewater. *J. Environ. Chem. Eng.* **2023**, *11*, 110944. [CrossRef]
31. Wang, X.; Chen, J.; Zhou, J.; Bao, L.; Zhang, L.; Yang, L.; Wu, J.; Hao, C. Novel carboxymethyl cellulose-based hydrogel embedded with metal organic framework for efficient cationic dye removal from water. *Int. J. Biol. Macromol.* **2024**, *282*, 137387. [CrossRef] [PubMed]
32. Nie, Z.; Guo, Q.; Xia, H.; Song, M.; Qiu, Z.; Fan, S.; Chen, Z.; Zhang, S.; Zhang, S.; Li, B. Cyclodextrin self-assembled graphene oxide aerogel microspheres as broad-spectrum adsorbent for removing dyes and organic micropollutants from water. *J. Environ. Chem. Eng.* **2021**, *9*, 104749. [CrossRef]

33. Zang, X.; Liang, L.; Wang, X.; Li, C.; Li, P.; Shao, Q.; Cao, N. β-Cyclodextrin anchor $NiCo_2S_4$ on graphene to enhance electrochemical performance of supercapacitor. *Energy Technol.* **2022**, *10*, 2200490. [CrossRef]
34. Qu, Y.; Li, H.; Yakub, I.; He, W.; Dong, W.; Barawi, M.H.; Wang, S.; Ma, H.; Zhu, Z. Synthesis and characterization of efficient adsorbents for methylene blue based on graphene oxide/β-cyclodextrin composites. *Water Air Soil Pollut.* **2025**, *56*, 236. [CrossRef]
35. Yang, X.; Wu, Y.; Qi, Y.; Zheng, M.; Xie, F.; Wang, Q.; Xiao, X.; Xiao, D. Multifunctional bionanohybrids of graphene oxide/β-cyclodextrin for applications in methylene blue adsorption, solvent response, and smart light-driven. *J. Mol. Liq.* **2024**, *404*, 124977. [CrossRef]
36. Wang, J.; Sha, X.; Chen, X.; Zhuo, H.; Xie, W.; Zhou, Z.; He, X.; Wu, L.; Li, B. Removal and distribution of antibiotics and resistance genes in conventional and advanced drinking water treatment processes. *J. Water Process Eng.* **2022**, *50*, 103217. [CrossRef]
37. Zhang, G.; Zhang, C.; Liu, J.; Zhang, Y.; Fu, W. Occurrence, fate, and risk assessment of antibiotics in conventional and advanced drinking water treatment systems: From source to tap. *J. Environ. Manag.* **2024**, *358*, 120746. [CrossRef]
38. Ashish, K.; Rajinder, K.; Shashikala, V.; Samer, S. Antimicrobials and antibiotic resistance genes in water bodies: Pollution, risk, and control. *Front. Environ. Sci.* **2022**, *10*, 830861. [CrossRef]
39. Jerry, O.; Timothy, A.; Damian, C. Mineralization of antibiotics in wastewater via photocatalysis. *Water Air Soil Pollut.* **2021**, *232*, 219. [CrossRef]
40. Xie, J.; Li, X.; Ma, W.; Wang, L.; Lu, Y.; Li, S.; Li, J.; Guo, X.; Li, Z.; Ding, C. Fate of dissolved organic matter and antibiotics in conventional treatment system and constructed wetlands system applied to source water pretreatment: Removal efficiency and risk assessment. *Water Air Soil Pollut.* **2024**, *235*, 24. [CrossRef]
41. Wu, N.; Zhang, X.; Zhang, X.; Yang, K.; Li, Y. Simultaneous degradation of trace antibiotics in water by adsorption and catalytic oxidation induced by N-doped reduced graphene oxide (N-rGO): Synergistic mechanism. *Mater. Res. Express* **2022**, *9*, 065601. [CrossRef]
42. Zhang, X.; Zhao, B.; Wu, N. Active sites and structure regulation of non-radical pathways in N-doped graphene-activated persulfate. *Catal. Commun.* **2024**, *187*, 106910. [CrossRef]
43. Li, J.; Zhao, L.; Feng, M.; Huang, C.; Sun, P. Abiotic transformation and ecotoxicity change of sulfonamide antibiotics in environmental and water treatment processes: A critical review. *Water Res.* **2021**, *202*, 117463. [CrossRef]
44. Guo, Y.; Guo, S.; Ren, J.; Zhai, Y.; Dong, S.; Wang, E. Cyclodextrin functionalized graphene nanosheets with high supramolecular recognition capability: Synthesis and host−Guest inclusion for enhanced electrochemical performance. *ACS Nano* **2010**, *4*, 4001–4010. [CrossRef]
45. Zhao, F.; Repo, E.; Yin, D.; Meng, Y.; Jafari, S.; Sillanpää, M. EDTA-cross-linked β-cyclodextrin: An environmentally friendly bifunctional adsorbent for simultaneous adsorption of metals and cationic dyes. *Environ. Sci. Technol.* **2015**, *49*, 10570–10580. [CrossRef]
46. Wang, J.; Chen, Z.; Chen, B. Adsorption of polycyclic aromatic hydrocarbons by graphene and graphene oxide nanosheets. *Environ. Sci. Technol.* **2014**, *48*, 4817–4825. [CrossRef] [PubMed]
47. Chen, Z.; Liu, Z.; Zhang, L.; Cai, Q.; Hu, J.; Wang, W.; Ju, X.; Xie, R.; Chu, L. Functional graphene oxide nanosheets modified with cyclodextrins for removal of Bisphenol A from water. *Chin. J. Chem. Eng.* **2021**, *39*, 79–87. [CrossRef]
48. Zhao, Z.; Long, X.; Millan, M.; Yuan, G.; Cui, Z.; Dong, Z.; Cong, Y.; Zhang, J.; Li, X. The influence of carbon supports and their surface modification on aqueous phase highly selective hydrogenation of phenol to cyclohexanol over different Ni/carbon catalysts. *Carbon* **2023**, *213*, 118227. [CrossRef]
49. Tan, Y.; Zhao, N.; Song, Q.; Ling, H. Alkali synergistic sulfide-modified nZVI activation of persulfate for phenanthrene removal. *Chem. Eng. J.* **2023**, *11*, 109923. [CrossRef]
50. Cheng, N.; Deng, F.; Wang, J.; Hu, L.; Yang, J.; Jiang, S.; Wang, H.; Ma, X.; Zhao, L.; Li, G.; et al. Simultaneous alkali recovery, coagulant recycling and organics removal from textile wastewater via membrane electrochemical system. *Sep. Purif. Technol.* **2025**, *354*, 129448. [CrossRef]
51. Wang, C.; Kang, J.; Sun, H.; Ang, H.M.; Tadé, M.O.; Wang, S. One-pot synthesis of N-doped graphene for metal-free advanced oxidation processes. *Carbon* **2016**, *102*, 279–287. [CrossRef]
52. Zhao, W.; Ye, C.; Li, J.; Yu, X. Increased risk of antibiotic resistance in surface water due to global warming. *Environ. Res.* **2024**, *263*, 120149. [CrossRef] [PubMed]
53. Zhang, X.; Wu, N. Adsorption characteristics of N-rGO for multiple representative trace antibiotics in water. *J. Environ. Qual.* **2022**, *51*, 1298–1309. [CrossRef]
54. Li, T.; Zhu, P.; Wang, D.; Zhang, Z.; Zhou, L. Efficient utilization of the electron energy of antibiotics to accelerate Fe(III)/Fe(II) cycle in heterogeneous Fenton reaction induced by bamboo biochar/schwertmannite. *Environ. Res.* **2022**, *209*, 112830. [CrossRef]
55. Luo, Y.; Su, R. Cobalt-based MOF material activates persulfate to degrade residual ciprofloxacin. *Water* **2024**, *16*, 2299. [CrossRef]
56. Carmen, S.; Luis, M. p-Nitrophenol degradation by activated persulfate. *Environ. Technol. Innov.* **2021**, *21*, 101265. [CrossRef]
57. Tan, C.; Gao, N.; Deng, Y.; Zhang, Y.; Sui, M.; Deng, J.; Zhou, S. Degradation of antipyrine by UV, UV/H_2O_2 and UV/PS. *J. Hazard. Mater.* **2013**, *260*, 1008–1016. [CrossRef]

58. Lou, X.; Wu, L.; Guo, Y.; Chen, C.; Wang, Z.; Xiao, D.; Fang, C.; Liu, J.; Zhao, J.; Lu, S. Peroxymonosulfate activation by phosphate anion for organics degradation in water. *Chemosphere* **2014**, *117*, 582–585. [CrossRef] [PubMed]
59. Gu, D.; Guo, C.; Hou, S.; Lv, J.; Zhang, Y.; Feng, Q.; Zhang, Y.; Xu, J. Kinetic and mechanistic investigation on the decomposition of ketamine by UV-254 nm activated persulfate. *Chem. Eng. J.* **2019**, *370*, 19–26. [CrossRef]
60. Dan, J.; Wang, Q.; Mu, K.; Rao, P.; Dong, L.; Zhang, X.; He, Z.; Gaod, N.; Wang, J. Degradation of sulfachloropyridazine by UV-C/persulfate: Kinetics, key factors, degradation pathway. *Environ. Sci. Water Res.* **2020**, *6*, 2510–2520. [CrossRef]
61. Fan, Z.; Feng, T.; Wu, S.; Wang, S.; Tan, Y.; Yu, Q.; Huang, R.; Zhang, X. Chitin-derived biochar with nitrogen doping to activate persulfate for phenol degradation: Application potential and electron transfer pathway in system. *Chemosphere* **2023**, *330*, 138641. [CrossRef] [PubMed]

catalysts

MDPI

Article

Effect of Alkaline Soil Solution on the Material Characteristics and Photocatalytic Activity of Strontium Titanate Nanomaterials

Badam Ariya [1,2], John Chagu [1], Karolina Solymos [3], Tamás Gyulavári [1], Zsejke-Réka Tóth [4], Ákos Kukovecz [1], Zoltán Kónya [1], Gábor Veréb [5,*] and Zsolt Pap [1,6,7,*]

1 Department of Applied and Environmental Chemistry, University of Szeged, Rerrich Tér 1, HU-6720 Szeged, Hungary; badama@mas.ac.mn (B.A.); johnchagu7@gmail.com (J.C.); gyulavarit@chem.u-szeged.hu (T.G.); kakos@chem.u-szeged.hu (Á.K.); konya@chem.u-szeged.hu (Z.K.)

2 Institute of Geography-Geoecology, Mongolian Academy of Science, Baruunselbe 15, Chingeltei District, Ulaanbaatar 15170, Mongolia

3 Department of Physical and Environmental Geography, University of Szeged, Egyetem Street 2-6, HU-6722 Szeged, Hungary; solymoskarolina@geo.u-szeged.hu

4 Centre of Nanostructured Materials and Bio-Nano Interfaces, Institute for Interdisciplinary Research on Bio-Nano-Sciences, RO-400271 Cluj-Napoca, Romania; zsejke.toth@ubbcluj.ro

5 Institute of Biosystem Engineering, Faculty of Engineering, University of Szeged, Moszkvai Blvd. 9, HU-6725 Szeged, Hungary

6 Hungarian Department of Biology and Geology, Babeş–Bolyai University, Clinicilor 5-7, RO-400006 Cluj-Napoca, Romania

7 STAR-UBB Institute, Babeş–Bolyai University, Mihail Kogălniceanu 1, RO-400084 Cluj-Napoca, Romania

* Correspondence: verebg@mk.u-szeged.hu (G.V.); pzsolt@chem.u-szeged.hu or zsolt.pap@ubbcluj.ro (Z.P.)

Abstract: The present study aimed to investigate the interaction between strontium titanate photocatalysts and alkaline soil (solonetz) soil solutions. For this purpose, one commercially available and several synthesized strontium titanates were considered. The photocatalytic activity and material characteristics were assessed before and after immersion in the soil solutions. All samples were characterized by X-ray diffractometry (XRD), infrared spectroscopy (IR), diffuse reflectance spectroscopy (DRS), and scanning electron microscopy (SEM). After interaction with the soil solution, most of the samples became more active for phenol degradation. It was found that the crystalline structure of each sample was preserved, while the primary crystallite sizes decreased after both phenol degradation and immersion in solonetz soil solutions. Moreover, the surface of all synthesized nanostructures was covered by organic residues from either the soil solution or the by-products of phenol degradation. This was also visible from the DR spectra, as an intensive color change was observed. The bandgaps of most samples were also changed, except for the commercial material. The results imply that it is important to investigate the ecofriendly nature of any photocatalytic material, as it tends to influence the surrounding environment. This is important, as solar photocatalysis is rising among the possible methods for water purification and disinfection.

Keywords: strontium titanate; photocatalysis; solonetz soil solution; environmental interaction

1. Introduction

Wastewater treatment is of great importance in providing clean water to the increasing population of Earth [1,2]. Due to the high costs required for the various water treatment and purification technologies, extensive research has been conducted into alternative methods such as photocatalysis. Numerous semiconductor materials have already been investigated, including TiO_2 [3–5], ZnO [6,7], WO_3 [8], Bi_2O_3 [9], etc.

Strontium titanate has been reported as one of the most promising photocatalysts due to its approachable bandgap value of 3.2 eV (UVA-excitable), high photocatalytic activity and quantum yield, and chemical stability [10,11], making it a strong candidate for environmental applications such as solar water treatment (at least 5% of solar energy is UV light). Moreover, strontium titanate is also utilized in dye-sensitized solar cells due to its ability to improve energy conversion efficiency by facilitating electron transport. Its stability and optoelectronic properties also make it valuable in gas sensors, where it detects changes in gas concentration through conductivity shifts [12]. These multiple applications reflect its versatility and importance in both renewable energy and environmental technologies. However, like many nanomaterials, its end of life poses challenges related to material recovery and environmental impact, highlighting the need for sustainable recycling technologies [13].

Research shows that nanomaterials end up in landfills. This means that they can enter the environment and potentially cause local changes in the soil, flora, and fauna [14]. The majority of engineered nanoparticles (ENPs) released into the environment tend to accumulate in the soil, reaching higher concentrations than in water or air [15,16]. Titanium-containing ENPs show the highest accumulation tendency in soil, water, and air, followed by iron- and zinc-based ENPs [15,16]. It is expected that Ti-based nanoparticles will remain in particulate form due to their low solubility, but are unlikely not to change in size [17].

Rainfall and running water cause the formation of soil solutions, which can interact with nanomaterials. The migration and deposition of ENPs are influenced by various parameters such as ionic strength and organic matter. ENPs have been investigated for their potential toxicity and environmental risks and have consequently been identified as emerging pollutants [18,19] endangering various levels of the food chain (bacteria, algae, fish, crustaceans, nematodes, etc.) [20].

Soil is a significant chapter of environmental research, and thus many studies focus on the chemical, biological, and physical changes that occur there [21]. Soil pH is an important soil characteristic that impacts many soil properties and processes, including nutrient availability, microbial activity, and plant growth. It can range from strongly acidic (pH < 5) to strongly alkaline (pH > 8), with most agricultural soils falling in the slightly acidic to slightly alkaline range (pH 6–8). Solonetz alkaline soil is characterized by its high pH and elevated sodium content, which can have a profound impact on soil chemistry and plant growth [22]. Solonetz soils are very common and cover a significant area of the Great Hungarian Plain [23]. This type of soil is abundant in arid and semiarid regions worldwide.

Considering the relevant information and the significance in evaluating soil characteristics, solonetz alkaline soil was selected for investigation. While past research mainly focused on the effects of ENPs in water, there is a lack of understanding regarding their interaction with soil solutions. The goal was to gain valuable insights into the environmental risks associated with ENPs, particularly focusing on the limited knowledge about how the physical and chemical properties of soil solutions affect the catalytic activity of $SrTiO_3$ ENPs. Further investigation in this area is crucial for developing a more comprehensive understanding of the environmental consequences of these ENPs.

Therefore, this study focused on investigating the properties of strontium titanates (structure, morphology, optical properties, and photocatalytic activity) and the changes in these characteristics after interaction with solonetz soil solution. These results give information on the role a specific soil type plays in the environmental distribution and fate of a specific nanomaterial.

2. Results

2.1. Assessment of Photocatalytic Activity

The abbreviations used in this study are listed at the end of the manuscript. Sample names and synthesis methods are described in Section 4.2.

By analyzing the photoactivity of the samples (REF, pristine STO, and soil solution-treated ones) the following questions can be formulated:

- Does the synthesis of the catalysts have any influence on the properties of the samples?
- Following the interaction with soil solutions, can the activity of a catalyst be preserved?
- If the answer to both the above questions is yes, then can it be stated that these nanoparticles are "safe" for the environment?

The photocatalytic degradation experiments revealed that the synthesized catalysts have higher photocatalytic activity than the reference sample (STO_REF). Among the synthesized catalysts, STO_I had the highest photocatalytic activity, as can be seen in Figure 1. The conversion values obtained were 3.8% for STO_REF, 32.1% for STO_I, 12.8% for STO_II, and 5.6% for STO_III.

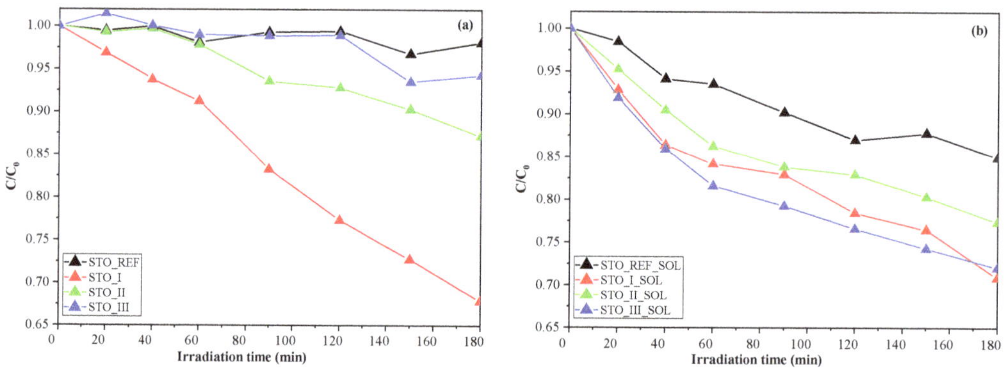

Figure 1. Phenol degradation efficiency of the STO photocatalysts (**a**) and their soil solution-treated counterparts (**b**).

The results show that the soil solution modified (detailed in Section 4.2) the photoactivity of the STO samples. The degradation efficiency for each photocatalyst (following treatment with the soil solution) was 15.0% for STO_REF_SOL, 29.1% for STO_I_SOL, 22.6% for STO_II_SOL, and 27.9% for STO_III_SOL. The immersion in soil solution increased the photocatalytic activity of samples STO_REF, STO_II, and STO_III, while in the case of STO_I, a slight decrease was observed (Figure 1). Usually, surface contamination yields a decreased activity of every photocatalysts; however, here seems to be an exception. There are reports in the scientific literature of an organic compound enhancing the photoactivity of a given nanostructure. In specific cases, surface-anchored compounds can act as charge separators via polar functional groups or aromatic rings, and can lock the photogenerated electrons, thus facilitating the charge separation mechanism and enhancing the overall photoactivity. In some reported situations, surface-anchored compounds can act as suspension stabilizers by enhancing the hydrophilicity of the particles. This way, the crystallites are in more intensive contact with the surrounding matrix, the water, permitting and facilitating the mass transfer between particles and aqueous media.

The photocatalytic degradation of phenol is influenced by many parameters, such as pH, initial concentration of phenol, primary crystallite size, suspension concentration, light intensity, and the presence of electron acceptors (NO_3^-, SO_4^{2-}, etc.) [20]. The alkaline pH (in our case, 9.4 ± 0.2) might lead to small particles (due to the slow solubilization of STO in alkaline media and subsequent recrystallization), which can contribute to the observed high photocatalytic activity.

Comparing $SrTiO_3$ photocatalysts that appear in different publications based on their performance may not be correct, due to variations in several experimental parameters, such as lamp emission spectra, reactor design, air supply flux differences, lamp power, lamp distance, different organic compounds, etc. Moreover, a standardized $SrTiO_3$ material should be available and used (which is not always the case) to discuss the performance of the material. However, photocatalysis has just one such major reference material, Evonik Aeroxide P25 TiO_2, which is not suitable in the present case. Additionally, there is also the issue of soil solution composition. The chemical composition of soil extracts varies depending on the geographical region. Hence, if there are similar experiments in the literature, comparison with any results will be difficult due to the variance in the chemical composition of the soil extract, which cannot be controlled.

Based on the results discussed above and visualized below (Figure 2a), it is clear that the semiconductor nature of nanoparticles is preserved even after interaction with the basic soil solutions. This was proven by the enhanced photoactivity values following the treatment. Hence, it is imperative to establish a relationship between the treated STO samples and their structural and morphological properties.

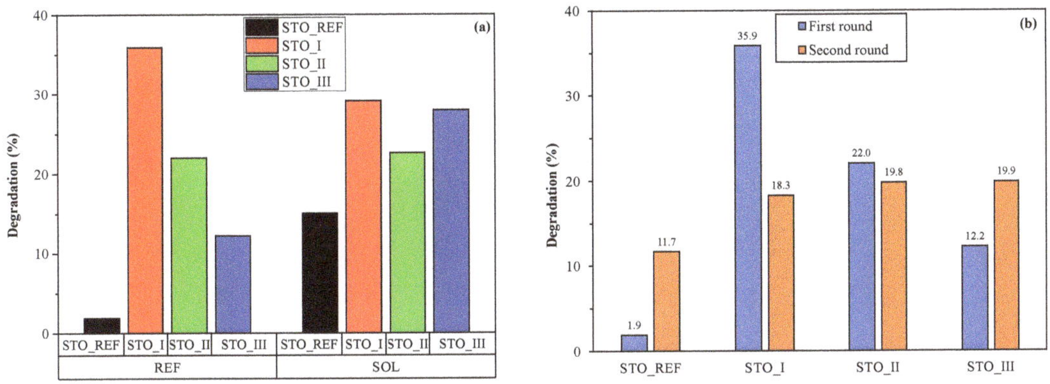

Figure 2. Comparison of degradation efficiencies before and after immersion in soil solutions (**a**) and the cyclic reutilization experiment (**b**).

However, before proceeding with the investigations, the reusability of each catalyst was tested within two cycles. It should be noted that no differences were observed in their XRD patterns (no new diffraction peaks, no widening of the diffraction signals), meaning that every activity change must be related to the surface quality of the particles. Figure 2b shows that the activity of the photocatalysts changed following the first degradation cycle. The most intriguing fact was that the activity changes followed nearly the same trend as what we observed after soil solution immersion, namely that the activity of STO_REF and STO_III increased (1.9% to 11.7%; 12.2 to 19.9%), while the performance of STO_I decreased (from 35.9% to 18.3%) and that of STO_II was nearly the same (22.0% vs. 19.8%). This means that the reason for the activity change after the recycling experiments is most likely identical to the one that will be identified for the soil-solution effect on the $SrTiO_3$ nanostructures.

2.2. *Structure and Morphology*

2.2.1. Primary Crystallite Size and Composition Analysis

The degradation results suggest that there might be differences in the crystallographic properties of the samples, as these can influence photoactivity. Hence, XRD was applied to verify the hypothesis, and this was complemented with the specific surface area values. First, the success of the synthesis procedure was investigated. All three samples and the reference contained cubic $SrTiO_3$ (JCPDS 35-0734), space group Pm$\bar{3}$m, lattice parameters: 3.90264 pm. Sample STO_I contained $SrCO_3$ (~17 wt.%; JCPDS 05-0418), while the other three materials (STO_REF, STO_II, and STO_III) were carbonate-free (Figures 3, S1 and S2). It should be noted that $SrCO_3$ may originate from two different sources, the first being absorption of CO_2 from air. However, the amount of $SrCO_3$ formed via this route should be low, as the exposure of the precursors and other reagents to air is short, in the order of a few minutes. The other formation route should be directly from the carbonated NaOH, as suggested in the experimental section.

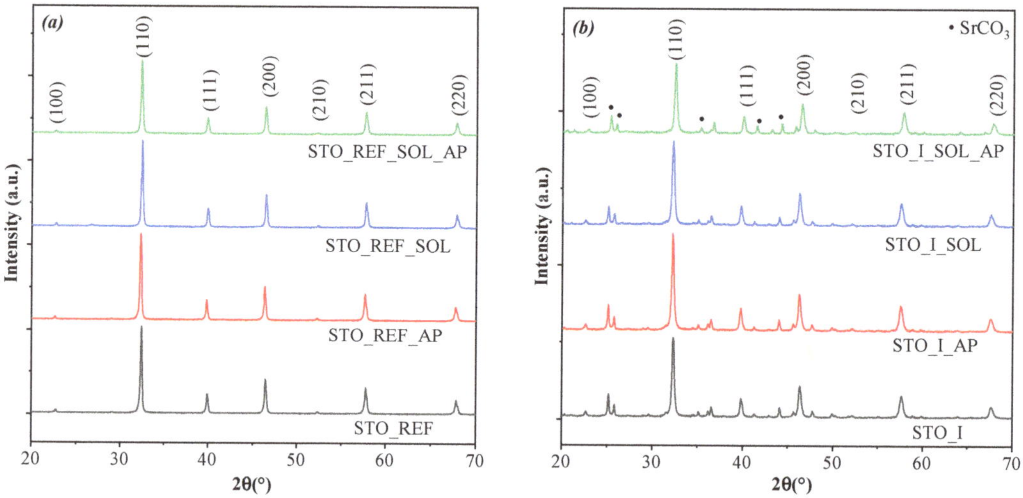

Figure 3. XRD patterns for (**a**) STO_REF and (**b**) STO_I in phenol, solonetz soil and solonetz soil with phenol.

Primary crystallite sizes were calculated for the pristine samples, which were between 42 and 45 nm (STO_I—38.1 m^2/g, STO_II—37.7 m^2/g, and STO_III—39.0 m^2/g), while for STO_REF, this value was 78 nm (12.1 m^2/g) (Table S2). Also, it should be mentioned that STO_III did not contain any other Sr or Ti compounds (even though this was expected due to the Sr:Ti = 1:1 ratio, Section 4.2, Figure S2).

The XRD patterns of the soil solution-treated samples were investigated before and after phenol degradation. They showed that the crystalline structure of the samples was preserved at each stage (Figures 3 and S1), except for sample STO_I, where an increase in $SrCO_3$ content was observed. However, after calculating the crystallite sizes, it was found that the interaction with the soil solution resulted in different values compared to the original ones (Table S2). In the case of STO_REF_SOL, it was 77.6 nm (11.8 m^2/g), which is lower than the original value (78 nm). For the synthesized samples, this change was as follows: from 44.9 nm to 42.9 nm (38.3 m^2/g) in STO_I; from 44.1 nm to 46.1 (36.2 m^2/g) in STO_II; and from 42.0 nm to 35.4 nm (42.4 m^2/g) in STO_III. After this, all the samples were investigated again following phenol degradation, and it was found that the crystallite sizes had changed again. For STO_REF_AP, 84.2 nm (9.0 m^2/g) was obtained, which is larger than the original value (78.3 nm). However, at this size, the Scherrer equation yields less accurate primary crystallite sizes (by 10%–15%) [24]. After phenol degradation, the size of the soil solution-treated sample (STO_REF_SOL_AP) decreased to 67.1 nm (19.2 m^2/g). For the synthesized samples, these changes were as follows: from 44.9 nm to 28.2 (49.4 m^2/g) nm in STO_I; from 46.1 nm to 26.9 (50.2 m^2/g) in STO_II; and from 42.0 nm to 41.4 nm (38.3 m^2/g) in STO_III (Figure S1).

2.2.2. Surface Chemistry of the Samples

All the samples mentioned until now were also analyzed using IR spectroscopy. IR results showed that there was a significant difference in specific cases at ~3000–3900 cm^{-1}, which is the so-called water band [25]. In the case of the STO_REF sample set (Figure 4) no changes were observed. In the case of the STO_I sample group (Figure 4), the water band diminished after soil solution immersion (STO_I_SOL, less intense signal diminishing) and also following phenol degradation (STO_I_AP, STO_I_SOL_AP). Sample set STO_II behaved like STO_REF, the only difference being in the case of STO_II_SOL, where a slight increase in this band was noticed. The only case where this band registered an increase after phenol degradation was in sample set STO_III (STO_III_SOL_AP). The other samples from this group did not show any changes in the water band (Figure S2). This phenomenon can be attributed to the fact that during phenol degradation, some intermediates may be adsorbed on the surface of the catalyst, covering the oxide layer. This was partially proven by the results shown below.

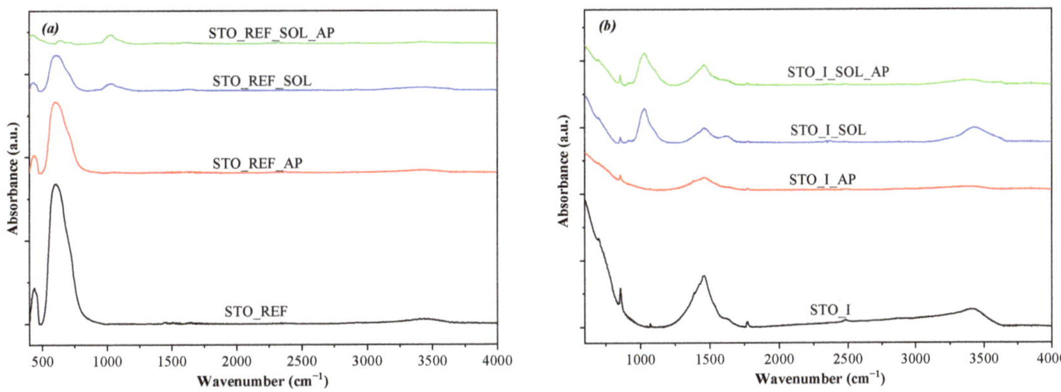

Figure 4. IR spectra recorded for (**a**) STO_REF and (**b**) STO_I after interaction with solonetz soil solution and/or phenol degradation.

One of the major bands is the metal–oxygen (Sr–O, Ti–O) band at ~300–750 cm^{-1} [26]. The intensity of this signal was stable for STO_II and STO_III (Figures 4 and S2), while for STO_REF, intense adsorption of some organic species may occur, covering the oxide surface and leading to the lower intensity of this band. The same behavior was observed for STO_I as well, with lower intensity.

A band centered at ~1000 cm^{-1} (C–O) [27] was present in the materials that were immersed in soil solutions. The intensity of this signal was stable for STO_REF, STO_I, and STO_II, while for STO_III, its intensity increased after phenol degradation (STO_III_SOL_AP). The signal found in the 1250–1500 cm^{-1} region was sharp and intense for the STO_I materials (Figure 4), which can be attributed to several organic compounds, including amines [28]. The other samples did not show this signal. Another intense signal, which behaved like the one located at ~1000 cm^{-1}, was detected at 1650 cm^{-1} (this signal is also characteristic of highly polar organic species, such as amides, carboxylic acids, etc. [27]). This signal appeared separately in STO_II_SOL and STO_III_SOL, while in the STO_I sample series, it overlapped with the band located at 1250–1500 cm^{-1}

2.2.3. Morphological Changes Induced by the Soil Solution and Phenol Degradation

The measurements revealed that all materials showed large particles with signs of aggregation. Specifically, for STO_REF, no major changes were observed in the size or shape of the particles (~0.5–1.5 μm). The aggregates were shapeless and could not be characterized by a well-defined morphology (Figures 5 and S3).

Figure 5. SEM micrographs of STO_REF and STO_I photocatalysts.

In STO_I, no major changes were observed in the size or shape of the particles. Signs of aggregation were observed for the soil solution-immersed sample STO_I_SOL. Aggregates were found also in STO_II (containing cube-shaped crystals), STO_II_SOL, and STO_II_AP, but after testing the soil solution-modified STO_II sample for phenol degradation (STO_II_SOL_AP), its morphology had changed (the cubic crystal morphology was partially destroyed). For STO_III, this observation was also valid (Figure S3).

2.2.4. Optical Properties and Bandgap Energy of the Samples

Based on the DRS results, STO_REF, STO_I, STO_II, and STO_III did not show any major color changes after phenol degradation. However, it should be mentioned that the reflectance values for STO_REF_AP, STO_I_AP, STO_II_AP, and STO_III_AP were lower compared to those for their pure counterparts, meaning that some phenol degradation intermediates might have adsorbed on their surface. Also, natural organic matter (e.g., humic acids) containing various functional groups can interact with the samples, contributing to changes in the optical properties [29]. When STO_REF was immersed in the soil solution, it became grayish (STO_REF_SOL), then brightened somewhat following phenol degradation (STO_REF_SOL_AP, Figure 6). This color change was valid for all other STO_II (Figure S4) and STO_III (Figure S5) soil solution-immersed samples. The same trend was not valid for STO_I, as the color of STO_I_SOL_AP (Figure 7) became darker after phenol degradation.

The bandgap energy of STO_REF (3.10 eV) did not change after phenol degradation (STO_REF_AP) or after immersion in the soil solution (STO_REF_SOL) followed by phenol degradation (STO_REF_SOL_AP). This was not surprising, because some organic compounds are capable of changing the color of semiconductors, but not their bandgap. This phenomenon is often confused with doping [30]. The bandgap of STO_I slightly decreased from 3.33 eV to 3.07 eV after phenol degradation (STO_I_AP). When the sample was immersed in soil solution, 2.93 eV was obtained (STO_I_SOL). When the previous material was applied in phenol degradation, the bandgap energy remained nearly unchanged (2.83 eV, STO_I_SOL_AP). These phenomena were similar for the STO_II and STO_III sample series, as shown by the values presented in Table S2. A decreasing bandgap value may be associated with decreased crystallite size (Section 2.2.1.) and the covering of the surface with compounds capable of intervening in charge transfer processes [31].

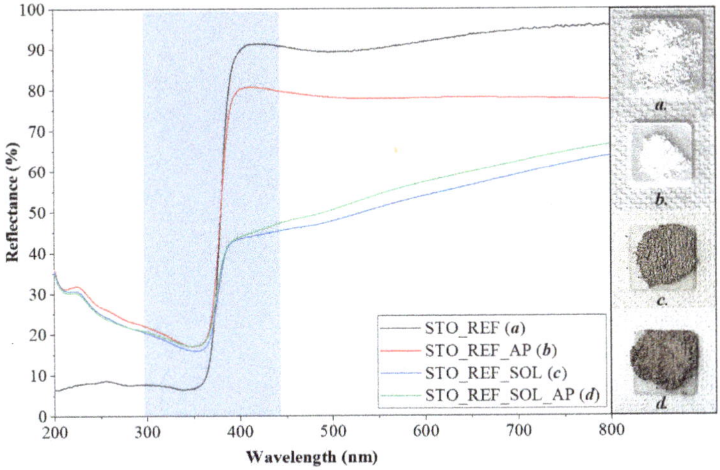

Figure 6. DR spectra of STO_REF samples (Blue area is the light absorption range).

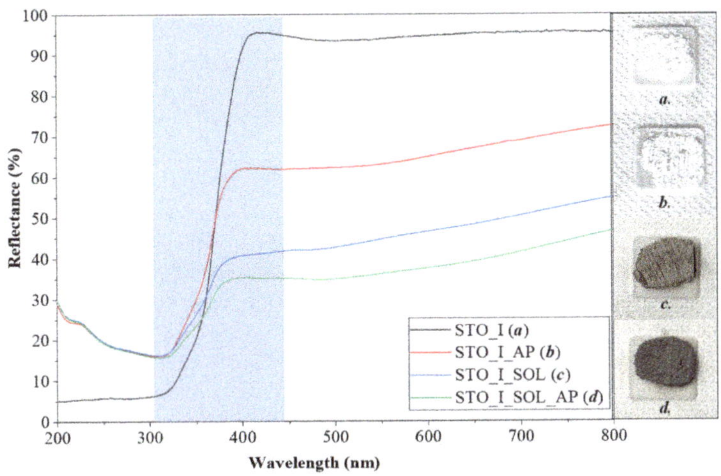

Figure 7. DRS spectra of STO_I sample series (Blue area is the light absorption range).

2.2.5. XPS Analysis

In the XP spectra (Figure S6) of $SrTiO_3$, the primary peaks correspond to the core levels of Sr 3d, Ti 2p, and O 1s [32]. The Sr 3d spectrum typically exhibits a doublet with Sr $3d_{5/2}$ and Sr $3d_{3/2}$ peaks, confirming the presence of Sr in the Sr^{2+} oxidation state. The Ti 2p region also presents a characteristic doublet with Ti $2p_{3/2}$ and Ti $2p_{1/2}$ peaks, indicating that titanium is in the Ti^{4+} oxidation state, consistent with the stoichiometry of $SrTiO_3$. No Ti^{3+} was detected in the investigated samples.

The O 1s peak appears around 529.5 eV, attributed to the lattice oxygen in $SrTiO_3$. Additional features in the O 1s region may indicate surface hydroxylation or adsorbed species. Surface-adsorbed species, such as organic compounds, were detected in the C 1s region, supporting the IR results. However, due to the complexity of organic compounds adsorbed, neither clear identification nor quantification could be carried out.

3. Discussion and Contextual Analysis of the Results

The first phase of this investigation was the evaluation of the photocatalytic performances of all the nanomaterials obtained at different stages, i.e., (i) pristine synthesized samples, (ii) materials immersed in soil solutions, and (iii) samples obtained after photocatalytic degradation and immersion in the soil solution. Two types of effects were observed:

- Activity increase: STO_REF, STO_II, and STO_III sample series.
- Activity decrease: STO_I sample series.
- Recycled catalyst activity: followed the trends listed above.

To explain the diverse activity changes, the main properties of the samples were analyzed.

- Primary crystallite size and composition: It was found that the crystalline structure of the samples was preserved at each stage, except for STO_I. In this sample, the amount of $SrCO_3$ increased following the soil solution treatment, meaning that $SrCO_3$ inhibited the activity of $SrTiO_3$. The primary crystallite sizes decreased more intensively following the photocatalytic degradation process, which was carried out for the samples immersed in the soil solution. This means that recrystallization occurs during the catalytic process, increasing the available surface area (lower crystallite sizes mean higher specific surface areas and in general higher photocatalytic activities), except for STO_I (data regarding the specific surface area can be found in the same section where the primary crystallite size values are discussed). Here, the primary crystallite size decrease effect was most probably balanced by the presence of $SrCO_3$ (manifested in activity decrease).
- Morphological changes: Not only structural but also morphological changes were observed as well. The only sample that did not show any changes in morphology was STO_REF, while aggregation and fragmentation were observable in the other cases. The latter can be associated with the changes registered in primary crystallite size values. This assumption was supported by the correlation between XRD-derived primary crystallite size reduction (Table S2) and the aggregation and fragmentation patterns observed in the SEM micrographs (Figure 5). For instance, sample STO_I showed a decrease in crystallite size from 44 nm to 27 nm, which coincided with visible particle fragmentation in SEM micrographs.
- Changes in the optical properties: Here, two types of changes were observed. The first was the decrease in reflectance in the visible region (above 450 nm) for all samples following their immersion in soil solutions. This was expected, as organic compounds were adsorbed on the surface. This was confirmed by the IR results. The second was the decrease of bandgap energy from 3.1 eV to 2.8 eV. This can be because the adsorbed organic compounds may participate in electron transport routes, hindering recombination (reinforced by the activity enhancements). This conclusion was drawn based on one of our earlier studies, where we observed similar behavior. The only difference was that the photocatalyst studied was titania, while the organic compounds were deposited on the surface of the nanoparticles via the calcination of the amorphous matter [30]. However, a clear mechanism would be challenging to propose because

of the wide variety of organic compounds present on the surface of the particles. The only exception was the STO_REF sample series, where no bandgap changes were observed.

- Surface chemistry: Due to the wide variety of organic compounds originating from the soil solution, only a general discussion can be had. It was observed that the concentration of these compounds increases if immersion in soil solution is followed by a photocatalytic degradation experiment. The complexity of the surface is well reflected in the width of the IR band at 1200–1500 cm^{-1}, which in some cases overlaps with the band centered at 1650 cm^{-1}. It is also evident that both soil solution immersion and photocatalysis contribute to the changes registered in the water band. Due to the complexity, XPS results could not be utilized for the explanation of photoactivity changes.

4. Materials and Methods

4.1. Chemicals

Commercial STO (Alfa Aesar, Kandel, Germany; 99%) was used as a reference. Phenol (Spektrum 3D, Debrecen, Hungary; analytical grade) was used as a model pollutant to investigate the photocatalytic activity of the nanomaterials under UVA light irradiation. Soil solutions were prepared using Milli-Q water. For the synthesis of STO, NaOH (VWR; Leuven, Belgium, ≥97.0%), titanium dioxide—P25 (Acros Organics, Waltham, MA, USA, ≥99.5%), urea (Molar Chemicals; Halásztelek, Hungary, ≥99.4%), and strontium nitrate Sr $(NO_3)_2$ (Alfa Aesar; Kandel, Germany; 99%) were used. Potassium bromide (Sigma Aldrich; St. Louis, MO, USA, ≥99.0%), (KBr) was applied as the base material for the pellets used during IR measurements.

4.2. Synthesis

In the first step, a 5 M NaOH (80 mL), 2 M urea (10 mL), and 3 M Sr $(NO_3)_2$ (10 mL) solution was prepared using Milli-Q water under continuous stirring separately for 10 min. In a second step, the prepared solutions were transferred slowly into a vessel. After the mixture had cooled, 15 mmol of Aeroxide P25 (2 g) was added (Ti:Sr molar ratio was 1:1.2). Subsequently, the mixture was further stirred for 10 min and sonicated for another 10 min. Afterwards, the resulting mixture was transferred into a 150 mL stainless-steel Teflon-lined autoclave and placed in an oven at 180 °C for 24 h. After cooling to room temperature, the content of the autoclave was neutralized until pH 7 was reached with a 5 M HCl solution. The excess liquid was removed through vacuum filtration using a microporous membrane (Pall, GH polypro, 0.2 µm, Pall Corporation, New York, NY, USA). Finally, the product was dried overnight in an oven at 75 °C. The samples obtained were as follows:

- STO_I—this is the sample that followed the synthesis description provided above.
- STO_II—this sample was prepared using the experimental approach listed above, the only difference being the use of Na_2CO_3-free NaOH, thus reducing the formation of $SrCO_3$. $SrCO_3$ is known to affect the photoactivity of STO, so a carbonate-free sample was also synthesized [33].
- STO_III—this is the sample that followed the synthesis description provided above, but the ratio of the Sr and Ti precursors was identical, and Na_2CO_3-free NaOH was applied. This was necessary to check whether residual titania clusters can form on the surface and if they contribute to the surface properties of STO, which is a critical aspect of nanoparticles' behavior in soil.

The samples used throughout this work were named as follows: STO_X_SOL_AP. "STO" stands for strontium titanate and "X" refers to the sample type described above (I, II, III, or "REF" standing for "reference"). "SOL" refers to the solonetz soil solution. Last, "AP" means that the samples were obtained by recollecting them after phenol degradation.

For example, "STO_I_SOL_AP" refers to the sample obtained based on the synthesis description provided above following interaction in solonetz soil solution and phenol degradation.

4.3. Characterization of Nanomaterials

X-ray diffraction (XRD) measurements were conducted using a Rigaku Miniflex II (Rigaku, Neu-Isenburg, Germany) diffractometer. The diffraction patterns were recorded over a $2\theta^\circ$ range of 20–70°, with a scan speed of $2^\circ \cdot min^{-1}$. The primary crystallite size values were determined using the Scherrer equation [24]:

$$L = K\lambda/(\beta \cos \theta)$$

where:

L is the crystallite size (nm).
K is the Scherrer/shape constant, typically around 0.9.
λ is the wavelength of the X-rays used in the experiment (nm).
β is the full width at half maximum (FWHM) of the diffraction peak, in radians.
θ is the Bragg angle (half of the diffraction angle).

The specific surface area of the catalysts was determined by nitrogen adsorption at 77 K with a Micromeritics gas adsorption analyzer (Gemini Type 2375, Micromeritics Instrument Corporation, Norcross, GA, USA). The specific surface area was calculated using the BET method.

The sample morphology was examined with a Hitachi S-4700 Type II scanning electron microscope (SEM, Tokyo, Japan), which was equipped with an Everhart–Thornley detector and operated with an electron beam accelerated to 10 kV.

Infrared (IR) spectroscopy was performed to analyze surface changes on the nanoparticles (NPs) using a Bruker Equinox 55 spectrometer (Billerica, MA, USA). The samples were prepared by pressing KBr powder into pellets, and IR spectra were recorded with a spectral resolution of 2 cm^{-1} over a wavelength range of 600–4000 cm^{-1}.

Diffuse reflectance spectroscopy (DRS) was employed to study the optical properties of the ENPs using a Jasco V-650 spectrophotometer (Kingsgrove, Australia) in a wavelength range of 300–800 nm. The bandgap energy was calculated using the Kubelka–Munk method [34].

XPS measurements were performed on a SPECS PHOIBOS 150 MCD instrument (Berlin, Germany) with monochromatized Al Kα radiation (1486.69 eV) at 14 kV and 20 mA and pressure of ~5 mbar. The samples were mounted on the sample holder using double-sided adhesive carbon tape. High-resolution Sr3d, Ti2p, and O1s spectra were recorded with a data acquisition rate of 0.05 eV. The analysis of the data was carried out with CasaXPS software version 2.3.25PR1.0. All peaks were deconvoluted using Shirley background and Lorentzian–Gaussian line shapes. The applied value of the Gaussian–Lorentzian ratio was 30.

4.4. Soil Sampling and Solution Preparation

Solonetz alkaline soil samples were taken near Szatymaz (Figure 8) in Hungary at a depth of 20 cm. The samples were dried for 10 days and then sieved with a 2 mm-pore sieve. The soil pH, electrical conductivity (EC), total salt, and organic matter (OM) content were determined following the ISO 10390:2021 standard [35].

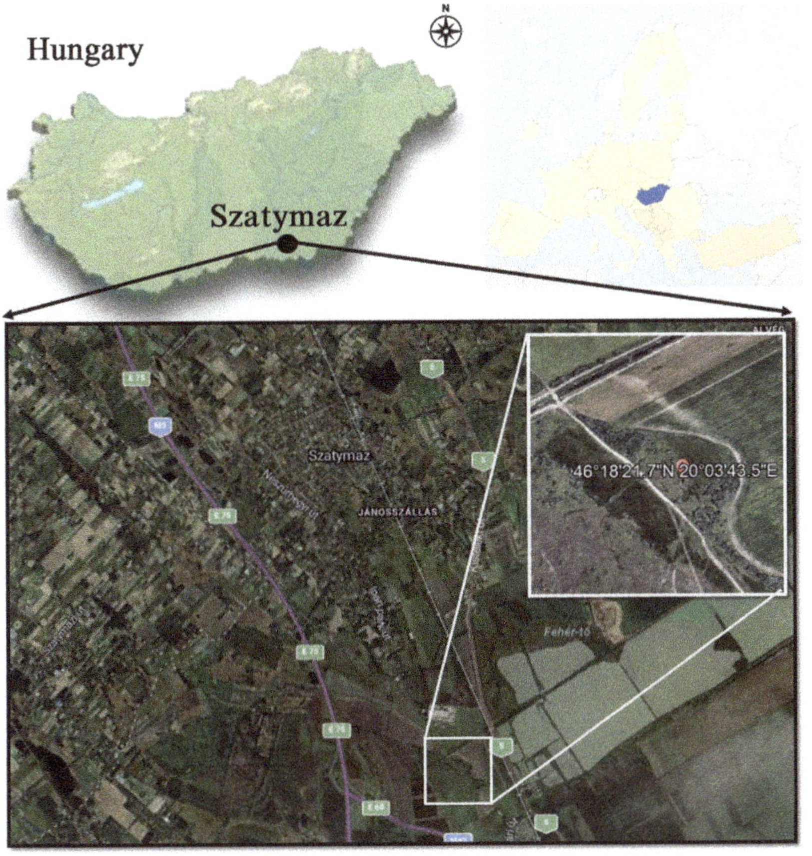

Figure 8. Localization of the sampling area in Hungary.

The pH was measured with a digital pH meter (Inolab pH 720) with an accuracy of ± 0.05 units in distilled water applying a 1:2.5 soil–water ratio (ISO 10390:2021). The determination of the soil texture was based on the plasticity index values according to Arany (Arany plasticity index). This is calculated from the amount of deionized water added to an air-dry soil sample (100 g) until reaching the upper limit of its plasticity. The electrical conductivity (with an accuracy of ± 10 mS cm^{-1}) and the total salt content were analyzed with an Orion 3-Star conductivity meter (Thermo Electron Corporation, Waltham, MA, USA) in saturated soil samples.

The organic matter (OM) content was determined with a UV-vis spectrophotometer (Spectronic Helios-γ, Thermo Fisher Scientific, Waltham, MA, USA). The OM was oxidized overnight with H_2SO_4 (95%) and 0.33 M $K_2Cr_2O_7$ (MSZ-21470–52 1983). The possible error of the measurement is considered to be ± 2%.

To determine the major elemental contents (Na, K, Ca, Mg, Al, Fe, Mn, Zn, Cu, Ni, Co, and As), 0.5 g of the samples were weighed, and 7 mL aqua regia (HNO_3/HCl = 1:3) was added. Soil samples were digested in a microwave oven (AntonPaar Multiwave 3000). Element concentrations in the digested samples were determined by an inductively coupled plasma optical emission spectrometer (ICP-OES; Optima 7000 DV, PerkinElmer) with an accuracy of ±10%. The soil parameters are summarized in the Supplementary Information (Table S1).

4.5. Soil Solution and Analysis

The soil solution of each sample was prepared as follows. Milli-Q water was added to the soil samples at a solid-to-solution ratio of 1:2.5. The suspension obtained was shaken for 18 h (40 rpm), centrifuged at 3700 rpm for 30 min, and then filtered with a 0.5 µm-pore cellulose nitrate membrane (VWR Chemicals, Leuven, Belgium).

The pH was measured as described in the previous section. The ionic strength was calculated from the EC values. Chemical oxygen consumption was determined according to the ISO 10390:2021 standards.

The major (Na, K, Ca, Mg, Al, Fe, Mn) and trace (Cu, Zn, Ni, Co, As) element concentrations were measured by ICP-OES, and the anion contents (F^-, Cl^-, NO_3^-, NO_2^-, PO_4^{3-}) were determined by ion chromatography (IC) and flow injection analysis (FIA).

4.6. Behavior of STO Nanoparticles in Soil Solutions

Suspensions containing 10 $g \cdot L^{-1}$ STO NPs were prepared using Milli-Q water. Before the adsorption experiments, the STO NP suspensions were ultrasonicated for 10 min.

For the adsorption experiments, 90 mL of soil solution and 10 mL of the stock NP suspension were combined in a volumetric flask and stirred for 3 h. During this time, the pH of the suspensions was monitored every 15 min for the first hour and every 20 min for the remaining two hours. Following the stirring period, the suspensions were centrifuged at 300 rpm for 10 min and then dried for 24 h at 36 °C.

4.7. pH Measurements

Before performing the photocatalytic experiments, the pH of the soil solution was determined as the catalysts were immersed in it. For this purpose, 0.2 g of catalyst was dispersed in 20 mL of H_2O, and the suspension was mixed with 180 mL of soil solution. The pH was measured for 3 h (every 15 min in the first hour and every 20 min in the second and third hours), and the results are shown in Figure 9.

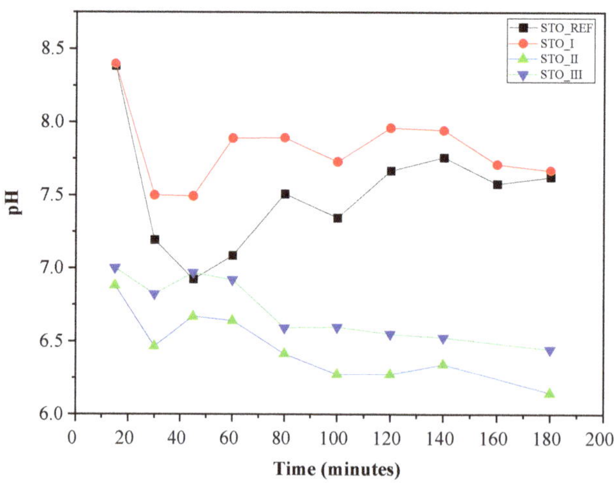

Figure 9. pH as a function of time following the addition of STO to the soil solution.

It can be observed that the pH changes are similar for STO_I and commercial STO, and stabilize around 7.5. For STO_III and STO_II, the pH stabilizes around 6.5. STO_I contains $SrCO_3$, which is why the pH value shifts to a more basic range. However, the results obtained for STO_II and STO_III are in the acidic region (pH 6.0–6.5), because the surface of nanoparticles can vary with 2.5 pH units while not having $SrCO_3$ [36]. Moreover, based on the literature [36], the isoelectric point variance of STO depends on the particle size (even as large as 7–8 pH), so the high pH of STO_REF can be explained.

4.8. Photocatalytic Activity

The photocatalytic activity of STO NPs was evaluated by phenol degradation. For this purpose, 1 $g \cdot L^{-1}$ suspension was prepared containing the model pollutant ($c_{0,phenol}$ = 0.1 mM). The suspension was ultrasonicated and kept for 10 min in the dark to achieve a high degree of homogenization and adsorption–desorption equilibrium (the adsorption of phenol was negligible). The investigation was carried out in a double-walled glass vessel (V = 100 mL) that was surrounded by six fluorescent tubes (Vilber-Lourmat T-6L UV-A, 6 W, Paris, France). Throughout the 3 h of measurements, air was continuously introduced into the suspension to maintain a constant level of dissolved oxygen. The suspensions were stirred in the dark for 10 min before turning on the lamps to allow the adsorption and desorption processes to reach equilibrium. The phenol concentration changes were assessed by high-performance liquid chromatography (HPLC), which consisted of a Merck-Hitachi L-4250 UV-vis detector and a Merck Hitachi L-7100 low-pressure pump (Merck, Darmstadt, Germany; Hitachi, Tokyo, Japan). The cyclic reusability of the base samples was also checked following the first degradation experiment as follows. All the powder was recovered after the experiment via centrifugation at 4400 rpm and the powders were directly introduced into a fresh phenol solution, with the same initial concentration (all the other parameters were also kept constant). The error of the photocatalytic activity assessment process is 2–3%, so in the case of low photoactivity samples, a fluctuation in phenol concentration values is expected.

5. Conclusions

Exposing semiconductor nanoparticles to environmental conditions using laboratory approaches is one way of studying potential scenarios regarding the behavior of nanoparticles in nature. In the present work, solonetz soil solutions were used for this purpose, and their effect on strontium titanate photocatalysts was investigated. It was found that the soil solution influenced the primary crystallite size, bandgap, surface composition, and morphology of the nanoparticles, changing their semiconductive properties. This resulted in enhanced photocatalytic activity in nearly all cases. This means that strontium titanates do not become inactivated under such environmental conditions, possibly leading to unwanted environmental effects. This finding warrants further investigation to better evaluate the risks associated with these perovskite nanomaterials. Moreover, the results here can point out a specific methodology that can be applied to verify the environmentally friendly nature of solar light-active photocatalysts.

Supplementary Materials: The following supporting information can be downloaded at https://www.mdpi.com/article/10.3390/catal15070608/s1. Figure S1: XRD patterns for the STO_II (a) and STO_III (b) in phenol, solonetz soil, and solonetz soil with phenol; Figure S2: IR spectra of STO_II and STO_III in solonetz soil and solonetz soil with phenol; Figure S3: SEM micrographs of STO_II and STO_III photocatalysts; Figure S4: DRS diffractometer of STO_II; Figure S5: DRS diffractometer of STO_III; Figure S6: XPS core spectra of sample STO_REF: Sr 3d (a), Ti 2p (b), O 1s (c) and C 1s (d). Table S1: Physicochemical properties and ionic content of the soil solution; Table S2: Properties of the samples.

Author Contributions: B.A.: conceptualization, investigation, writing, editing; J.C.: investigation; K.S.: conceptualization, investigation, writing, editing; T.G.: investigation, writing—review and editing; Z.-R.T.: investigation; Z.K.: resources, funding acquisition; Á.K.: resources, funding acquisition; G.V.: conceptualization, writing—review and editing; Z.P.: conceptualization, supervision, writing—review and editing, resources, funding acquisition. All authors have read and agreed to the published version of the manuscript.

Funding: T. Gyulavári acknowledges the Bolyai János scholarship (BO/00447/23) provided by the Hungarian Academy of Sciences. The authors are grateful for the financial support of 2021-1.2.6-TÉT-IPARI-MA-2022-00009 from the National Research, Development and Innovation Fund.

Data Availability Statement: Data are contained within the article or Supplementary Material.

Conflicts of Interest: The authors declare no conflicts of interest.

Abbreviations

The following abbreviations are used in this manuscript:

STO	strontium titanate ($SrTiO_3$)
STO_REF	pure/reference STO
STO_AP	STO after phenol degradation
STO_SOL	STO with solonetz soil solution
STO_SOL_AP	STO with solonetz soil solution and phenol degradation
STO_I	pure/reference STO_I
STO_I_AP	STO_I after phenol degradation
STO_I_SOL	STO_I with solonetz soil solution
STO_I_SOL_AP	STO_I with solonetz soil solution and phenol degradation
STO_II	pure/reference STO_II
STO_II_AP	STO_II after phenol degradation
STO_II_SOL	STO_II with solonetz soil solution
STO_II_SOL_AP	STO_II with solonetz soil solution and phenol degradation
STO_III	pure/reference STO_III
STO_III_AP	STO_III after phenol degradation
STO_III_SOL	STO_III with solonetz soil solution
STO_III_SOL_AP	STO_III with solonetz soil solution and phenol degradation

References

1. Qian, Y.; Qin, C.; Chen, M.; Lin, S. Nanotechnology in soil remediation—Applications vs. implications. *Ecotoxicol. Environ. Safety* **2020**, *201*, 110815. [CrossRef]
2. Kurade, M.B.; Ha, Y.-H.; Xiong, J.-Q.; Govindwar, S.P.; Jang, M.; Jeon, B.-H. Phytoremediation as a green biotechnology tool for emerging environmental pollution: A step forward towards sustainable rehabilitation of the environment. *Chem. Eng. J.* **2021**, *415*, 129040. [CrossRef]
3. Phoon, B.L.; Lai, C.W.; Juan, J.C.; Show, P.L.; Chen, W.H. A review of synthesis and morphology of $SrTiO_3$ for energy and other applications. *Int. J. Energy Res.* **2019**, *43*, 5151–5174. [CrossRef]
4. Zhang, Q.; Fan, W.; Gao, L. Anatase TiO_2 nanoparticles immobilized on ZnO tetrapods as a highly efficient and easily recyclable photocatalyst. *Appl. Catal. B Environ.* **2007**, *76*, 168–173. [CrossRef]
5. Solymos, K.; Babcsányi, I.; Ariya, B.; Gyulavári, T.; Ágoston, Á.; Szamosvölgyi, Á.; Kukovecz, Á.; Kónya, Z.; Farsang, A.; Pap, Z. Photocatalytic and surface properties of titanium dioxide nanoparticles in soil solutions. *Environ. Sci. Nano* **2024**, *11*, 1204–1216. [CrossRef]
6. Simonin, M.; Martins, J.M.; Uzu, G.; Spadini, L.; Navel, A.; Richaume, A. Low mobility of CuO and TiO_2 nanoparticles in agricultural soils of contrasting texture and organic matter content. *Sci. Total Environ.* **2021**, *783*, 146952. [CrossRef] [PubMed]
7. Solymos, K.; Kanász, E.; Ágoston, Á.; Gyulavári, T.; Pálffy, B.; Szamosvölgyi, Á.; Kukovecz, Á.; Kónya, Z.; Pap, Z. Impact of different soil solutions on the stability and photocatalytic activity of commercial zinc oxide nanoparticles. *Environ. Sci. Nano* **2025**, *12*, 1328–1339. [CrossRef]
8. Qamar, M.; Gondal, M.; Yamani, Z. Synthesis of highly active nanocrystalline WO_3 and its application in laser-induced photocatalytic removal of a dye from water. *Catal. Commun.* **2009**, *10*, 1980–1984. [CrossRef]
9. Dong, W.; Zhu, C. Optical properties of surface-modified Bi_2O_3 nanoparticles. *J. Phys. Chem. Solids* **2003**, *64*, 265–271. [CrossRef]
10. Cornelis, G.; Pang, L.; Doolette, C.; Kirby, J.K.; McLaughlin, M.J. Transport of silver nanoparticles in saturated columns of natural soils. *Sci. Total Environ.* **2013**, *463*, 120–130. [CrossRef]

11. Moss, B.; Wan, Q.; Butler, K.T.; Grau-Crespo, R.; Selim, S.; Regoutz, A.; Hisatomi, T.; Godin, R.; Payne, D.J.; Kafizas, A.; et al. Linking in situ charge accumulation to electronic structure in doped $SrTiO_3$ reveals design principles for hydrogen-evolving photocatalysts. *Nat. Mater.* **2021**, *20*, 511–517. [CrossRef] [PubMed]
12. Bhan, C.; Kumar, N.; Elangovan, V. Microplastics pollution in the rivers, its source, and impact on aquatic life: A review. *Int. J. Environ. Sci. Technol.* **2024**, *22*, 1937–1952. [CrossRef]
13. Sarijan, S.; Azman, S.; Said, M.I.M.; Jamal, M.H. Microplastics in freshwater ecosystems: A recent review of occurrence, analysis, potential impacts, and research needs. *Environ. Sci. Pollut. Res.* **2021**, *28*, 1341–1356. [CrossRef]
14. Siddiqi, K.S.; Husen, A. Plant response to engineered metal oxide nanoparticles. *Nanoscale Res. Lett.* **2017**, *12*, 92. [CrossRef]
15. Fazeli Sangani, M.; Owens, G.; Fotovat, A. Transport of engineered nanoparticles in soils and aquifers. *Environ. Rev.* **2019**, *27*, 43–70. [CrossRef]
16. Peralta-Videa, J.R.; Zhao, L.; Lopez-Moreno, M.L.; de la Rosa, G.; Hong, J.; Gardea-Torresdey, J.L. Nanomaterials and the environment: A review for the biennium 2008–2010. *J. Hazard. Mater.* **2011**, *186*, 1–15. [CrossRef] [PubMed]
17. Rizwan, M.; Ali, S.; Qayyum, M.F.; Ok, Y.S.; Adrees, M.; Ibrahim, M.; Zia-ur-Rehman, M.; Farid, M.; Abbas, F. Effect of metal and metal oxide nanoparticles on growth and physiology of globally important food crops: A critical review. *J. Hazard. Mater.* **2017**, *322*, 2–16. [CrossRef]
18. Cañas, J.E.; Qi, B.; Li, S.; Maul, J.D.; Cox, S.B.; Das, S.; Green, M.J. Acute and reproductive toxicity of nano-sized metal oxides (ZnO and TiO_2) to earthworms (*Eisenia fetida*). *J. Environ. Monit.* **2011**, *13*, 3351–3357. [CrossRef]
19. Chai, H.; Yao, J.; Sun, J.; Zhang, C.; Liu, W.; Zhu, M.; Ceccanti, B. The effect of metal oxide nanoparticles on functional bacteria and metabolic profiles in agricultural soil. *Bull. Environ. Contam. Toxicol.* **2015**, *94*, 490–495. [CrossRef]
20. Smita, S.; Gupta, S.K.; Bartonova, A.; Dusinska, M.; Gutleb, A.C.; Rahman, Q. Nanoparticles in the environment: Assessment using the causal diagram approach. *Environ. Health* **2012**, *11*, S13. [CrossRef]
21. Lee, W.-M.; Kim, S.W.; Kwak, J.I.; Nam, S.-H.; Shin, Y.-J.; An, Y.-J. Research trends of ecotoxicity of nanoparticles in soil environment. *Toxicol. Res.* **2010**, *26*, 253–259. [CrossRef] [PubMed]
22. Joseph, B.; Gladis, R.; Aparna, B.; JS, B. Amelioration of Acidity and Iron Toxicity in Lateritic Lowland Rice Soil using Non-Conventional Sources of Calcium. *Int. J. Chem. Stud.* **2020**, *8*, 2041–2044. [CrossRef]
23. Rossel, R.V.; Walvoort, D.; McBratney, A.; Janik, L.J.; Skjemstad, J. Visible, near infrared, mid infrared or combined diffuse reflectance spectroscopy for simultaneous assessment of various soil properties. *Geoderma* **2006**, *131*, 59–75. [CrossRef]
24. Holzwart, U.; Gibson, N. The Scherrer equation versus the 'Debye-Scherrer equation'. *Nat. Nanotechnol.* **2011**, *6*, 534. [CrossRef]
25. Libnau, F.O.; Kvalheim, O.M.; Christy, A.A.; Toft, J. Spectra of water in the near- and mid-infrared region. *Vib. Spectrosc.* **1994**, *7*, 243–254. [CrossRef]
26. Seo, S.S.A.; Lee, H.N.; Noh, T.W. Infrared spectroscopy of $CaTiO_3$, $SrTiO_3$, $BaTiO_3$, $Ba_{0.5}Sr_{0.5}TiO_3$ thin films, and $(BaTiO_3)_5/(SrTiO_3)_5$ superlattice grown on $SrRuO_3/SrTiO_3$(001) substrates. *Thin Solid Film.* **2005**, *486*, 94–97. [CrossRef]
27. Smith, B.C. The C=O Bond, Part VI: Esters and the Rule of Three. *Spectroscopy* **2018**, *33*, 20–23.
28. Smith, B.C. Organic Nitrogen Compounds III: Secondary and Tertiary Amines. *Spectroscopy* **2019**, *34*, 22–26.
29. Oades, J.M. An introduction to organic matter in mineral soils. *Miner. Soil Environ.* **1989**, *1*, 89–159.
30. Pap, Z.; Karácsonyi, É.; Cegléd, Z.; Dombi, A.; Danciu, V.; Popescu, I.; Baia, L.; Oszkó, A.; Mogyorósi, K. Dynamic changes on the surface during the calcination of rapid heat treated TiO_2 photocatalysts. *Appl. Catal. B Environ.* **2012**, *111*, 595–604. [CrossRef]
31. Mylsamy, S.; Karazhanov, S.; Subramanian, B. Lattice distortion-driven band gap engineering and enhanced electrocatalytic activity of Mn-substituted nanostructured $SrTiO_3$ materials: A comprehensive investigation. *Chemosphere* **2024**, *346*, 140577. [CrossRef] [PubMed]
32. Vasquez, R.P. $SrTiO_3$ by XPS. *Surf. Sci. Spectra* **1992**, *1*, 129–135. [CrossRef]
33. Boga, B.; Steinfeldt, N.; Moustakas, N.G.; Peppel, T.; Lund, H.; Rabeah, J.; Pap, Z.; Cristea, V.M.; Strunk, J. Role of $SrCO_3$ on Photocatalytic Performance of $SrTiO_3$-$SrCO_3$ Composites. *Catalysts* **2022**, *12*, 978. [CrossRef]
34. Alcaraz de la Osa, R.; Iparragirre, I.; Ortiz, D.; Saiz, J. The extended Kubelka–Munk theory and its application to spectroscopy. *ChemTexts* **2020**, *6*, 2. [CrossRef]
35. *ISO 10390:2021*; Soil, Treated Biowaste and Sludge—Determination of pH. ISO: Geneva, Switzerland, 2021.
36. Zhang, M.; Salvador, P.A.; Rohrer, G.S. Influence of pH and surface orientation on the photochemical reactivity of $SrTiO_3$. *ACS Appl. Mater. Interfaces* **2020**, *12*, 23617–23626. [CrossRef]

MDPI AG
Grosspeteranlage 5
4052 Basel
Switzerland
Tel.: +41 61 683 77 34

Catalysts Editorial Office
E-mail: catalysts@mdpi.com
www.mdpi.com/journal/catalysts

www.ingramcontent.com/pod-product-compliance
Lightning Source LLC
Chambersburg PA
CBHW040821250926
51850CB00007BA/1232
9783725858019